PARTIAL DIFFERENTIAL EQUATIONS

Theory and Technique

PARTIAL DIFFERENTIAL EQUATIONS

Theory and Technique

GEORGE F. CARRIER

Harvard University
Cambridge, Massachusetts

CARL E. PEARSON

University of Washington
Seattle, Washington

MA PING HANG
UBC SEPT 81

ACADEMIC PRESS   New York   San Francisco   London   1976

A Subsidiary of Harcourt Brace Jovanovich, Publishers

ACADEMIC PRESS, INC.
111 Fifth Avenue, New York, New York 10003

United Kingdom Edition published by
ACADEMIC PRESS, INC. (LONDON) LTD.
24/28 Oval Road, London NW1

Library of Congress Cataloging in Publication Data

Carrier, George F
    Partial differential equations.

    Bibliography: p.
    Includes index.
    1.   Differential equations, Partial—Numerical
solutions.   I.   Pearson, Carl E., joint author.
II.   Title.
QA374.C36   515'.353   75-13107
ISBN 0−12−160450−0

AMS (MOS) 1970 Subject Classifications: 35-01, 35-02

PRINTED IN THE UNITED STATES OF AMERICA
    82   9 8 7 6 5

CONTENTS

PREFACE   xi

Introduction   1

    I.1   Definitions and Examples   1

Chapter 1   THE DIFFUSION EQUATION   7

    1.1   Derivation   7
    1.2   Problems   9
    1.3   Simple Solutions   10
    1.4   Problems   12
    1.5   Series Solutions   13
    1.6   Problems   16
    1.7   Nonhomogeneous End Conditions   17
    1.8   Problems   18
    1.9   The Maximum Principle   19
    1.10 Problems   20

Chapter 2   LAPLACE TRANSFORM METHODS   23

    2.1   Introductory Example   24
    2.2   Problems   25
    2.3   A Finite Interval Problem   27
    2.4   Problems   29
    2.5   Delta Function   30
    2.6   Problems   33
    2.7   Supplementary Problems   34

Chapter 3   THE WAVE EQUATION   35

    3.1   Derivation   35
    3.2   Problems   36
    3.3   An Infinite-Interval Problem   38
    3.4   Problems   42
    3.5   Series Solutions   45
    3.6   Problems   46
    3.7   A Problem with Radial Symmetry   48
    3.8   Problems   49
    3.9   Transforms   50
    3.10 Problems   51
    3.11 Uniqueness   52
    3.12 Supplementary Problems   53

Chapter 4   THE POTENTIAL EQUATION   55

    4.1   Laplace's and Poisson's Equations   55
    4.2   Problems   58
    4.3   Simple Properties of Harmonic Functions   59
    4.4   Some Special Solutions—Series   62
    4.5   Problems   64
    4.6   Discontinuous Boundary Data   66
    4.7   Complex Variables and Conformal Mapping   68
    4.8   Problems   72

Chapter 5   CLASSIFICATION OF SECOND-ORDER
                EQUATIONS   75

    5.1   Cauchy Data on y-Axis   75
    5.2   Cauchy Data on Arbitrary Curve   77
    5.3   Problems   78
    5.4   Case I: $B^2 - AC > 0$ 79
    5.5   Case II: $B^2 - AC = 0$ 81
    5.6   Case III: $B^2 - AC < 0$ 82
    5.7   Problems   83
    5.8   Discontinuities; Signal Propagation   86
    5.9   Problems   87
    5.10 Some Remarks   89

Chapter 6   FIRST-ORDER EQUATIONS   91

    6.1   Linear Equation Examples   91
    6.2   Problems   94
    6.3   Quasi-Linear Case   95
    6.4   Problems   97

6.5   Further Properties of Characteristics   99
6.6   Problems   101
6.7   More Variables   101

Chapter 7   EXTENSIONS   103

7.1   More Variables   103
7.2   Problems   107
7.3   Series and Transforms   112
7.4   Problems   115
7.5   Legendre Functions   116
7.6   Problems   121
7.7   Spherical Harmonics   121
7.8   Problems   125

Chapter 8   PERTURBATIONS   127

8.1   A Nonlinear Problem   127
8.2   Problems   130
8.3   Two Examples from Fluid Mechanics   130
8.4   Boundary Perturbations   134
8.5   Problems   135

Chapter 9   GREEN'S FUNCTIONS   139

9.1   Some Consequences of the Divergence Theorem   139
9.2   The Laplacian Operator   142
9.3   Problems   144
9.4   Potentials of Volume and Surface Distributions   147
9.5   Problems   150
9.6   Modified Laplacian   152
9.7   Problems   154
9.8   Wave Equation   156
9.9   Problems   158

Chapter 10   VARIATIONAL METHODS   161

10.1   A Minimization Problem   161
10.2   Problems   164
10.3   Natural Boundary Conditions   167
10.4   Subsidiary Conditions   168
10.5   Problems   172
10.6   Approximate Methods   174
10.7   Problems   178
10.8   Finite-Element Method   180
10.9   Supplementary Problems   183

Chapter 11   EIGENVALUE PROBLEMS   189

    11.1   A Prototype Problem   189
    11.2   Some Eigenvalue Properties   190
    11 3   Problems   193
    11.4   Perturbations   197
    11.5   Approximations   199
    11.6   Problems   201

Chapter 12   MORE ON FIRST-ORDER EQUATIONS   203

    12.1   Envelopes   203
    12.2   Characteristic Strips   205
    12.3   Complete Integral   209
    12.4   Problems   212
    12.5   Legendre Transformation   214
    12.6   Problems   215
    12.7   Propagation of a Disturbance   217
    12.8   Complete Integral and Eikonal Function   220
    12.9   Hamilton–Jacobi Equation   222
    12.10   Problems   224

Chapter 13   MORE ON CHARACTERISTICS   227

    13.1   Discontinuities—A Preliminary Example   227
    13.2   Weak Solutions   230
    13.3   Burgers' Equation   233
    13.4   Problems   236
    13.5   A Compressible Flow Problem   239
    13.6   A Numerical Approach   242
    13.7   Problems   246
    13.8   More Dependent Variables   250
    13.9   More Independent Variables   253
    13.10   Problems   255

Chapter 14   FINITE-DIFFERENCE EQUATIONS AND
                   NUMERICAL METHODS   257

    14.1   Accuracy and Stability; A Diffusion Equation Example   257
    14.2   Error Analysis   259
    14.3   Problems   262
    14.4   More Dimensions, or Other Complications   264
    14.5   Series Expansions   267
    14.6   Problems   269
    14.7   Wave Equation   270
    14.8   A Nonlinear Equation   272

14.9   Problems   273
14.10 Boundary Value Problems   274
14.11 Problems   278
14.12 Series; Fast Fourier Transform   278
14.13 Problems   281

Chapter 15   SINGULAR PERTURBATION METHODS   285

15.1   A Boundary Layer Problem   285
15.2   A More General Procedure   287
15.3   Problems   289
15.4   A Transition Situation   292
15.5   Problems   295
15.6   Asymptotic Analysis of Wave Motion   297
15.7   Boundary Layer near a Caustic   300
15.8   Problems   303
15.9   Multiple Scaling   303
15.10 Problems   306

References   309

INDEX   313

PREFACE

This book reflects the authors' experience in teaching partial differential equations, over several years, and at several institutions. The viewpoint is that of the user of mathematics; the emphasis is on the development of perspective and on the acquisition of practical technique.

Illustrative examples chosen from a number of fields serve to motivate the discussion and to suggest directions for generalization. We have provided a large number of exercises (some with answers) in order to consolidate and extend the text material.

The reader is assumed to have some familiarity with ordinary differential equations of the kind provided by the references listed in the Introduction. Some background in the physical sciences is also assumed, although we have tried to choose examples that are common to a number of fields and which in any event are intuitively straightforward.

Although the attitudes and approaches in this book are solely the responsibility of the authors, we are indebted to a number of our colleagues for useful suggestions and ideas. A note of particular appreciation is due to Carolyn Smith, who patiently and meticulously prepared the successive versions of the manuscript, and to Graham Carey, who critically proofread most of the final text.

INTRODUCTION

We collect here some formal definitions and notational conventions. Also, we analyze a preliminary example of a partial differential equation in order to point up some of the differences between ordinary and partial differential equations.

The systematic discussion of partial differential equations begins in Chapter 1. We start with the classical second-order equations of diffusion, wave motion, and potential theory and examine the features of each. We then use the ideas of characteristics and canonical forms to show that any second-order linear equation must be one of these three kinds. First-order linear and quasi-linear equations are considered next, and the first half of the book ends with a generalization of previous results to the case of a larger number of dependent or independent variables, and to sets of equations.

Included in the second half of the book are separate chapters on Green's functions, eigenvalue problems, and a more extensive survey of the theory of characteristics. Much of the emphasis, however, is on practical approximation techniques; attention is directed toward variational methods, perturbations (regular and singular), difference equations, and numerical methods.

I.1 DEFINITIONS AND EXAMPLES

A *partial differential equation* is one in which there appear partial derivatives of an unknown function with respect to two or more independent variables. A simple example of such an equation is

$$\frac{\partial u}{\partial x} + \frac{\partial u}{\partial y} - \sigma u = 0 \tag{I.1}$$

where σ is a constant. By a *solution* of this equation in a region R of the (x, y) plane we mean a function $u(x, y)$ for which u, $\partial u/\partial x$, and $\partial u/\partial y$ are defined at each point (x, y) in R and for which the equation reduces to an identity at each such point. Such a function u is said to *satisfy* the equation in R.

We denote partial derivatives by subscripts, so that $u_x = \partial u/\partial x$, $u_{xx} = \partial^2 u/\partial x^2$, $u_{xy} = \partial^2 u/\partial x \, \partial y$, etc. Other examples of partial differential equations are

$$x^2 u_{xx} + u_{xy} - \pi^2 u_{yy} + 3u_x - u = e^{x+y} \tag{I.2}$$

$$u_{xxy} - uu_x + \sin(xu^2) = 0 \tag{I.3}$$

$$u_{xyzz} + 2u_{zz} - u = \sin(x^2 + yz) \tag{I.4}$$

[In Eq. (I.4), u is a function of the three variables x, y, z.] Since the highest-order partial derivative that occurs in Eq. (I.1) is the first, Eq. (I.1) is said to be a *first-order* equation. Similarly, Eqs. (I.2), (I.3), and (I.4) are of the second, third, and fourth orders, respectively.

An important property that a partial differential equation may or may not possess is that of *linearity*. By definition, a linear partial differential equation for $u(x, y)$ has the form

$$\sum_{n=0}^{N} \sum_{m=0}^{M} a_{nm}(x, y) \frac{\partial^{n+m} u}{\partial x^n \, \partial y^m} = g(x, y) \tag{I.5}$$

where $a_{nm}(x, y)$ and $g(x, y)$ are given functions of x and y, and where N, M are fixed positive integers. (We define $\partial^0 u/\partial x^0 \, \partial y^0$ to equal u.) If $g(x, y) \equiv 0$, we say that Eq. (I.5) is *homogeneous*. As with ordinary differential equations, the applicability of the principle of superposition is what makes linearity a useful property. Let $U(x, y)$ be one solution of Eq. (I.5), and let each of a set of functions $u^{(1)}(x, y)$, $u^{(2)}(x, y)$, . . . , $u^{(p)}(x, y)$ be solutions of the homogeneous counterpart of Eq. (I.5); i.e.,

$$\sum_{n=0}^{N} \sum_{m=0}^{M} a_{nm}(x, y) \frac{\partial^{n+m} u^{(j)}}{\partial x^n \, \partial y^m} = 0, \qquad j = 1, 2, \ldots, p$$

Then if $a^{(1)}, a^{(2)}, \ldots, a^{(p)}$ are any p chosen constants, direct substitution into Eq. (I.5) shows that

$$u = U + a^{(1)} u^{(1)} + a^{(2)} u^{(2)} + \cdots + a^{(p)} u^{(p)}$$

is also a solution of Eq. (I.5).

Thus, Eqs. (I.1), (I.2), and (I.4) are linear, whereas Eq. (I.3) is nonlinear. Only rarely can one make much formal progress with nonlinear equations; fortunately, many equations of practical interest turn out to be linear (or almost linear).

Just as with an ordinary differential equation, many questions can be asked in connection with an equation such as (I.1). For example, (1) what function or functions, if any, satisfy Eq. (I.1) when $\sigma = 1$? (2) For what values of σ does a function $u(x, y)$ exist that satisfies Eq. (I.1)? (3) How many functions satisfy Eq. (I.1) in $y > 0$, $-\infty < x < \infty$ if we also require that $u(x, 0) = x^2$ for x in the interval $(0, 1)$?

In contrast to most of the partial differential equations we will encounter, Eq. (I.1) is rather easy to solve explicitly. In fact, for any value of σ we can define a new function $\phi(x, y)$ via

$$u = \phi \cdot \exp[\tfrac{1}{2}\sigma(x + y)]$$

(noting that the exponential factor is always nonzero); then $\phi(x, y)$ satisfies the equation

$$\phi_x + \phi_y = 0 \qquad (I.6)$$

With the change in variables $\xi = x + y$, $\eta = x - y$, and with

$$\psi(\xi, \eta) = \phi\left(\frac{\xi + \eta}{2}, \frac{\xi - \eta}{2}\right)$$

[i.e., $\phi(x, y) = \psi(\xi, \eta)$ at corresponding points (x, y) and (ξ, η)], Eq. (I.6) becomes

$$2\psi_\xi = 0$$

so that ψ is a function of η alone, say $f(\eta)$. Since $\eta = x - y$, we can say equivalently that ϕ must be a function of $(x - y)$ alone. Thus u must have the form

$$u = f(x - y) \cdot \exp[\tfrac{1}{2}\sigma(x + y)] \qquad (I.7)$$

where f is an as-yet-undetermined function of the argument $(x - y)$. Conversely, the reader should show that if we choose any continuously differentiable function f and define a function u by Eq. (I.7), then u will satisfy Eq. (I.1).

The reader may now answer such questions as those posed above. In particular, the answer to question (3) can be found by use of Eq. (I.7). At $y = 0$, we have

$$u(x, 0) = f(x) \cdot \exp[\tfrac{1}{2}\sigma x]$$

and if this is to equal x^2 for x in $(0, 1)$, we must choose $f(x)$ such that

$f(x) = x^2 \exp[-\tfrac{1}{2}\sigma x]$ for x in $(0, 1)$. Replacing the argument x by $x - y$, it follows that

$$f(x - y) = (x - y)^2 \exp[-\tfrac{1}{2}\sigma(x - y)] \tag{I.8}$$

for $0 < x - y < 1$. Thus in that region of the (x, y) plane lying between the lines $y = x$ and $y = x - 1$, Eq. (I.7) yields

$$u = (x - y)^2 e^{\sigma y} \tag{I.9}$$

Outside the strip $0 < x - y < 1$, $f(x - y)$ can be any continuously differentiable function of $x - y$ that, at $x - y = 0$ and $x - y = 1$, joins continuously and with continuous first derivatives onto the function described by Eq. (I.8). The answer to question (3) is therefore that there are infinitely many solutions.

Notice that the general solution (I.7) involves an undetermined *function*, rather than simply an undetermined *constant*, as would be the case for a typical first-order ordinary differential equation. We can therefore anticipate that to determine completely the solution to an equation such as (I.1) we will have to specify u along some curve, rather than merely at a single point. Moreover, even such a specification of the solution along a curve may determine the solution only within a region determined by that curve, as in the example just discussed.

The above discussion can be generalized in several ways. Instead of only two independent variables x and y, we may have a number of such variables, and instead of only one dependent variable u, there may be a number of such functions to be determined. We will let a single example suffice. If u and v are each functions of (x, y, z), then the equations

$$u u_x + u u_y + u_z v_z = 1$$

$$x^2 u_{xx} + u_y + v_{zz} = \sin(x + u)$$

would form a coupled pair of nonlinear equations for u and v.

As a different kind of generalization, we can weaken the term "solution" as defined in the first paragraph. It may be physically reasonable to permit a particular derivative, for example, to be discontinuous at a certain point or along a certain curve in the (x, y) plane, and perhaps even greater liberties with the idea of a "solution" can be taken when they are consistent with the context in which a problem arises. We shall encounter such situations later in this book, but in the early sections the given definition is to apply unless an alternative is explicitly stated.

The subject of partial differential equations is a broad one, and it seems useful to begin by acquiring experience with certain frequently encountered special equations. This we will do in the next few chapters.

In so doing we shall focus attention primarily on the techniques by which equations are generated (as a result of model-building) and by which solutions are found, and on the features that characterize these equations and their solutions.

Throughout, it will be assumed that the reader is familiar with, or can easily refer to, such properties of ordinary differential equations as are discussed in standard texts.† The abbreviations ODE and PDE will sometimes be used for "ordinary differential equation" and "partial differential equation," respectively. When particular attention is to be directed to a continuity property, the notation $C^{(n)}$ may be used to indicate continuity of nth derivatives. When no specification to the contrary is made, it is to be understood that boundary curves or surfaces are smooth, in the sense of having continuously turning tangents, and that functional data specified on such boundaries are continuous.

The problems are considered to be an integral part of the text. The reader who evades them will miss 72% of the value of the book.

† A representative selection follows: Kreyszig (1967, Chaps. 1–4); Boyce and DiPrima (1969); Coddington (1961); Birkhoff and Rota (1962); Carrier and Pearson (1968); Ince (1956); Kamke (1948; this text contains a dictionary of solutions).

1

THE DIFFUSION EQUATION

1.1 DERIVATION

One of the more common partial differential equations of practical interest is that governing diffusion in a homogeneous medium; it arises in many physical, biological, social, and other phenomena. A simple example of such an equation is

$$\phi_t = a^2 \phi_{xx} \tag{1.1}$$

Here x is position, t time, a a positive constant,† and we seek a function $\phi(x, t)$ satisfying this equation for a certain range of x and t values. In addition, ϕ is usually required to satisfy certain auxiliary conditions.

Much of our attention in this chapter will be directed toward Eq. (1.1)—the one-dimensional diffusion equation with constant coefficients. However, before considering properties of the equation itself, it seems worthwhile to derive it (with reasonable care) in at least one context in which it arises. We choose the problem of heat flow along a thin rod with insulated sides, since the associated physics is rudimentary.

Let the rod be oriented along the x-axis; denote its cross-sectional area by A, its density by ρ, its specific heat by c, and its thermal conductivity by k. We take the temperature ϕ (measured relative to some chosen reference level) as being a function of x and t only, i.e., ϕ has the same value at each point in any chosen cross section. To start with, we restrict ourselves

---

† We write the coefficient as a^2 for future convenience.

to the case in which each of A, ρ, c, and k is a constant (and so is independent of x, t, or ϕ).

Let us single out for consideration a portion of the rod lying between any two points α and β, with $\beta > \alpha$. From the definition of specific heat the rate at which thermal energy is accumulating within this portion of the rod is

$$R_1 = \int_\alpha^\beta \phi_t(x, t) c\rho A \, dx \tag{1.2}$$

However, heat is transported by diffusion in the direction of, and at a rate proportional to, the negative of the temperature gradient. Thus, the net rate R_2 at which heat enters the segment $\alpha < x < \beta$ is

$$R_2 = kA\phi_x(\beta, t) - kA\phi_x(\alpha, t) \tag{1.3}$$

Clearly, $R_1 = R_2$. Hence

$$\int_\alpha^\beta \phi_t(x, t) c\rho A \, dx = kA\phi_x(\beta, t) - kA\phi_x(\alpha, t)$$

$$= kA \int_\alpha^\beta \phi_{xx}(x, t) \, dx$$

It follows that

$$\int_\alpha^\beta [\phi_t c\rho A - kA\phi_{xx}] \, dx = 0 \tag{1.4}$$

But α and β were arbitrarily chosen positions along the rod. Equation (1.4) can hold for *any* choice of α and β only if the integrand is identically zero.†
This implies then that

$$\phi_t c\rho A - kA\phi_{xx} = 0$$

or

$$\phi_t = a^2\phi_{xx} \tag{1.5}$$

where $a^2 = k/(c\rho)$ is termed the *thermal diffusivity*.

---

† The basic theorem to which we appeal here is as follows. Let $\psi(x)$ be a continuous function of x satisfying the condition that $\int_\alpha^\beta \psi(x) \, dx = 0$ for all choices of α and β in some interval. Then $\psi(x) \equiv 0$ in that interval. For otherwise, $\psi(x)$ would be nonzero at some point x_0 in the interval, and in consequence of continuity, $\psi(x)$ would be nonzero and would have the same sign as $\psi(x_0)$ in some small subinterval around x_0; a choice of α and β within this subinterval would then lead to a nonzero value of the integral, which provides a contradiction.

To use this theorem, we require that the integrand of Eq. (1.4) be continuous, and for temperature this is a reasonable physical expectation.

An alternative derivation of Eq. (1.5) might proceed as follows. By use of the mean value theorem of the integral calculus, Eq. (1.2) can be re-written as

$$R_1 = \phi_t(\xi, t) c \rho A \cdot (\beta - \alpha)$$

where $\alpha < \xi < \beta$. Setting $R_1 = R_2$ and dividing by $(\beta - \alpha)$, we obtain

$$\phi_t(\xi, t) c \rho A = \frac{k A \phi_x(\beta, t) - k A \phi_x(\alpha, t)}{\beta - \alpha}$$

Holding α fixed and letting $\beta \to \alpha$ (and noting that then $\xi \to \alpha$ also), we again obtain Eq. (1.5)

These two derivations are, of course, closely related. In complicated multidimensional situations, the first method—in which a conservation law is applied to a finite portion of the material by use of definite integrals, and the result manipulated so as to imply the vanishing of an integrand—is frequently the more straightforward.

For copper at room temperature, approximate values for the thermal constants are $\rho \doteq 8.9 \, \text{g/cm}^3$, $c = 0.09 \, \text{cal/(g °C)}$, $k = 0.93 \, \text{cal/(cm sec °C)}$, so that $a^2 \cong 1.1 \, \text{cm}^2/\text{sec}$. For steel, rock, glass, and water, typical values for a^2 are 0.1, 0.01, 0.006, and 0.0015 cm^2/sec, respectively.

1.2 PROBLEMS

1.2.1 As a result of runoff from the surrounding land following a rain-storm, a large lake of uniform depth h receives a sudden influx of a soluble phosphate compound in its upper layers. The phosphate then diffuses downward into the rest of the lake. Assuming that the phosphate concen-tration c is a function of depth z (measured positively downward from the lake surface) and time t only, and that the rate at which phosphate diffuses across any horizontal cross section is proportional to $\partial c/\partial z$, show that c satisfies an equation of diffusion type. Check that the equation is dimen-sionally consistent. Explain why reasonable additional conditions to be satisfied by c are that $c_z = 0$ at $z = 0$ and at $z = h$.

1.2.2 Modify the derivation leading to Eq. (1.5) so that it is valid for diffusion in a nonhomogeneous medium for which c and k are functions of x and ϕ, and so that it is valid for a geometry in which A is a function of x. Show that Eq. (1.5) is now replaced by

$$c \rho A \phi_t = (k A \phi_x)_x$$

1.2.3 In the derivation of Eq. (1.5), no heat was permitted to enter or leave the sides of the rod. Remove this restriction (i.e., remove the insulation), but retain the assumption that the rod is sufficiently thin that ϕ is a function of x and t only. Let the heat loss per unit length of rod per unit of time be given by $\beta(\phi - \phi_0)$, where β and ϕ_0 are constants, ϕ_0 being the temperature of the environment. Also, let heat be generated inside the bar (perhaps via an electrical current or chemical reaction) at a rate of $h(x, t)$ heat units per unit volume per unit of time. With c, ρ, k, A constant, show that Eq. (1.5) is replaced by

$$c\rho\phi_t = k\phi_{xx} - \beta(\phi - \phi_0)/A + h \tag{1.6}$$

Next, let $h = 0$ and show that the change in variables $\phi - \phi_0 = \exp[-\beta t/(c\rho A)] \cdot \psi(x, t)$ reduces Eq. (1.6) to an equation for $\psi(x, t)$ that has the form of Eq. (1.5). Is a similar reduction feasible for the equation

$$\phi_t = a^2\phi_{xx} + \lambda_1\phi_x + \lambda_2\phi_t$$

where a^2, λ_1, and λ_2 are constants?

1.2.4 Let the temperature ϕ inside a solid sphere (or spherical annulus) be a function only of radial distance r from the center and of time t. Show that the equation corresponding to Eq. (1.5) is now

$$\phi_t = a^2(\phi_{rr} + 2\phi_r/r)$$

and that a transformation of the form $\phi = r^\alpha\psi$, for a suitable choice of the constant α, reduces this equation to one having the form of Eq. (1.5). Discuss also the corresponding problem of one-dimensional heat flow in a cylinder (consider here the transformation $\xi = \ln r$).

1.3 SIMPLE SOLUTIONS

The examples of the Introduction suggest that some kind of auxiliary data (such as the specification of solution values along some curve) must be given in order to complete a problem statement involving a partial differential equation. For an equation such as Eq. (1.1), we can examine one of the physical contexts in which it arises in order to see what kind of auxiliary conditions would be reasonable. In the case of temperature distribution along a thin rod of finite length l, we would expect the solution to be physically determined by a specification of the initial temperature distribution $\phi(x, 0)$ and by a specification of the two end temperatures $\phi(0, t)$ and $\phi(l, t)$ for all values of $t > 0$. Thus if Eq. (1.1) adequately represents the physical situation, the corresponding mathematical problem should be well posed—i.e., the combination of Eq. (1.1) together with the above initial and end conditions should enable us to determine the solution $\phi(x, t)$ for $0 < x < l$ and for $t > 0$ (and uniquely so).

Most of the problems we will discuss will include information concerning such initial or boundary conditions as part of the problem statement, and unless this is done the problem statement is not complete. However, our immediate purpose is to explore the properties of Eq. (1.1), and as a first step in this direction it is useful to try to construct simple solutions of the equation, without specifying beforehand what the initial or boundary conditions satisfied by these solutions will be.

A frequently useful technique is that of "separation of variables." We try

$$\phi(x, t) = X(x) \cdot T(t) \tag{1.7}$$

where $X(x)$ is a function of x alone, and $T(t)$ is a function of t alone, and ask whether functions $X(x)$ and $T(t)$ can be found so that Eq. (1.7) provides a solution of Eq. (1.1). Substitution of Eq. (1.7) into Eq. (1.1) yields

$$T'(t)/T(t) = a^2 X''(x)/X(x) \tag{1.8}$$

The left-hand side of Eq. (1.8) is a function of t alone, and the right-hand side is a function of x alone. However, the two sides of the equation are to be equal for *any* values of x and t in our range of interest. It follows that each side must be a constant—for otherwise, one of x or t can be held fixed and the other varied in order to obtain a contradiction to Eq. (1.8). Thus,

$$T'(t)/T(t) = C \quad \text{and} \quad X''(x)/X(x) = C/a^2 \tag{1.9}$$

Since Eq. (1.8) is equivalent to the condition that $\phi(x, t)$ as defined by Eq. (1.7) be a solution of Eq. (1.1), it follows that if we choose any value for the constant C and solve the two ordinary differential equations (1.9), then ϕ will indeed be a solution of Eq. (1.1).

At this point, we are free to choose C in any way we wish—real positive, real negative, purely imaginary, or complex†; different choices will lead to different solution functions ϕ. We try first $C = \lambda^2$, where λ is a real positive constant. Then Eqs. (1.7) and (1.9) lead to

$$\phi = \exp[\lambda^2 t + \lambda x/a] \quad \text{and} \quad \phi = \exp[\lambda^2 t - \lambda x/a] \tag{1.10}$$

as special solutions of Eq. (1.1) (any linear combination is of course also a solution). We note that each involves an exponential growth in time.

---

† If the constant C is complex, the resulting solution function ϕ will also be complex. However, each of the real and imaginary parts of ϕ will then also be a solution, and in physical problems requiring real-valued solutions, these real and imaginary parts will provide the solutions of primary interest.

Second, set $C = -\lambda^2$, again with the real constant $\lambda > 0$; then

$$\phi = \exp[-\lambda^2 t] \sin(\lambda x/a) \qquad \text{and} \qquad \phi = \exp[-\lambda^2 t] \cos(\lambda x/a) \qquad (1.11)$$

are solutions of Eq. (1.1). Each decays exponentially in time.

We can easily devise problems whose boundary and initial conditions are compatible with the functions ϕ given by Eqs. (1.10) or (1.11). For example, choosing $\lambda = n\pi a/l$ in the first of Eqs. (1.11), where n is a positive integer and l a positive constant, we see that

$$\phi = A \exp[-(n^2\pi^2 a^2/l^2)t] \sin(n\pi x/l) \qquad (1.12)$$

(with A constant) is a solution of the following problem: find $\phi(x, t)$ in the region $0 < x < l$, $t > 0$, such that $\phi(x, 0) = A \sin(n\pi x/l)$, $\phi(0, t) = \phi(l, t) = 0$, and such that Eq. (1.1) is satisfied. (As a result of a uniqueness theorem to be proved subsequently, we will be able to conclude that there is no other solution to this problem.)

Equation (1.12) is rather interesting. It represents a function that is always sinusoidal in x, i.e., that has a ripple in x. It decays at the same rate for all values of x and, moreover, decays more rapidly if n or a is increased or if l is decreased—and if we think again of the rod temperature problem that led to Eq. (1.5), this is in accordance with physical expectation.

Finally, let us set $C = i\omega$ in Eq. (1.9), where $\omega > 0$ and $i = \sqrt{-1}$. Then

$$\phi = \exp\left[i\omega t \pm \frac{1+i}{a\sqrt{2}} \sqrt{\omega x}\right]$$

(or, alternatively, its real part or its imaginary part) is a solution of Eq. (1.1), so that we can form a linear combination to obtain special solutions of the form

$$\phi = \exp(\sqrt{\omega x}/a\sqrt{2}) \cdot \cos[\omega t + (\sqrt{\omega x}/a\sqrt{2}) + \theta_1]$$

$$\phi = \exp(-\sqrt{\omega x}/a\sqrt{2}) \cdot \cos[\omega t - (\sqrt{\omega x}/a\sqrt{2}) + \theta_2] \qquad (1.13)$$

where θ_1 and θ_2 are arbitrary phase constants. In particular, the second of these, with $\theta_2 = 0$, would provide a solution of the following problem: find $\phi(x, t)$ in the region $0 < x < \infty$, $t > 0$, such that ϕ is periodic in time with angular frequency ω, $\phi(x, t) \to 0$ as $x \to \infty$, $\phi(0, t) = \cos \omega t$, and such that ϕ satisfies Eq. (1.1).

1.4 PROBLEMS

1.4.1 Investigate solutions of Eq. (1.1) obtained by setting $C = (\alpha + i\beta)^2$ in Eq. (1.9), where α and β are real.

1.4.2 Suppose that we want to use the function

$$\Phi(x, t) = \sum_{n=1}^{N} c_n \exp[-n^2\pi^2 a^2 t/l^2] \sin(n\pi x/l)$$

where N is a chosen integer and the c_n constants, to approximate the solution of the following problem: find $\phi(x, t)$ satisfying Eq. (1.1) in $0 < x < l$, $t > 0$, with $\phi(0, t) = \phi(l, t) = 0$ and $\phi(x, 0) = $ *a prescribed function* $g(x)$. What would be a good way to determine the constants c_n? If we permit $N \to \infty$, what feature of the series would appear to ensure convergence for $t > 0$? [*Hint:* consider $\int_0^l \{\phi(x, 0) - \Phi(x, 0)\}^2 dx$.]

1.4.3 Use the function $\phi(x, t)$ given in Eq. (1.13) to discuss the problem of the penetration into the earth's crust of the daily and seasonal heating cycle to which the sun subjects the surface of the earth. Make numerical estimates.

1.4.4 With $\xi = x^\alpha t$, find α such that, for some function f, $\phi = f(\xi)$ is a solution of Eq. (1.1). Find f.

1.4.5 Since $\phi = \exp[-\lambda^2 t] \cos(\lambda x/a)$, as given by Eq. (1.11), is a solution of Eq. (1.1) for any value of λ, it may be possible to construct other solutions by integration with respect to λ. Thus,

$$\psi(x, t) = \int_0^\infty f(\lambda) \cdot \exp[-\lambda^2 t] \cos(\lambda x/a) \, d\lambda \qquad (1.14)$$

could provide a new special solution of Eq. (1.1) for appropriate choices of the function $f(\lambda)$. Intuitively, the right-hand side of Eq. (1.14) can be thought of as a superposition of solutions of the form of Eq. (1.11) (in the sense that an integral is defined to be the limit of a certain sum). Set $f(\lambda) = 1$ and carry out the integration. [*Hint:* with $f = 1$, differentiation of Eq. (1.14) with respect to x, followed by an integration by parts, yields the ODE $\psi_x = -(x/2a^2 t)\psi$.] Verify that the function $\psi(x, t)$ so obtained does indeed satisfy Eq. (1.1). Plot $\psi(x, t)$ for various values of $t > 0$ for the region $-\infty < x < \infty$. For the case of temperature distribution along an infinitely long thin rod, what initial value problem would this function $\psi(x, t)$ solve?

1.5 SERIES SOLUTIONS

Problem 1.4.2 suggests the possibility of obtaining a solution for a suitable problem involving Eq. (1.1) by means of a series expansion of the form

$$\phi(x, t) = \sum_{n=1}^{\infty} \alpha_n(t)\beta_n(x) \qquad (1.15)$$

where $\alpha_n(t)$ and $\beta_n(x)$ are functions of t alone and of x alone, respectively. Instead of Eq. (1.1), we will consider the slightly more general equation

$$\phi_t = a^2\phi_{xx} + w(x, t) \tag{1.16}$$

where $w(x, t)$ is a specified function of x and t (in the problem of heat conduction along a rod, w represents internally generated heat; cf. Problem 1.2.3). To complete the problem statement, let

$$\phi(x, 0) = f(x) \quad \text{for} \quad 0 < x < l$$
$$\phi(0, t) = \phi(l, t) = 0 \tag{1.17}$$

where $f(x)$ is a specified function.

We hope to solve Eq. (1.16), subject to conditions (1.17), by means of an expansion of the form (1.15). Now, for any choice of t, the solution function $\phi(x, t)$ will be some function of x, and Eq. (1.15) represents an expansion of this function in terms of certain functions $\beta_n(x)$ [the coefficients of the expansion being the values of the functions $\alpha_n(t)$ for that particular choice of t]. In order that the right-hand side of Eq. (1.15) be capable of representing, within broad limits, any such function of x, it seems very reasonable to choose for the functions $\beta_n(x)$ a *complete set* over the interval $(0, l)$—i.e., a set of functions in terms of which any function of interest can be expanded. One way to do this is to choose the functions $\{\beta_n(x)\}$ to be the eigenfunctions of a suitable Sturm–Liouville problem.† A natural Sturm–Liouville problem is that which arises via the separation-of-variables technique as applied to our basic differential operator [cf. Eq.

---

† As discussed in texts on ordinary differential equations, a Sturm–Liouville problem has the form

$$(py')' + (q + \lambda r)y = 0 \quad \text{in} \quad a < x < b$$
$$c_1 y(a) + c_2 y'(a) = 0, \qquad c_3 y(b) + c_4 y'(b) = 0$$

where p, q, r are suitable functions of x, and where (c_i) are constants. Only for certain values of λ (termed eigenvalues) will such a problem have nontrivial solutions (termed eigenfunctions). Denoting eigenvalues by λ_j and corresponding eigenfunctions by $y_j(x)$, the expansion theorem states that, given any function $h(x)$ defined in (a, b) and satisfying certain mild conditions, then there exist constants (k_j) such that $h(x) = k_1 y_1(x) + k_2 y_2(x) + \cdots$. To be precise, the convergence may be only in the mean square sense—i.e., $\int_a^b [h(x) - \Sigma_1^N k_j y_j(x)]^2\, dx \to 0$ as $N \to \infty$. The k_i are easily determined by use of the orthogonality condition satisfied by the y_i, which requires that $\int_a^b r(x) y_i(x) y_j(x)\, dx = 0$ for $\lambda_i \neq \lambda_j$. Use of this result leads to

$$k_i = \left[\int_a^b rhy_i\, dx \right] \Big/ \left[\int_a^b ry_i^2\, dx \right]$$

For details, see for example Ince (1956, Section 11.5).

(1.9)]. Thus we choose the $\{\beta_n(x)\}$ to be the eigenfunctions of the problem

$$\beta'' + \lambda\beta = 0 \qquad \text{for} \quad 0 < x < l \tag{1.18}$$

$$\beta(0) = \beta(l) = 0. \tag{1.19}$$

At this point it is worth emphasizing that we could have chosen any eigenvalue-type equation in place of Eq. (1.18) and the existence of an expansion of the form (1.15) would still have been assured (in the usual mean square sense). However, the functions β_n would not have "fitted" Eq. (1.16) as well, and the manipulations that follow would have been less simple. Also, we could have chosen any homogeneous boundary conditions in place of those of Eq. (1.19), but the advantage of choosing these particular ones is that they happen to be the ones satisfied by the desired solution function $\phi(x, t)$, and we anticipate that this fact will "strengthen" the convergence of Eq. (1.15). [Otherwise, although Eq. (1.15) would still be valid in the mean square sense, we would not necessarily have equality between the two sides of the equation at the points $x = 0$ and $x = l$; such a nonuniformity of convergence could well lead to divergence when the series is differentiated—cf. Section 1.7.]

From Eqs. (1.18) and (1.19) we find

$$\beta_n(x) = \sin(n\pi x/l), \qquad n = 1, 2, 3, \ldots \tag{1.20}$$

so that expansion (1.15) becomes

$$\phi(x, t) = \sum_{n=1}^{\infty} \alpha_n(t) \sin(n\pi x/l) \tag{1.21}$$

(which happens in this special case to have the form of a conventional Fourier series).

Substitution of Eq. (1.21) into (1.16) leads to

$$\alpha_n'(t) = -a^2(n^2\pi^2/l^2)\alpha_n(t) + w_n(t) \tag{1.22}$$

where $w_n(t)$ is the coefficient of the expansion of $w(x, t)$ in terms of the $\beta_n(x)$; i.e.,

$$w(x, t) = \sum_{n=1}^{\infty} w_n(t) \sin(n\pi x/l)$$

The solution of Eq. (1.22) is

$$\alpha_n(t) = \exp\left(-\frac{n^2\pi^2 a^2}{l^2} t\right)\left[\int_0^t \exp\left(\frac{n^2\pi^2 a^2}{l^2} \tau\right) w_n(\tau) \, d\tau + c_n\right]$$

To determine the constants of integration c_n, we now use the fact that

$\phi(x, 0) = f(x)$; i.e.,

$$\sum_{n=1}^{\infty} \alpha_n(0) \sin(n\pi x/l) = f(x)$$

so that the $c_n = \alpha_n(0)$ are simply the expansion coefficients of the given intitial temperature distribution $f(x)$. Thus, the $\alpha_n(t)$ are completely determined, and Eq. (1.21) now gives us the desired expansion for $\phi(x, t)$.

In this derivation we have implicitly assumed that series (1.21) converges even after the differentiations to which we subjected it. This can be justified on an a posteriori basis by examining the uniformity of convergence of the differentiated series made up of the actual functions $\alpha_n(t)$ obtained above; here and in future we will leave such formal justifications to the interested reader. We will, however, discuss further the matter of convergence of a differentiated series in Section 1.7.

1.6   PROBLEMS

1.6.1   In the problem solved by Eq. (1.21), let $w(x, t) = 0$ and $f(x) = 1$.†
Choose various numerical values for a^2, x, l, and t, and decide how many terms of the series expansion are necessary to represent $\phi(x, t)$ adequately. Consider also the case in which, for $0 < \varepsilon \ll l$,

$$f(x) = \begin{cases} 0 & \text{for} \quad |x - l/2| > \varepsilon/2 \\ 1/\varepsilon & \text{for} \quad |x - l/2| < \varepsilon/2 \end{cases}$$

1.6.2   (a)   For $0 < x < l$, solve the problem

$$\phi_t = a^2\phi_{xx} + w(x, t)$$

$$\phi(0, t) = 0, \qquad \phi_x(l, t) = 0, \qquad \phi(x, 0) = f(x)$$

by means of a series expansion involving the eigenfunctions of $\beta'' + \lambda\beta = 0$, $\beta(0) = \beta'(l) = 0$. As before, w and f are prescribed functions.

(b)   If the end conditions are altered to read

$$\phi(0, t) = 0, \qquad \phi(l, t) + c\phi_x(l, t) = 0$$

---

† Notice that there is a discontinuity between the end conditions $\phi(0, t) = \phi(l, t) = 0$ and the initial condition $f(x) = 1$, at the points $x = 0$, $t = 0$ and $x = l$, $t = 0$. To be meticulous, we should phrase the boundary and initial conditions for this case as

$$\phi(0, t) = \phi(l, t) = 0 \qquad \text{for} \quad t > 0$$

$$\lim_{t \to 0} \phi(x, t) = 1 \qquad \text{for} \quad 0 < x < l$$

where $c > 0$ is a constant, find a set of appropriate eigenfunctions and obtain a series solution to the problem.

1.7 NONHOMOGENEOUS END CONDITIONS

The boundary conditions associated with the problem of Eq. (1.16) were $\phi(0, t) = \phi(l, t) = 0$, and the eigenfunctions $\beta_n(x)$ satisfied similar conditions; viz., $\beta(0) = \beta(l) = 0$. Suppose, however, that the problem to be solved is

$$\phi_t = a^2\phi_{xx} + w(x, t)$$

$$\phi(x, 0) = f(x), \qquad \phi(0, t) = g(t), \qquad \phi(l, t) = h(t)$$

(1.23)

where f, g, h are given functions. In attempting a series expansion of the form

$$\phi(x, t) = \sum_{n=1}^{\infty} a_n(t)\beta_n(x)$$

(1.24)

we could try to use the same set of $\beta_n(x)$ functions as before, i.e., $\beta_n(x) = \sin(n\pi x/l)$. Sturm–Liouville theory guarantees that there exists a representation of the function we seek in the form of Eq. (1.24), despite the fact that it would sum to zero at $x = 0$ or at $x = L$. If, however, we substitute Eq. (1.24) into the differential equation (1.23) and proceed as before, we are led to exactly the same functions $a_n(t)$ as were obtained in Section 1.5, and this is not surprising since in this procedure we have *nowhere* an opportunity to use the boundary constraints $\phi(0, t) = g$, $\phi(l, t) = h$. The conclusion, of course, is that term-by-term differentiation of Eq. (1.24) is not permissible in this case.

A simple device will permit us to evade this difficulty. Define an auxiliary function $\psi(x, t)$ via

$$\psi(x, t) = \phi(x, t) - [g(t) + (x/l)\{h(t) - g(t)\}]$$

(1.25)

so that $\psi(x, t)$ satisfies the equation

$$\psi_t = a^2\psi_{xx} + w_1(x, t)$$

and the conditions

$$\psi(0, t) = \psi(l, t) = 0, \qquad \psi(x, 0) = f_1(x)$$

where

$$w_1(x, t) = w(x, t) - g'(t) - (x/l)\{h'(t) - g'(t)\}$$

$$f_1(x) = f(x) - g(0) - (x/l)\{h(0) - g(0)\}$$

Thus, $\psi(x, t)$ satisfies exactly the same kind of conditions as those satisfied by the function $\phi(x, t)$ in Section 1.5, and so may be expanded without difficulty in a series of the form of the right-hand side of Eq. (1.24). Equation (1.25) then yields the desired representation for $\phi(x, t)$.

Let us summarize for the general case. If the end conditions in Eq. (1.23) are homogeneous, say

$$c_1\phi(0, t) + c_2\phi_x(0, t) = 0, \qquad c_3\phi(l, t) + c_4\phi_x(l, t) = 0$$

where the c_j are constants, then we choose similar boundary conditions for the $\beta_n(x)$; i.e.,

$$c_1\beta(0) + c_2\beta'(0) = 0, \qquad c_3\beta(l) + c_4\beta'(l) = 0$$

If the boundary conditions for $\phi(x, t)$ are not homogeneous, then we introduce an auxiliary function $\psi(x, t)$ that does satisfy homogeneous boundary conditions and carry out an expansion for the $\psi(x, t)$ function.

1.8 PROBLEMS

1.8.1   Use a series expansion technique to solve the problem

$$\phi_t = a^2\phi_{xx} + 1 \qquad \text{for} \quad t > 0, \quad 0 < x < l,$$

$$\phi(x, 0) = 0, \qquad \phi(0, t) = t, \qquad \phi_x(l, t) = -c\phi(l, t)$$

where $c > 0$ is a constant.

1.8.2   A rod occupies the portion $1 < x < 2$ of the x-axis. The thermal conductivity depends upon x in such a manner that the temperature $\phi(x, t)$ satisfies the equation

$$\phi_t = A^2(x^2\phi_x)_x$$

where A is a constant. For $\phi(1, t) = \phi(2, t) = 0$ for $t > 0$, with $\phi(x, 0) = f(x)$ for $1 < x < 2$, show that appropriate eigenfunctions $\beta_n(x)$ are

$$\beta_n(x) = \frac{1}{\sqrt{x}}\sin\left(\frac{\pi n \ln x}{\ln 2}\right)$$

and work out the series expansion solution of this problem.

1.8.3   Show that the following technique enables us to obtain an expansion (1.24) for the solution of the problem (1.23), without the necessity of introducing an auxiliary function ψ. Choosing $\beta_n = \sin(n\pi x/l)$, Eq. (1.24) becomes

$$\phi(x, t) = \sum_{n=1}^{\infty} \alpha_n(t) \sin(n\pi x/l)$$

Multiplication of each side by $\sin(n\pi x/l)$ and integration between 0 and l shows that

$$\alpha_n(t) = 2/l \int_0^l \phi(x, t) \sin(n\pi x/l) \, dx \qquad (1.26)$$

Now multiply each side of the equation

$$\phi_t = a^2\phi_{xx} + w(x, t)$$

by $\sin(n\pi x/l)$ and integrate between 0 and l (using integration by parts if needed) to obtain a differential equation for $\alpha_n(t)$ [involving $g(t)$ and $h(t)$], and solve this problem. Note that this method does not require term-by-term differentiation of a series. Can you generalize it to problems involving more complicated functions $\beta_n(x)$? Because of the form of Eq. (1.26), the method is often referred to as utilizing a "finite transform." Do the coefficients of the series obtained here decrease more or less rapidly than those of the function ψ in Eq. (1.25)?

1.8.4 The temperature ϕ inside a long, thick-walled hollow cylinder depends only on r and t, where r is the radial distance from the axis. Use a series solution to determine $\phi(r, t)$ satisfying the equation

$$\phi_t = a^2[\phi_{rr} + (1/r)\phi_r]$$

for $\phi(r, 0) = f(r)$ in $\alpha < r < \beta$. Let $\phi_r(\alpha, t) = \phi(\beta, t) = 0$ for $t > 0$. What happens as $\alpha \to 0$?

1.9 THE MAXIMUM PRINCIPLE

Let the temperature $\phi(x, t)$ in a finite bar $(0 < x < l)$ satisfy the equation $\phi_t = a^2\phi_{xx}$, where $\phi(x, 0)$ for $0 < x < l$, $\phi(0, t)$ for $t > 0$, and $\phi(l, t)$ for $t > 0$ are all given. Consider the region $0 < x < l$, $0 < t < T$ of the (x, t) plane, where T is some chosen time. Then our region of interest is the shaded region of Fig. 1.1, where $\phi(x, t)$ is given on the darker horizontal and vertical lines shown in the figure; this darker portion of the boundary we will denote by Γ. Because the direction of heat flow between two bodies (or parts of the same body) in contact is always from a higher to a lower temperature, we expect that the value of $\phi(x, t)$ in the shaded region should attain its maximum on Γ; our purpose is to prove this mathematically.

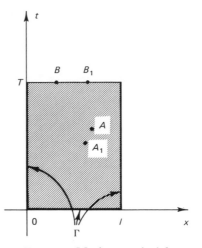

Fig. 1.1 Maximum principle.

Suppose, on the contrary, that we can find a point (x_0, t_0)—either an interior point A or an upper boundary point B as shown in the figure—such that $\phi(x_0, t_0)$ is greater than the least upper bound of $\phi(x, t)$ on Γ. Define an auxiliary function $\psi(x, t)$ via

$$\psi(x, t) = \phi(x, t) - \varepsilon(t - t_0)$$

where $\varepsilon > 0$ is a constant. Since $\psi(x_0, t_0) = \phi(x_0, t_0)$, which exceeds by some definite amount the greatest value of $\phi(x, t)$ on Γ, it follows that we can choose ε so small that $\psi(x_0, t_0)$ is greater than the maximum value attained by $\psi(x, t)$ on Γ. Thus, $\psi(x, t)$ attains its maximum at some point not on Γ—say A_1 or B_1. At this maximum point we have the conditions

$$\psi_{xx} \le 0, \qquad \psi_t \ge 0$$

which imply

$$\phi_{xx} \le 0, \qquad \phi_t > 0$$

But this result contradicts the requirement that $\phi_t = a^2 \phi_{xx}$ at this point; hence, the maximum principle is proved.

1.10 PROBLEMS

1.10.1   (a)   Show that $\phi(x, t)$ also attains its minimum on the boundary Γ of the shaded region in Fig. 1.1. Show also that if $\phi(x, t) = 0$ on Γ, then $\phi(x, t) = 0$ throughout the shaded region.

(b) By considering the difference between any two solutions of a problem in which $\phi(x, t)$ is given on Γ, show that the solution to such a problem is unique. Does it matter whether or not the given values of $\phi(x, t)$ on Γ are continuous at the point $(0, 0)$?

1.10.2 Multiply each side of $\phi_t = a^2\phi_{xx}$ by ϕ and integrate with respect to x from 0 to l to show that

$$\frac{\partial}{\partial t}\left(\frac{1}{2}\int_0^l \phi^2 \, dx\right) = a^2[\phi\phi_x]_0^l - a^2\int_0^l (\phi_x)^2 \, dx$$

Use this result to discuss uniqueness—for what kinds of initial or boundary value problems?

2

LAPLACE TRANSFORM METHODS

A suitable ordinary differential equation can be transformed into an equivalent algebraic equation by use of the Laplace transform method. In an analogous way, it may be possible to use Laplace transforms to reduce a partial differential equation to an ordinary one, or in any event, to reduce the number of independent variables involved in the differentiation. In this chapter we will consider some examples arising from the one-dimensional diffusion equation of Chapter 1.

We assume the reader to be familiar with the basic properties of Laplace transforms.† Capital letters will be used to denote transforms of functions represented by corresponding lowercase letters; thus, if $f(t)$ is defined for $t > 0$,

$$F(s) = \int_0^\infty e^{-st} f(t) \, dt \tag{2.1}$$

represents the Laplace transform of $f(t)$ for those values of s for which the integral exists.

In dealing with the diffusion equation, one family of transform pairs is often encountered, and for convenience we list three members of this family ($\lambda > 0$ is a constant, and $s > 0$):

$$f(t) = (1/\sqrt{t}) \exp(-\lambda^2/4t), \quad F(s) = (\pi/s)^{1/2} \exp(-\lambda\sqrt{s})$$
$$f(t) = (\lambda/2t^{3/2}) \exp(-\lambda^2/4t), \quad F(s) = \sqrt{\pi} \exp(-\lambda\sqrt{s}) \tag{2.2}$$
$$f(t) = \operatorname{erfc}(\lambda/2\sqrt{t}), \quad F(s) = (1/s) \exp(-\lambda\sqrt{s})$$

† As presented for example by Kreyszig (1967, Chap. 4).

23

The complementary error function appearing in the last of these is defined by

$$\text{erfc}(x) \ = \ 2/\sqrt{\pi} \int_{x}^{\infty} \exp(-\alpha^2) \, d\alpha \qquad (2.3)$$

2.1 INTRODUCTORY EXAMPLE

Consider the problem of determining $\phi(x, t)$ in the region $0 < x < \infty$, $0 < t < \infty$, such that

$$\phi_t = a^2 \phi_{xx}$$

$$\phi(x, 0) = 0, \qquad \text{for all} \quad x > 0 \qquad (2.4)$$

$$\phi(0, t) = 1, \qquad \text{for all} \quad t > 0$$

If we take the Laplace transform with respect to time of this equation (i.e., multiply each side by e^{-st} and integrate from $t = 0$ to $t = \infty$), we obtain

$$-\phi(x, 0) + s\Phi(x, s) = a^2 \Phi_{xx} \qquad (2.5)$$

where

$$\Phi(x, s) \ = \ \int_{0}^{\infty} e^{-st} \phi(x, t) \, dt$$

and where we assume $\phi(x, t)$ to be sufficiently well behaved that the processes of integration with respect to t and differentiation with respect to x are interchangeable, so that

$$\int_{0}^{\infty} e^{-st} \phi_{xx}(x, t) \, dt = \left(\int_{0}^{\infty} e^{-st} \phi(x, t) \, dt \right)_{xx}$$

Since $\phi(x, 0) = 0$ in this case, the solution of the ordinary differential equation (2.5) is

$$\Phi = A \, \exp(-\sqrt{s}x/a) + B \, \exp(\sqrt{s}x/a)$$

where A and B are as yet undetermined constants (i.e., constants insofar as the variable x is concerned; they may, and in fact will, depend on the parameter s).

For large x, we anticipate that $\phi(x, t)$ will be a bounded function of t, so that its Laplace transform must approach zero as $s \to \infty$. Thus, the coefficient B must be set equal to zero. To determine A, we note that

$\phi(0, t) = 1$, so that $\Phi(0, s) = 1/s$; consequently,

$$\Phi(x, s) = (1/s) \exp(-\sqrt{sx}/a) \qquad (2.6)$$

Consulting a table of transforms†—here one of the pairs (2.2)—we can invert both sides of Eq. (2.6) to obtain

$$\phi(x, t) = \text{erfc}(x/2a\sqrt{t}) \qquad (2.7)$$

as the solution to the problem of Eq. (2.4). The reader should verify that this expression does indeed satisfy the partial differential equation (2.4) and that the general behavior of the erfc function for small or large values of x or t is in accordance with one's physical expectation for the solution of the problem.

An interesting feature of Eq. (2.7) is that, no matter how large a value we assign to x, $\phi(x, t)$ will be nonzero for *all* values of $t > 0$; thus the "signal speed" is infinite. Of course, as a practical matter, the erfc function decays so rapidly as its argument becomes large that $\phi(x, t)$ as given by Eq. (2.7) is exceedingly small for large x and small t. This is evident from the asymptotic expression for the function:

$$\text{erfc}(z) \sim \frac{1}{\sqrt{\pi}} \exp(-z^2) \left[\frac{1}{z} - \frac{1}{2z^3} + \frac{3}{4z^5} - \cdots \right] \qquad \text{as} \quad z \to +\infty$$

2.2 PROBLEMS

2.2.1 (a) In the problem statement (2.4), alter the condition $\phi(0, t) = 1$ to read $\phi(0, t) = f(t)$ for all $t > 0$, where $f(t)$ is a given function. Obtain the formula for $\Phi(x, s)$ that replaces Eq. (2.6) and use the convolution theorem of Laplace transform theory to show that

$$\phi(x, t) = \frac{x}{2a\sqrt{\pi}} \int_0^t f(t - \tau) \frac{1}{\tau^{3/2}} \exp\left(-\frac{x^2}{4a^2\tau}\right) d\tau \qquad (2.8)$$

(b) Discuss the special case of Eq. (2.8) obtained by setting $f(t) = 1$, and also that in which $f(t) = 1$ for $0 < t < T$, with $f(t) = 0$ for $t > T$. Here T is some positive constant.

2.2.2 Let $\phi(x, t)$ satisfy the equation $\phi_t = a^2\phi_{xx}$ for $-\infty < x < \infty$,

---

† A useful book is "Tables of Integral Transforms" (Erdélyi, 1954, Vol. 1). A shorter table will be found in "Handbook of Mathematical Functions" (Nat. Bur. Std., 1964).

$t > 0$, with $\phi(x, 0) = f(x)$ in $-\infty < x < \infty$. Show† that the Laplace transform of $\phi(x, t)$ is given by

$$\Phi(x, s) = \frac{1}{2a\sqrt{s}} \int_{-\infty}^{\infty} f(\xi) \exp\left(-\frac{\sqrt{s}}{a}\,|x - \xi|\right) d\xi$$

and hence deduce that

$$\phi(x, t) = \frac{1}{2a(\pi t)^{1/2}} \int_{-\infty}^{\infty} f(\xi) \exp\left[-\frac{(x - \xi)^2}{4a^2 t}\right] d\xi$$

2.2.3   In Problem 2.2.2, let $f(-x) = -f(x)$ for all x, and verify that $\phi(x, t)$ so obtained is the solution, for $x > 0$, of the following problem: find $\phi(x, t)$ satisfying $\phi_t = a^2 \phi_{xx}$ in $0 < x < \infty$, $0 < t$, with $\phi(0, t) = 0$ and $\phi(x, 0) = f(x)$. This technique, in which the solution for a semi-infinite interval is obtained from that for an infinite interval, is an example of the *method of images*.

As a second example of the method of images, set $f(x) = 0$ for $x > 0$, $f(x) = 2$ for $x < 0$, in the result of Problem 2.2.2, so as to re-derive the solution to the problem of Eq. (2.4).

2.2.4   Show that if $f(x) = 1$ for $|x| < \beta$, $f(x) = 0$ for $|x| > \beta$, where β is a given constant, then Problem 2.2.2 implies

$$\phi(x, t) = \frac{1}{2}\left[\operatorname{erf}\left(\frac{\beta - x}{2a\sqrt{t}}\right) + \operatorname{erf}\left(\frac{\beta + x}{2a\sqrt{t}}\right)\right]$$

2.2.5   Use the method of images (cf. Problem 2.2.3) to solve for $\phi(x, t)$ in $0 < x < L$, satisfying $\phi_t = a^2 \phi_{xx}$, $\phi(x, 0) = f(x)$ (prescribed) for $0 < x < L$, $\phi(0, t) = \phi_x(L, t) = 0$ for $t > 0$.

2.2.6   In the problem of Eq. (2.4), the range of x is from 0 to ∞, and the Laplace transform of ϕ with respect to x should therefore also exist. Why would a transform of Eq. (2.4) with respect to x be less useful for the problem as posed?

2.2.7   At time $t = 0$, an initially very thick snowfield starts accreting at

† *Hint:* write the solution of Eq. (2.5) as

$$\Phi = \left\{B - (1/2a\sqrt{s}) \int_0^x \exp(-\sqrt{s}\xi/a) f(\xi)\, d\xi\right\} \exp(\sqrt{s}x/a)$$

$$+ \left\{A + (1/2a\sqrt{s}) \int_0^x \exp(\sqrt{s}\xi/a) f(\xi)\, d\xi\right\} \exp(-\sqrt{s}x/a)$$

and observe that the first bracketed expression must approach zero as $x \to \infty$ to avoid uncontrolled growth in $\Phi(x, s)$. This condition determines B; A may be determined similarly by considering $x \to -\infty$.

a constant rate V. The initial temperature T_0 was constant throughout the snowfield; the temperature of the accreting snow is T_1 (also a constant, different from T_0). Measuring x downward from the moving surface, show that the temperature $\phi(x, t)$ satisfies the equation $\phi_t + V\phi_x = a^2\phi_{xx}$, and solve for ϕ by use of Laplace transforms. Take the initial thickness of the snowfield as infinite. (This kind of problem also arises in situations involving erosion or sublimation, or in the burning of a cigarette.)

2.2.8 In an ideal heat exchanger, two fluids flow parallel to one another in opposite directions; they are separated by a thin wall across which the rate of heat transfer is proportional to the temperature difference between the two fluids. Neglecting diffusion in the axial (flow) direction, formulate and solve an appropriate transient problem.

2.3 A FINITE INTERVAL PROBLEM

Let it be required to find $\phi(x, t)$ in $-L < x < L$, for $t > 0$, such that

$$\phi_t = a^2\phi_{xx}$$

$$\phi(x, 0) = 1 \qquad \text{for} \quad -L < x < L \qquad (2.9)$$

$$\phi(-L, t) = \phi(L, t) = 0 \qquad \text{for} \quad t > 0$$

Use of a Laplace transform in t yields

$$-1 + s\Phi(x, s) = a^2\phi_{xx}(x, s)$$

whence

$$\Phi(x, s) = 1/s + A \cosh(\sqrt{s}x/a) + B \sinh(\sqrt{s}x/a)$$

Use of the end conditions implies that $\Phi(-L, s) = \Phi(L, s) = 0$, so that A and B can be determined; the final result is

$$\Phi(x, s) = \frac{1}{s} - \frac{\cosh(\sqrt{s}x/a)}{s \cosh(\sqrt{s}L/a)} \qquad (2.10)$$

(As a check on the calculation, note the symmetry with respect to x.)

We have thus reduced the problem to one of inverting the right-hand side of Eq. (2.10). In principle, any such Laplace transform expression can be inverted by a formal process involving contour integration in the complex s plane. However, in this book we will restrict ourselves to methods that do not make use of this general inversion integral (which might well require numerical integration in any event).† In the present case we

---

† A discussion will be found in Carrier *et al.* (1966, Chapter 7).

expand in series to obtain

$$\frac{\cosh(\sqrt{s}x/a)}{s\cosh(\sqrt{s}L/a)} = \frac{\exp(\sqrt{s}x/a) + \exp(-\sqrt{s}x/a)}{s\exp(\sqrt{s}L/a)[1 + \exp(-2\sqrt{s}L/a)]}$$

$$= \frac{1}{s}\exp\left(-\frac{\sqrt{s}}{a}L\right)\left[\exp\left(\frac{\sqrt{s}}{a}x\right) + \exp\left(-\frac{\sqrt{s}}{a}x\right)\right]$$

$$\times\left[1 - \exp\left(-2\frac{\sqrt{s}}{a}L\right) + \exp\left(-4\frac{\sqrt{s}}{a}L\right)\right.$$

$$\left. - \exp\left(-6\frac{\sqrt{s}}{a}L\right) + \cdots\right]$$

$$= \frac{1}{s}\left\{\exp\left[-\frac{\sqrt{s}}{a}(L-x)\right] - \exp\left[-\frac{\sqrt{s}}{a}(3L-x)\right]\right.$$

$$\left. + \exp\left[-\frac{\sqrt{s}}{a}(5L-x)\right] - \cdots\right\}$$

$$+ \frac{1}{s}\left\{\exp\left[-\frac{\sqrt{s}}{a}(L+x)\right] - \exp\left[-\frac{\sqrt{s}}{a}(3L+x)\right]\right.$$

$$\left. + \exp\left[-\frac{\sqrt{s}}{a}(5L+x)\right] - \cdots\right\} \qquad (2.11)$$

Using this result in Eq. (2.10) and inverting term by term, we obtain

$$\phi(x,t) = 1 - \left\{\text{erfc}\left(\frac{L-x}{2a\sqrt{t}}\right) - \text{erfc}\left(\frac{3L-x}{2a\sqrt{t}}\right) + \text{erfc}\left(\frac{5L-x}{2a\sqrt{t}}\right) - \cdots\right\}$$

$$- \left\{\text{erfc}\left(\frac{L+x}{2a\sqrt{t}}\right) - \text{erfc}\left(\frac{3L+x}{2a\sqrt{t}}\right) + \text{erfc}\left(\frac{5L+x}{2a\sqrt{t}}\right) - \cdots\right\}$$

$$(2.12)$$

We have previously obtained a solution for this kind of problem in the form of a product series. With the appropriate modification in Eq. (1.21), this series solution becomes, for the present problem,

$$\phi(x,t) = \sum_{n=0}^{\infty}\frac{4(-1)^n}{(2n+1)\pi}\exp\left(-\frac{(2n+1)^2\pi^2a^2}{4L^2}t\right)\cos\left(\frac{(2n+1)\pi x}{2L}\right)$$

$$(2.13)$$

We conclude that the series of Eqs. (2.12) and (2.13) are equal to one another; they provide two representations of the same function. The reader should verify that the series (2.13) is particularly useful for *large* values of t, whereas the series (2.12) is particularly useful for *small* values of t.

2.4 PROBLEMS

2.4.1 Obtain result (2.12) by use of the method of images (cf. Problem 2.2.3).

2.4.2 Let $\phi(x, t)$ satisfy the equation $\phi_t = a^2\phi_{xx} + h(x, t)$ for $0 < x < l$, $t > 0$, with

$$\phi(x, 0) = f(x) \qquad \text{for} \quad 0 < x < l$$

$$\phi(0, t) = g_1(t), \qquad \phi(l, t) = g_2(t) \qquad \text{for} \quad t > 0.$$

Use Laplace transforms to show that

$$\Phi(x, s) = [\sinh(\sqrt{s}x/a)/\sinh(\sqrt{s}l/a)][G_2(s) - G_1(s)\cosh(\sqrt{s}l/a)$$

$$+ (1/a\sqrt{s}) \int_0^l \{f(\xi) + H(\xi, s)\} \sinh(\sqrt{s}(l - \xi)/a) \, d\xi]$$

$$+ G_1(s)\cosh(\sqrt{s}x/a) - (1/a\sqrt{s}) \int_0^x \{f(\xi) + H(\xi, s)\}$$

$$\times \sinh(\sqrt{s}(x - \xi)/a) \, d\xi \tag{2.14}$$

In what sense does Eq. (2.14) represent a superposition of effects? How could the convolution theorem now be used to simplify the inversion process?

2.4.3 Let $\phi(x, t)$ satisfy the equation $\phi_t = a^2\phi_{xx}$ for $0 < x < l$, $t > 0$, with $\phi(x, 0) = 0$ for $0 < x < l$, with $\phi(0, t) = 0$ for $t > 0$, and with $\phi(l, t) + \phi_x(l, t) = 1$ for $t > 0$. Obtain two series solutions, one useful for large t and one useful for small t.

2.4.4 Solve Problems 1.8.2 and 1.8.4 with $f(x) = 1$ and $f(r) = 1$, respectively, by a Laplace transform method. Determine the character of the solutions for small t.

2.4.5 A heated slab of uniform thickness is suddenly cooled by placing it between two infinitely thick cold slabs. Determine, for small values of t, the temperature profile in the hot slab. Can you think of a geophysical problem analogous to this situation?

2.4.6   In some of the foregoing problems, use has been made of the fact that the larges-s behavior of $F(s)$ is determined† by the small-t behavior of $f(t)$, and vice versa. As a further example, use Laplace transforms to show that if $\phi(r, t)$ satisfies

$$\phi_t = a^2(\phi_{rr} + (1/r)\phi_r) \qquad \text{in} \quad R < r < \infty, \quad t > 0$$

$$\phi(r, 0) = 0, \qquad \phi(R, t) = 1$$

then

$$\phi(r, s) = K_0(r\sqrt{s}/a)/sK_0(R\sqrt{s}/a)$$

and use the asymptotic expansion for the modified Bessel function (for large argument) to determine the behavior of $\phi(r, t)$ for small values of t.

2.4.7   For how large a range of values of x and t do the first two terms of Eq. (2.12), or of Eq. (2.13), provide a useful description of the solution of the problem of Section 2.3?

2.5 DELTA FUNCTION

In Problem 2.2.2 the solution of the problem $\phi_t = a^2\phi_{xx}$, $\phi(x, 0) = f(x)$, for $-\infty < x < \infty$, $t > 0$, was found to be

$$\phi(x, t) = \int_{-\infty}^{\infty} f(\xi)g(x - \xi, t) \, d\xi \tag{2.15}$$

where

$$g(x, t) = \frac{1}{2a(\pi t)^{1/2}} \exp\left(-\frac{x^2}{4a^2t}\right) \tag{2.16}$$

By the definition of an integral, Eq. (2.15) can be replaced by a sum over $\Delta\xi$ intervals:

$$\phi(x, t) = \lim_{\Delta\xi \to 0} \sum_{j=-\infty}^{\infty} f(\xi_j)g(x - \xi_j, t) \, \Delta\xi$$

where, say, $\xi_j = j \, \Delta\xi$. In terms of the temperature distribution problem of Section 1.1, it is clear that the term $f(\xi_j) \, \Delta\xi \, g(x - \xi_j, t)$ can be interpreted as providing the effect at any point x and time t of an initial temperature distribution that is zero everywhere except in an interval $\Delta\xi$

---

† This follows directly from the definition (2.1), for as s increases, that neighborhood of the origin in which the factor e^{-st} is of significant size decreases. Thus, for large s, only the behavior of $f(t)$ near the origin can appreciably affect $F(s)$.

about the point ξ_j, where it has the value $f(\xi_j)$. The heat content of this $\Delta\xi$ portion of the bar is proportional to $f(\xi_j)\ \Delta\xi$; as $\Delta\xi \to 0$, we approach closer and closer to a situation in which a heat "impulse" of this amount at the point ξ_j is released at time zero.

Thus the function $g(x, t)$ in Eq. (2.16) represents the temperature distribution at time t resulting from the release of an initially concentrated unit thermal impulse at the origin. The reader can easily verify two results that are compatible with this interpretation:

$$\int_{-\infty}^{\infty} g(x, t)\ dx = 1, \qquad t > 0$$

$$\lim_{t \to 0} g(x, t) = \begin{cases} 0, & x \neq 0 \\ \infty, & x = 0 \end{cases}$$

A plot of values of $g(x, t)$ for various values of t would look something like Fig. 2.1.

The limiting value of a function like $g(x, t)$, as $t \to 0$, is often written symbolically as

$$\lim_{t \to 0} g(x, t) \stackrel{?}{=} \delta(x)$$

and is referred to as the "delta function." However, $\delta(x)$ is not a conventional mathematical function (this is why we write $\stackrel{?}{=}$ rather than $=$), and to avoid difficulties, it must always be interpreted as a symbolic representation for a limiting process. Thus when we write

$$\int_{-\infty}^{\infty} w(x)\ \delta(x)\ dx = w(0) \qquad (2.17)$$

[where $w(x)$ is some given function], what we really mean is something

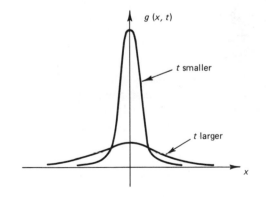

Fig. 2.1 Plot of g versus x.

like

$$\lim_{\beta \to 0} \int_{-\infty}^{\infty} w(x) \left\{ \frac{1}{2(\pi\beta)^{1/2}} \exp\left(-\frac{x^2}{4\beta}\right) \right\} dx = w(0) \qquad (2.18)$$

The truth of Eq. (2.18) follows from the properties of $g(x, t)$ (with $a^2t = \beta$) as described above and as depicted in Fig. 2.1. More generally, Eq. (2.17) can be replaced by

$$\int_a^b w(x) \, \delta(x - x_0) \, dx = w(x_0) \qquad (2.19)$$

where x_0 is any chosen point satisfying $a < x_0 < b$.

The function

$$\frac{1}{2(\pi\beta)^{1/2}} \exp\left(-\frac{x^2}{4\beta}\right)$$

(with $\beta > 0$) is not the only function whose approach to the limit as $\beta \to 0$ serves as a convenient replacement for the symbol $\delta(x)$. The reader should verify that another such function is

$$\delta(x) \stackrel{?}{=} \lim_{\beta \to 0} \frac{\beta}{\pi(x^2 + \beta^2)} \qquad (2.20)$$

The important thing is that the delta function is never used in all of its singular glory; whenever it occurs, we replace it by a function like

$$\frac{1}{2(\pi\beta)^{1/2}} \exp\left(-\frac{x^2}{4\beta}\right) \quad \text{or} \quad \frac{\beta}{\pi(x^2 + \beta^2)} \qquad (2.21)$$

with $\beta > 0$ small but nonzero, carry out (at least conceptually) whatever integration or other process that is to be applied, and only then permit $\beta \to 0$.

It is in this sense that we write, for $t_0 > 0$,

$$\int_0^{\infty} e^{-st} \delta(t - t_0) = \exp(-st_0)$$

and thus conclude that the Laplace transform of $\delta(t - t_0)$ is $\exp(-st_0)$ for $t_0 > 0$. When using Laplace transforms, it is conventional to interpret $\delta(t)$ as $\delta(t - \varepsilon)$, where $\varepsilon > 0$ is arbitrarily small; this avoids the uncertainty as to whether or not the "impulse" is included in the range of integration from 0 to ∞. Thus

$$\int_0^{\infty} e^{-st} \delta(t) \, dt = 1$$

by this convention.

A meaning can also be attached to the derivative of the delta function. We merely differentiate with respect to x either expression in (2.21) (or any similar one), perform any integrations that are called for, and then proceed to the limit $\beta \to 0$. Thus, the reader may show that

$$\int_{-\infty}^{\infty} w(x)\,\delta'(x - x_0)\,dx = \left[w(x)\,\delta(x - x_0) \right]_{-\infty}^{\infty} - \int_{-\infty}^{\infty} w'(x)\,\delta(x - x_0)\,dx$$

$$= -w'(x_0)$$

and also that the Laplace transform of $\delta'(t)$ [i.e., of $\delta'(t - [0+])$] is s.

2.6 PROBLEMS

2.6.1   Show that the convolution theorem of Laplace transform theory is applicable to the inversion of transform expressions containing a factor that is the transform of a member of the delta function family.

2.6.2   Referring to Eq. (1.6), interpret the equation

$$\phi_t = a^2\phi_{xx} + m(x)\cdot\delta(t) \tag{2.22}$$

for $-\infty < x < \infty$, $t > 0$, where $m(x)$ is a given function, in terms of the release at $t = 0$ of heat energy distributed along the rod with an intensity proportional to $m(x)$. Use Laplace transforms to find $\phi(x, t)$ if $\phi(x, 0) = 0$ for all x. Next, replace $m(x)$ by $\delta(x - \xi)$ (where ξ is a chosen point such that $-\infty < \xi < \infty$) and discuss the meaning of your results for this case.

2.6.3   At $t = 0$, a concentrated amount of heat is released uniformly over the cylindrical surface at $r = \xi$ inside a long cylinder whose outer surface $(r = R)$ is kept at the constant temperature T_0. Initially the interior temperature is also T_0. Determine $\phi(r, t)$ for all $0 \leq r < R$, $t > 0$.

2.6.4   Use Laplace transforms to solve the problem of Eq. 2.4 with the last condition altered to read $\phi(0, t) = \delta(t - t_0)$, where $t_0 > 0$. If, instead, the last condition reads $\phi(0, t) = f(t)$, where $f(t)$ is some given function, then use the fact that

$$f(t) = \int_0^{\infty} f(t_0)\,\delta(t - t_0)\,dt_0$$

to obtain the solution by superposition of solutions of the form just obtained. Compare your result with that of Problem 2.2.1.

2.7 SUPPLEMENTARY PROBLEMS

2.7.1 As shown in Problem 1.10.1, the solution to an appropriately posed problem involving the diffusion equation is unique. However, spurious solutions (i.e., physically unacceptable ones) sometimes appear when an implied constraint is neglected. To illustrate, it is easily verified that each of ϕ_t, ϕ_{tt}, ϕ_{ttt}, ... is a solution of $\phi_t = a^2\phi_{xx}$ when ϕ itself is. Moreover, in the problem of Eq. (2.4), in which $\phi = \operatorname{erfc}(x/2a\sqrt{t})$, the function ϕ_t obeys the boundary conditions $\phi_t(x, 0) = 0$ for $x > 0$, $\phi_t(0, t) = 0$ for $t > 0$. Each of the functions ϕ_{tt}, ϕ_{ttt}, ... obeys similar boundary conditions. Thus any linear combination

$$\phi + a_1\phi_t + a_2\phi_{tt} + \cdots$$

where the a_j are constants, also obeys the boundary conditions associated with the derivation of Eq. (2.7).

Examine the behavior of any such combination near $x = 0$, $t = 0$ to explain why an acceptable solution requires that all $a_j = 0$.

Why cannot one of these spurious solutions arise when the Laplace transform method is used?

2.7.2 In the upper layers of the ocean, microscopic plant life, phytoplankton, carries on photosynthesis. It tends to sink at a certain rate, is replenished by a combination of eddy diffusion and reproduction, and is diminished as a result of "grazing" by zooplankton. Consider only the sinking and diffusion processes and show that the equation governing the concentration c of phytoplankton is given by $c_t = (kc_z)_z - wc_z$ where k is the z-dependent coefficient of eddy diffusivity, z the distance from the surface, and w the sinking rate. Formulate the boundary condition at $z = 0$. For a constant $k = 6$ m²/day, $w = 0.5$ m/day, what would happen, and in how long a time, to an initially uniform concentration? If, instead, $k = 90 \exp(-z/10)$ m²/day (where z is in meters), what qualitative difference would you expect?

3

THE WAVE EQUATION

3.1 DERIVATION

In this chapter we will deal with the problem of finding a function $u(x, t)$ satisfying the wave equation. A simple form of this equation is

$$u_{tt} = c^2 u_{xx} \qquad (3.1)$$

where $c(x)$ is some positive function of x. In physical aplications, x frequently represents position, and t time. In comparison with the diffusion equation, we now have a second derivative with respect to t instead of a first derivative; it turns out that this change markedly alters the solution behavior.

As with the diffusion equation, we begin by deriving Eq. (3.1) for a simple physical situation in order to help provide an intuitive feeling for expected properties of solutions as well as for what constitutes an appropriate set of initial or boundary conditions. Other physical problems leading to Eq. (3.1) will be found in Problems 3.2.3 and 3.2.4.

Consider a tightly stretched string of negligible thickness that in an undisturbed state is coincident with the x-axis. In response to external disturbances, the string may deflect in the (x, y) plane. Denote the displacement of a string particle in the y direction by $u(x, t)$. We restrict ourselves to the case of small displacements, in the sense that the displacement of any string particle in the x direction may be disregarded, and that $| u_x | \ll 1$. By definition, a "string" has the property that the force vector in the string always has the direction of the tangent vector to the string; denote the horizontal component of the string tension by T (cf. Fig. 3.1). Since an

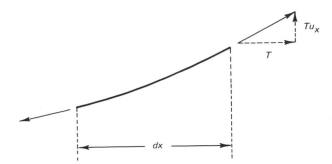

Fig. 3.1 Portion of string (vertical scale magnified).

elemental portion of the string, of length dx, does not accelerate in the x direction, T must be constant along the string. The vertical component of the force acting at either end of the free body of length dx equals Tu_x, so that Newton's law applied to the vertical acceleration of the free body yields

$$(Tu_x)_x\, dx + f\, dx = (\rho\, dx)u_{tt}$$

where $\rho(x)$ is the mass per unit length of the string, and where $f(x, t)$ is the y-direction force per unit length applied to the string by any external loading. Thus

$$u_{tt} = c^2 u_{xx} + g \tag{3.2}$$

where $c^2 = T/\rho$, $g(x, t) = f/\rho$. For the special case in which $g = 0$, Eq. (3.2) reduces to Eq. (3.1). Note that, either from a comparison of the dimensions of the two sides of Eq. (3.2) or from the formula $c^2 = T/\rho$, c itself has the dimensions of velocity.

3.2 PROBLEMS

3.2.1 Maintaining the small-displacement simplification made in the preceding derivation, derive Eq. (3.2) more meticulously using the kinds of arguments employed in Section 1.1.

3.2.2 Generalize Eq. (3.2) so as to apply to a situation in which the string is subjected to an elastic restraint (with a force per unit length ku, acting to return the string to its undeformed position) and to a damping force (of magnitude σu_t per unit length, acting so as to oppose the direction of motion). Here k and σ are constants. Show that Eq. (3.2) becomes

$$u_{tt} = c^2 u_{xx} - ku/\rho - \sigma u_t/\rho + g(x, t) \tag{3.3}$$

Let ρ be constant; determine to what extent this equation can now be simplified by a transformation of the form

$$u(x, t) = v(x, t) e^{\alpha x + \beta t}$$

3.2.3 An electric cable has resistance, capacitance, and inductance per unit length of R, C, and L, respectively. Using the cable element sketched in Fig. 3.2, derive the "telegraph equation" for the potential $e(x, t)$ and current $i(x, t)$:

$$e_x = -Ri - Li_t, \qquad i_x = -Ce_t$$

whence

$$e_{tt} = (1/LC) e_{xx} - (R/L) e_t$$

Verify the dimensional homogeneity. How would this equation be modified by the presence of a leakage resistance σ per unit length (in parallel to the capacitance)?

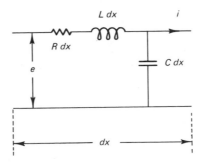

Fig. 3.2 Electric cable element.

3.2.4 Consider long waves in shallow water of constant density ρ (Fig. 3.3). Let the depth of the water be $d(x) > 0$, and let the wave height above the undisturbed surface be $w(x, t)$. Assume the particle acceleration to be sufficiently small that the pressure p at any point beneath the surface is essentially hydrostatic, so that $p = \rho g(w - y)$, where g is the acceleration of gravity. Then the accelerating force in the x direction is proportional to $p_x = \rho g w_x$, and since this is independent of y, it is reasonable to write the x component of the particle velocity as $u = u(x, t)$. Show that the horizontal mass-acceleration† equation becomes

$$u_t + u u_x = -g w_x$$

† Since $u = u(x, t)$, the acceleration of a fluid particle is given by the chain rule as

$$\frac{du}{dt} = \frac{\partial u}{\partial t} + \frac{\partial u}{\partial x}\frac{dx}{dt} = u_t + u_x u$$

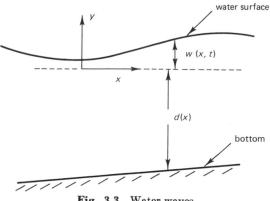

Fig. 3.3 Water waves.

(we neglect any viscous forces). Next, by considering conservation of mass for the fluid portion lying between x and $x + dx$, show that

$$[u(w + d)]_x = -w_t$$

Finally, consider the case in which $d = \text{const}$, and in which u, w, and their derivatives are small enough that the nonlinear terms can be neglected, so as to obtain

$$w_{tt} = (gd)w_{xx}$$

Verify dimensional homogeneity.

3.2.5 An elastic rod is placed along the x-axis. As a result of longitudinal stresses in the rod, a particle may experience a displacement u along the x-axis in such a way that a particle whose rest position is at x finds itself at $x + u(x, t)$ at time t. The local stress τ in the bar (force/unit area) is proportional to the local fractional elongation u_x; we write $\tau = Eu_x$, where E is the elastic modulus. The stress difference across two ends of an elemental portion of the rod will result in an acceleration of that portion; use this fact to derive the equation of propagation of longitudinal elastic waves, viz., $u_{tt} = (E/\rho)u_{xx}$, where ρ is density. List any assumptions. What would the boundary condition be at a free end?

3.3 AN INFINITE-INTERVAL PROBLEM

For the string problem of Section 3.1 a physically reasonable (and therefore, we expect, mathematically well-posed) problem would be one in which a finite string ($0 < x < l$, say) is fixed at each end and in which its initial configuration and velocity are given. Mathematically, we would have

$$u_{tt} = c^2 u_{xx} + f(x, t) \qquad \text{for} \quad 0 < x < l, \quad 0 < t$$

$$u(0, t) = u(l, t) = 0 \qquad \text{for all} \quad t > 0$$

$$u(x, 0) = h(x) \qquad \text{prescribed for} \quad 0 < x < l$$

$$u_t(x, 0) = p(x) \qquad \text{prescribed for} \quad 0 < x < l$$

(3.4)

and this exemplifies the kind of problem statement we can expect to encounter. Of course, alternative end conditions could be imposed; for example, we could require $u(0, t)$ to be some prescribed function of t—a sort of whip-cracking situation. Alternatively, if the end of the string at $x = 0$ were attached to a ring free to slide up and down a frictionless vertical wire support, the end condition would be $u_x(0, t) = 0$. Still another possibility is that there is an elastic restraint at the $x = 0$ end, in which case we would require a linear combination of $u(0, t)$ and $u_x(0, t)$ to vanish.

We will consider problems like that of (3.4) in the next section. To start with, however, we want to concentrate attention on Eq. (3.1) itself, and for that purpose it is convenient to make the end conditions unimportant by the simple device of considering an infinite interval (e.g., an infinitely long string). Let $c > 0$ be constant; we seek $u(x, t)$ such that

$$u_{tt} = c^2 u_{xx} \qquad \text{for} \quad -\infty < x < \infty, \quad t > 0$$

$$u(x, 0) = h(x) \qquad \text{prescribed for} \quad -\infty < x < \infty \qquad (3.5)$$

$$u_t(x, 0) = p(x) \qquad \text{prescribed for} \quad -\infty < x < \infty$$

For constant c, Eq. (3.5) is a very exceptional PDE, in that its general solution can easily be obtained. If we replace x and t by the new pair of independent variables

$$\xi = x + ct, \qquad \eta = x - ct$$

and define

$$\phi(\xi, \eta) = u\left(\frac{\xi + \eta}{2}, \frac{\xi - \eta}{2c}\right)$$

[i.e., $\phi(\xi, \eta)$ and $u(x, t)$ are equal for corresponding values of the variables], then Eq. (3.5) becomes

$$\phi_{\xi\eta} = 0$$

so that

$$\phi(\xi, \eta) = \alpha(\xi) + \beta(\eta)$$

where α and β are arbitrary (but, of course, adequately differentiable) functions of ξ and η, respectively. In terms of u, this result becomes

$$u(x, t) = \alpha(x + ct) + \beta(x - ct) \tag{3.6}$$

so that any solution of Eq. (3.5) must have this form. Conversely, substitution into Eq. (3.5) shows that any function of the form (3.6) satisfies the equation, so that Eqs. (3.5) and (3.6) are completely equivalent to one another.

Our problem then is to find α and β—each a function of the single variable $(x + ct)$ or $(x - ct)$, respectively—such that the given initial conditions on $u(x, 0)$ and $u_t(x, 0)$ are satisfied. Setting $t = 0$ in the expressions for $u(x, t)$ and $u_t(x, t)$ as obtained from Eq. (3.6), we find (where a prime indicates differentiation)

$$\alpha(x) + \beta(x) = h(x), \qquad c\alpha'(x) - c\beta'(x) = p(x)$$

If $P(x)$ is any function such that $P'(x) = p(x)$, these equations are easily solved to give

$$\alpha(x) = \frac{1}{2} h(x) + \frac{1}{2c} P(x) + K, \qquad \beta(x) = \frac{1}{2} h(x) - \frac{1}{2c} P(x) - K$$

where K is a constant of integration. When these results are substituted into Eq. (3.6) we find that the solution u is given by

$$u(x, t) = \alpha(x + ct) + \beta(x - ct)$$

$$= \frac{1}{2} [h(x + ct) + h(x - ct)] + \frac{1}{2c} [P(x + ct) - P(x - ct)]$$

(notice that K gets canceled out). Since $P(x)$ is any integral of $p(x)$, we have

$$P(x + ct) - P(x - ct) = \int_{x-ct}^{x+ct} p(\tau) \, d\tau$$

so that an alternative form for the solution of the problem associated with Eq. (3.5) is

$$u(x, t) = \frac{1}{2} [h(x + ct) + h(x - ct)] + \frac{1}{2c} \int_{x-ct}^{x+ct} p(\tau) \, d\tau$$

$$\tag{3.7}$$

$$= \frac{1}{2} [u(x + ct, 0) + u(x - ct, 0)] + \frac{1}{2c} \int_{x-ct}^{x+ct} u_t(\tau, 0) \, d\tau$$

This solution satisfies all the conditions of our problem; as we will prove later (and as we could in any event anticipate on physical grounds) it is unique. What does it look like? The simplest case is that in which $u_t(x, 0) = 0$ for all x (in the string problem of Section 3.1, the string is given some initial displacement and then released from rest). Equation (3.7) then shows that

$$u(x, t) = \tfrac{1}{2}[h(x + ct) + h(x - ct)] \qquad (3.8)$$

Consider first the function $h(x + ct)$. If an observer were to travel along the x-axis with constant velocity $(-c)$, the argument $(x + ct)$ would maintain a constant value for him, so that he would not observe any change in the function $h(x + ct)$. Thus the contribution $h(x + ct)$ represents a pattern moving to the left with constant velocity and without change in shape. Similarly, $h(x - ct)$ represents a pattern moving to the right with constant velocity c and without change in shape. Taking account of the factor $\tfrac{1}{2}$ in Eq. (3.8), it follows then that if $u_t(x, 0) = 0$, the initial displacement pattern breaks up into two patterns propagating with velocity c, one to the left and one to the right; each has the same shape as the initial pattern, but is only half as large. At time $t = 0$, the two patterns superimpose so as to reproduce the original displacement pattern $h(x)$. See Fig. 3.4.

An interesting fact emerges from this result. Suppose that we wanted to use a very long string as a signaling device. We impose some sort of local deformation on the string and then release the string (from rest) at $t = 0$. An observer some distance away will eventually see our pattern pass by him, and so will receive the transmitted signal. The velocity with which our signal moves along the string is exactly c, so this is the speed of signal transmission.

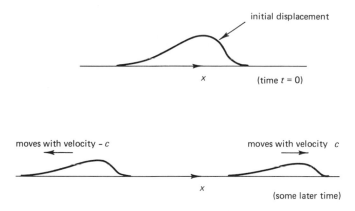

Fig. 3.4 Displacement patterns.

Next, suppose that we start with $u(x, 0) = 0$ and prescribe $u_t(x, 0) = p(x)$ as some given function. Then, from Eq. (3.7), we have

$$u(x, t) = \frac{1}{2c} \int_{x-ct}^{x+ct} p(\tau) \, d\tau \tag{3.9}$$

which may be interpreted geometrically as follows. In the (x, t) plane, draw two straight lines through a chosen observation point (x_1, t_1) such that $x + ct$ is constant on one line and $x - ct$ is constant on the other line. These lines intersect the x-axis at $x_A = x_1 - ct_1$ and at $x_B = x_1 + ct_1$; Eq. (3.9) corresponds to an integration of the $p(x)$ function between x_A and x_B—i.e., to a calculation of the area under that function. But this means that $u(x_1, t_1)$ is affected only by those portions of the x-axis from which signals traveling no faster than with a velocity c can reach the point x_1. Thus again there is a maximum velocity c for signal propagation.

The general case, in which both $h(x)$ and $p(x)$ may be nonzero, is a superposition of these two cases, so that the signal propagation velocity of c must hold for the general case also. In studying the diffusion equation (1.1), we observed that initial disturbances were felt at once throughout the region, although rapidly decaying in size as the distance between source and observation points increases. We have now seen that solutions to Eq. (3.5) behave very differently, in that disturbances travel with the finite speed c, and in that they need not progressively decay in size as they travel.

3.4 PROBLEMS

3.4.1 Sketch the solution (3.7) for various values of t corresponding to the initial conditions

(a) $u(x, 0) = \exp(-x^2)$, $u_t(x, 0) = 0$

(b) $u(x, 0) = 0$, $u_t(x, 0) = \exp(-x^2)$

(c) $u(x, 0) = 0$, $u_t(x, 0) = \sin \omega x$ (with $\omega = $ const)

(d)† $u(x, 0) = 0$, $u_t(x, 0) = \begin{cases} 1 & \text{for} \ \ 0 < x < 1 \\ -1 & \text{for} \ -1 < x < 0 \\ 0 & \text{for} \ \ |x| > 1 \end{cases}$

† Note that $u_t(x, 0)$ is discontinuous; the appropriate interpretation of the problem is that we consider a sequence of initial conditions, described by well-behaved functions, that approach the given initial condition in the limit.

3.4.2 A very thin steel wire—effectively a string—is held horizontally under 1000 kg/cm² tension. The specific gravity of steel is 7. Compute the speed of transverse waves.

With $E = 2 \times 10^6$ kg/cm², compute also the speed of a longitudinal wave (cf. Problem 3.2.5).

3.4.3 (a) Consider an infinite-interval problem $(-\infty < x < \infty)$ for which

$$u(x, 0) = \begin{cases} h(x) & \text{for} \quad x > 0 \\ -h(-x) & \text{for} \quad x < 0 \end{cases}$$

$$u_t(x, 0) = 0$$

Show that the solution of Eq. (3.5) satisfying these initial conditions also solves the following semi-infinite problem: find $u(x, t)$ satisfying $u_{tt} = c^2 u_{xx}$, in $0 < x < \infty$, $t > 0$, with initial conditions $u(x, 0) = h(x)$, $u_t(x, 0) = 0$, and with the fixed end condition $u(0, t) = 0$. [Here $h(x)$ is any given function, with $h(0) = 0$.] Sketch the solution for the case $h(x) = \frac{1}{2} - |x - \frac{3}{2}|$ for $1 < x < 2$, $h(x) = 0$ elsewhere.

(b) Use a similar idea to explain how you could use Eq. (3.7) to solve the finite-interval problem in which $u(0, t) = u(l, t) = 0$ for all t, with $u(x, 0) = h(x)$ and $u_t(x, 0) = 0$ for $0 < x < l$. [We take $h(0) = h(l) = 0$.]

(c) Reconsider parts (a) and (b) for situations in which $u_t(x, 0)$ is prescribed, with $u(x, 0) = 0$. Sketch the solution for a simple case.

3.4.4 In the problem associated with Eq. (3.5), the interval was chosen as infinite in order to simplify the evaluation of the functions α and β of Eq. (3.6); Eq. (3.6) will itself be valid for any interval, finite or infinite. Consider the "whip-cracking" problem, with c constant:

$$u_{tt} = c^2 u_{xx} \qquad \text{in} \quad x > 0, t > 0$$

$$u(x, 0) = u_t(x, 0) = 0 \qquad \text{for all} \quad x > 0$$

$$u(0, t) = \gamma(t) \qquad \text{prescribed for} \quad t > 0, \text{where } \gamma(0) = 0$$

Find α and β in Eq. (3.6) so as to determine u for $x > 0$. [*Hint:* since $h = p = 0$ for $x > 0$, it follows that $\alpha(\xi) = -\beta(\xi) = K$ for $\xi > 0$. Also, $\gamma(t) = \alpha(ct) + \beta(-ct)$, so $\beta(-\xi) = \gamma(\xi/c) - K$ for $\xi > 0$.]

3.4.5 (a) The deflection $u(x, T) = \phi(x)$ and velocity $u_t(x, T) = \psi(x)$ for an infinite string [governed by Eq. (3.5)] are measured at time T, and we are asked to determine what the initial displacement and velocity profiles $u(x, 0)$ and $u_t(x, 0)$ must have been. A student suggests that this problem is equivalent to that of determining the solution of Eq. (3.5) at time T when initial conditions $u(x, 0) = \phi(x)$, $u_t(x, 0) = -\psi(x)$ are prescribed. Is he correct? If not, can you rescue his idea?

(b) In the problem associated with Eq. (3.5), what conditions must $h(x)$ and $p(x)$ satisfy if each particle of the string is to be simultaneously at rest for some future value of t—i.e., if $u_t(x, T) = 0$ for some T?

(c) Suppose that in the problem of Eq. (3.5) $h(x)$ and $p(x)$ are unknown. However, $u(0, t)$ is observed for all values of t. Does this information determine $u(x, t)$ for all other values of x?

3.4.6 A semi-infinite string occupies the $x > 0$ portion of the x-axis. Its density varies with x in such a way that $c = (A + Bx)^2$, where A and B are positive constants. Obtain the general solution of the resulting equation

$$u_{tt} = (A + Bx)^4 u_{xx}$$

by means of the transformation $u(x, t) = (A + Bx)\psi(x, t)$ followed by the change in variables

$$\xi = x/[A(A + Bx)]$$

and discuss the speed of signal propagation.

3.4.7 (a) The density of an infinite string is such that $c = c_-$ for $x < 0$, $c = c_+$ for $x > 0$. Here both c_- and c_+ are constant, in general with $c_- \neq c_+$. The initial conditions are that $u_t(x, 0) = 0$ for all x, and

$$u(x, 0) = \begin{cases} h(x), & x < 0 \\ 0, & x > 0 \end{cases}$$

where $h(x)$ is prescribed (e.g., a pulse). Use the fact that the solution of the coupled wave problem must have the form of Eq. (3.6) (with different values for c, of course) in each half of the string, together with the requirement that u and u_x be continuous at $x = 0$, to discuss the character of the reflected and transmitted waves for each of the cases $c_+ < c_-$, $c_+ > c_-$. Consider also the limit as $c_+ \to c_-$.

(b) Solve a similar problem in which c has the same value for all x, but in which there is a concentrated mass attached to the string at $x = 0$.

(c) A student suggests that a problem with variable $c(x)$ could be solved, approximately, by dividing the length of the string into intervals in each of which c is constant, and using results like those of part (a). How practical is this approach?

3.5 SERIES SOLUTIONS

The methods of Chapter 1 carry over directly. For example, consider the problem

$$u_{tt} = c^2 u_{xx} + f(x, t) \qquad \text{in } 0 < x < l, \quad 0 < t$$

$$u(0, t) = u(l, t) = 0 \qquad \text{for all } t > 0$$

$$u(x, 0) = h(x) \qquad \text{for } 0 < x < l$$

$$u_t(x, 0) = p(x) \qquad \text{for } 0 < x < l$$

$$(3.10)$$

where c and l are positive constants, and where $f(x, t)$, $h(x)$, and $p(x)$ are prescribed functions.

We want to solve problem (3.10) by means of an expansion

$$u(x, t) = \sum_{1}^{\infty} \alpha_n(t) \beta_n(x) \tag{3.11}$$

and for this purpose we need an appropriate set of expansion functions $\beta_n(x)$. To find such a complete set, we construct an associated Sturm–Liouville problem by using separation of variables in the differential operator; we obtain

$$[\alpha(t)\beta(x)]_{tt} = c^2[\alpha(t)\beta(x)]_{xx}$$

from which

$$\beta'' + \lambda^2 \beta = 0 \tag{3.12}$$

where $\lambda = \text{const}$. Suitable homogeneous boundary conditions for $\beta(x)$ are suggested by the given end conditions on u; we therefore adjoin to Eq. (3.12) the boundary conditions

$$\beta(0) = \beta(l) = 0$$

and this completes the statement of the desired problem. The eigenfunctions are easily found to be

$$\beta_n(x) = \sin(n\pi x/l)$$

where n is a positive integer.

If we assume that the series (3.11) continues to converge after differentiation (cf. Problem 3.6.1), substitution of (3.11) into (3.10) yields the ordinary differential equation

$$\alpha_n''(t) = -(n^2\pi^2 c^2/l^2)\alpha_n(t) + f_n(t) \tag{3.13}$$

where $f_n(t)$ is the coefficient of $f(x, t)$ in the expansion

$$f(x, t) = \sum_1^\infty f_n(t) \sin(n\pi x/l)$$

The solution of Eq. (3.13), with the constants of integration being determined from the initial conditions, may be verified to be

$$\alpha_n(t) = (l/n\pi c) \int_0^t \sin(n\pi c(t - \tau)/l) \cdot f_n(\tau) \, d\tau$$

$$+ h_n \cos(n\pi ct/l) + (l/n\pi c) p_n \sin(n\pi ct/l) \tag{3.14}$$

where h_n and p_n are defined by

$$h(x) = \sum_1^\infty h_n \sin(n\pi x/l), \qquad p(x) = \sum_1^\infty p_n \sin(n\pi x/l)$$

In terms of the string problem of Section 3.1, the series constructed above solves the problem of forced vibration of a string with fixed ends. If $f(x, t) = 0$, the situation is said to be one of free vibration; each term in the series is then of the form of a constant times

$$\sin(n\pi c(t - t_n)/l) \cdot \sin(n\pi x/l) \tag{3.15}$$

where t_n is a phase constant. Each such term individually satisfies the equation $u_{tt} = c^2 u_{xx}$ and represents a simple harmonic motion in which all string particles move in phase with one another. Such a free vibrational mode is called a *normal mode*; our general result above thus shows that any free vibration of the string may be thought of as a superposition of normal modes. Incidentally, the normal modes described by (3.15) are the only ones possible, as is easily seen by substituting $u = \phi(x) \sin \omega t$ into $u_{tt} = c^2 u_{xx}$ and applying the end conditions.

Returning to the forced-vibration case, we can anticipate the occurrence of resonance if the applied force is periodic in time and has a frequency component that exactly coincides with a normal mode frequency. This matter, as well as the generation of series solutions for more complicated cases, we leave for exercises.

3.6 PROBLEMS

3.6.1 Use the finite-transform method of Problem 1.8.3 to solve problem (3.10), and make appropriate remarks.

3.6.2 Explain, with the use of examples, how series expansions could be used to solve problem (3.10) when different homogeneous or, alternatively, nonhomogeneous boundary conditions are imposed. [How could a boundary condition like $u_x(0, t) = 0$ be achieved physically?]

3.6.3 Solve problem (3.10) for the case

$$h(x) = p(x) = 0, \qquad f(x, t) = \sin(N\pi x/l) \sin(N\pi ct/l)$$

where N is some chosen integer. Discuss the relevance of this example to the discussion of the last paragraph of the preceding section. Consider also the effect of a concentrated periodic force; how does the point of application affect the nature of the solution?

3.6.4 Obtain a series solution for problem (3.10) as modified by the inclusion of a damping term $-ku_t$ (with k a positive constant) on the right-hand side of the equation. Discuss also the analog, if any, to a normal mode.

3.6.5 The density of a string varies with x in such a way that, for $0 < x < l$, c^2 is well approximated by the expression $c^2 = c_0^2(1 + \varepsilon x)^2$, where ε is a positive constant. The end conditions are $u(0, t) = u(l, t) = 0$. Obtain the normal modes and show how to find a series solution for the case of arbitrary initial conditions. As $\varepsilon \to 0$, do previous results reappear? [*Note:* Solutions of the equation $\xi^2 \Phi_{\xi\xi} + \lambda(1 - \lambda)\Phi = 0$, with $\lambda = $ const, are given by $\Phi = A\xi^\lambda + B\xi^{1-\lambda}$. If the range of interest is $0 < a < \xi < b$, and if $\Phi(a) = \Phi(b) = 0$, then λ must equal $\frac{1}{2} \pm \pi n i/\ln(b/a)$; this leads to eigenfunctions like

$$\left(\frac{\xi}{a}\right)^{1/2} \sin\left(\frac{\pi n \ln(\xi/a)}{\ln(b/a)}\right), \qquad n = 1, 2, 3, \ldots]$$

3.6.6 Use Eq. (3.6) to solve (3.10), and make appropriate comparisons (cf. Problem 3.4.4). Take $f(x, t) = 0$.

3.6.7 A hanging chain of length L undergoes small oscillations in a plane. Assuming that the tensile force in the chain does not differ appreciably from that required to withstand gravity, derive the governing equation

$$g(xu_x)_x = u_{tt}$$

for small lateral displacements $u(x, t)$, where x is measured upward from the free end, and where g is the acceleration of gravity. Obtain a series solution for the case of prescribed initial displacement with zero initial velocity. [*Hint:* the transformation $\xi = \sqrt{x}$ may be useful.]

3.7 A PROBLEM WITH RADIAL SYMMETRY

This problem has musical as well as mathematical interest; we are going to explore the radially symmetric vibrational modes of a drum. The mathematical problem is that of Eq. (3.16) below; we pause to derive this equation for the physical problem of membrane vibrations.

Consider a tightly stretched thin membrane that, when at rest, lies in the (x, y) plane. The deflection of any membrane particle out of this plane, $w(x, y, t)$, is assumed small (we require $w_x^2 + w_y^2 \ll 1$). Moreover, the membrane is supposed incapable of sustaining shear, so that the membrane force exerted across any line element is purely tensile and perpendicular in direction to the direction of the line element. Because of the small-deflection hypothesis, we will not distinguish between this membrane tensile force and its horizontal component. Let an external force in the z direction, of magnitude $f(x, y, t)$ per unit area, be applied to the membrane.

Let the magnitude of the membrane tensile force, per unit of length of line element across which it acts, be T. We observe first that the value of T is the same for all line elements through a particular membrane point; to see this, consider the horizontal equilibrium of a triangular membrane element (two of the sides, dx and dy, are parallel to the coordinate axes) centered at the point. With the notation of Fig. 3.5, we have $(T \, ds) \sin \theta = T_1 \, dy$, whence $T = T_1$; similarly $T = T_2$. Since this result holds for any θ, it follows that T is independent of θ.

Next, the quantity T has the same value at all points in the membrane surface—as is easily seen by considering the horizontal equilibrium of a rectangular membrane element of sides dx and dy. Now let the problem be radially symmetric, in the sense that the loading function f and the displacement w depend only on the distance r from some origin. We write these functions as $f(r, t)$ and $w(r, t)$. Let the uniform area density of the membrane be ρ. If we consider a portion of the membrane between r and

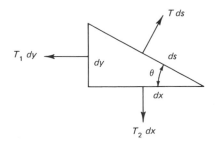

Fig. 3.5   Membrane element.

$r + dr$, the equation of mass acceleration in the vertical direction is clearly

$$(2\pi r\ dr)f + (2\pi r T w_r)_r\ dr = (2\pi r\ dr\ \rho) w_{tt}$$

whence

$$r w_{tt} = c^2 (r w_r)_r + rf/\rho \qquad (3.16)$$

where $c^2 = T/\rho$, a constant.

In the drum problem, we take $0 < r < R$ with $w(R, t) = 0$ and $f(r, t) = 0$. To find a normal mode, set $w = \phi(r) \sin \omega t$ in Eq. (3.16) to give

$$c^2 (r\phi')' + r\omega^2 \phi = 0$$

so that

$$\phi = AJ_0(\omega r/c) + BY_0(\omega r/c) \qquad (3.17)$$

where J_0 and Y_0 are the Bessel functions of first and second kinds, of order zero, and where A and B are constants. Since Y_0 becomes infinite at $r = 0$, we must set $B = 0$ to keep the drum displacement finite at the origin; this leaves only the J_0 term. To satisfy the condition $w(R) = 0$, we need

$$J_0(\omega R/c) = 0$$

so that $\omega R/c$ must equal one of the numbers ζ_i for which $J_0(\zeta_i) = 0$ ($\zeta_1 \cong 2.4048$, $\zeta_2 \cong 5.5201$, $\zeta_3 \cong 8.6537$, etc.—see any table of values of J_0). Thus ω can have only one of the values

$$\omega_i = c\zeta_i/R$$

The final result is then that the radially symmetric normal modes of a drum have the form

$$w_i(r, t) = J_0(r\zeta_i/R) \cdot \sin(c\zeta_i t/R) \qquad (3.18)$$

3.8 PROBLEMS

3.8.1   Obtain a formal series solution for Eq. (3.16) for $0 < r < R$, where $f(r, t)$, $w(r, 0)$, and $w_t(r, 0)$ are arbitrarily prescribed, and where $w(R, t) = 0$ for all t. Notice that, as in the string problem, the free-vibration solution is a superposition of normal modes.

3.8.2   Repeat Problem 3.8.1 for the case of an annular region, where $R_0 < r < R_1$, with $w(R_0, t) = w(R_1, t) = 0$.

3.9 TRANSFORMS

To solve the problem of Eq. (3.5) by Laplace transforms define

$$U(x, s) = \int_0^\infty e^{-st} u(x, t) \, dt$$

Taking the transform of each side of Eq. (3.5), we obtain

$$-u_t(x, 0) - s u(x, 0) + s^2 U = c^2 U_{xx}$$

or

$$-p(x) - s h(x) + s^2 U = c^2 U_{xx} \tag{3.19}$$

We require boundedness at $\pm\infty$; the resulting solution of Eq. (3.19) (cf. Problem 2.2.2) is

$$U = \frac{1}{2c} \int_{-\infty}^\infty \exp\left[-\frac{s}{c} | x - \xi |\right]\left[\frac{1}{s} p(\xi) + h(\xi)\right] d\xi \tag{3.20}$$

Now with $\alpha > 0$ the inverse transform of $e^{-\alpha s}$ is $\delta(t - \alpha)$, as shown in Section 2.5, and that of $(1/s)e^{-\alpha s}$ is 0 for $t < \alpha$ and 1 for $t > \alpha$. Using these results, the reader should verify that he can recover Eq. (3.7).

As a second example, let $f = 0$ in Eq. (3.16) and consider the problem in which an infinite membrane with a hole at its center moves in accord with a prescribed motion of the particles on the boundary of the hole.

Mathematically, let a be a positive constant, with

$$r w_{tt} = c^2 (r w_r)_r \quad \text{in} \quad a < r < \infty, \quad t > 0$$

$$w(r, 0) = w_t(r, 0) = 0 \quad \text{for all} \quad r > a \tag{3.21}$$

$$w(a, t) = 1 \quad \text{for} \quad t > 0$$

Taking transforms, we obtain

$$r W_{rr} + W_r - (r s^2/c^2) W = 0$$

With the change in variable $rs/c = z$, the general solution is found to be

$$W = A I_0(rs/c) + B K_0(rs/c)$$

where I_0 and K_0 are the modified Bessel functions of the first and second kinds. Referring to a handbook,[†] we observe that as $\lambda \to \infty$, $K_0(\lambda) \to 0$ but $I_0(\lambda) \to \infty$. This fact forces us to set $A = 0$ to avoid W becoming

† For example, "Handbook of Mathematical Functions" (Nat. Bur. Std., 1964, p. 374).

unboundedly large for large r (note that, physically, w is expected to decrease as r increases; its transform should therefore not become unbounded as $r \to \infty$). The boundary condition at $r = a$ is that $W = 1/s$, so that our final result is

$$W = K_0(rs/c)/sK_0(as/c) \tag{3.22}$$

Unfortunately, we do not easily find this expression in tables of Laplace transforms, so a continuation of our straightforward attack would force us either to expand the quotient in terms of a series or to use the general inversion formula for Laplace transforms together with, perhaps, numerical integration. Instead, we ask if we can at least determine w for small values of t by use of the fact that the large-s behavior of $W(r, s)$ is coupled to the small-t behavior of $w(r, t)$. Now, for large λ, our handbook shows that

$$K_0(\lambda) \sim (\pi/2\lambda)^{1/2}e^{-\lambda} \tag{3.23}$$

so that, for large s,

$$W(r, s) \sim (1/s)\,(a/r)^{1/2} \exp[-s(r-a)/c]$$

Consequently, for small t,

$$w(r, t) = \begin{cases} (a/r)^{1/2} & \text{for} \quad t > (r-a)/c \\ 0 & \text{otherwise} \end{cases} \tag{3.24}$$

Again we observe that signals do not travel more rapidly than with a velocity c. Note the decay factor $(a/r)^{1/2}$; cf. Problem 3.10.3.

3.10 PROBLEMS

3.10.1 Obtain the Laplace transform of the solution to the problem of Eq. (3.10). In the special case $f = 0$, invert the transform to obtain the solution. [*Hint:* use a series expansion, as in Chapter 2; for what t values is this solution particularly useful?]

3.10.2 An improved version of Eq. (3.23) is

$$K_0(\lambda) \sim (\pi/2\lambda)^{1/2}e^{-\lambda}(1 - 1/8\lambda)$$

for large λ. How would the use of this result modify Eq. (3.24)?

3.10.3 The decay rate with increasing r observed in Eq. (3.24) suggests that the transformation $w = (a/r)^{1/2}\phi$ might be interesting. Show that the first equation in (3.21) becomes

$$\phi_{tt} = c^2[\phi_{rr} + \phi/4r^2]$$

The term $\phi/4r^2$ is now, hopefully, small compared to ϕ_{rr}; as a first approximation, we could neglect it entirely to obtain a solution $\phi^{(1)}$ for the problem. A second approximation $\phi^{(2)}$ would then ensue from the solution of

$$\phi_{tt}^{(2)} = c^2[\phi_{rr}^{(2)} + \phi^{(1)}/4r^2]$$

Follow this idea as far as you can. Does it seem to work?

3.10.4   Solve Problem 3.6.4 by Laplace transform methods.

3.10.5   Let $w(r, t)$ satisfy the equation $rw_{tt} = c^2(rw_r)_r$ for $0 < r < \infty$, $t > 0$, and let $w(r, 0) = 1$ for $r < a$, $w(r, 0) = 0$ for $r > a$, with $w_t(r, 0) = 0$ for all r. Use Laplace transforms to find $w(r, t)$ in the form of a convolution integral. [*Hint:* W will be proportional to an I_0 function plus $1/s$ for $r < a$, and to a K_0 function for $r > a$; the coefficients can be determined by matching values and slopes at $r = a$. A Wronskian will appear in the expression, and this Wronskian can be simplified. Finally, transforms like $K_0(\lambda s)$ can be inverted by the use of standard tables.] Show that the effect of the initial displacement is not propagated outward more rapidly than with a velocity c.

3.11 UNIQUENESS

As an illustrative proof that the solution to a well-posed problem is unique, consider the problem

$$u_{tt} = [c(x)]^2 u_{xx} + f(x, t) \qquad \text{in}\ \ 0 < x < l, \quad t > 0$$

$$u(0, t) = \alpha(t), \qquad u(l, t) = \beta(t) \qquad \text{for all}\ \ t > 0 \qquad (3.25)$$

$$u(x, 0) = h(x), \qquad u_t(x, 0) = p(x) \qquad \text{in}\ \ 0 < x < l$$

where $c(x) > 0$ and where $c(x), h(x), p(x), \alpha(t), \beta(t)$, are given functions.

We first observe that if ϕ is any solution of $\phi_{tt} = [c(x)]^2\phi_{xx}$, for which $\phi(0, t) = \phi(l, t) = 0$, then with

$$I(t) = \tfrac{1}{2} \int_0^l [\phi_t^2/c^2 + \phi_x^2]\, dx$$

we have

$$dI/dt = \int_0^l \left[\phi_t \phi_{tt}/c^2 + \phi_x \phi_{xt} \right] dx$$

$$= \int_0^l \left[\phi_t \phi_{xx} + \phi_x \phi_{xt} \right] dx$$

$$= \int_0^l \left[\phi_t \phi_x \right]_x dx = \left[\phi_t \phi_x \right]_0^l = 0$$

since ϕ_t vanishes at each endpoint by hypothesis.

Now let u and v be two solutions to (3.25) and define $\phi = u - v$. It follows that $I = 0$ at $t = 0$, since $\phi_t = \phi_x = 0$ for all x in $(0, l)$ at time $t = 0$. Moreover, since u and v satisfy the same end conditions, we have $\phi = 0$ at $x = 0$ and at $x = l$; we conclude from the result just obtained that $dI/dt = 0$ for all $t > 0$. We conclude that $I \equiv 0$. But the integral is a sum of squares, and consequently I can vanish only if $\phi_x \equiv \phi_t \equiv 0$. Thus $u - v$ can be at most a constant, and since $u = v$ at one boundary point this constant must be zero. This proves uniqueness for (3.25), and the reader may construct similar proofs for other situations.

In the string problem of Section 1.1, the quantity I may be interpreted physically as being proportional to the energy; many existence proofs in partial differential equation theory are based on such energy considerations.

3.12 SUPPLEMENTARY PROBLEMS

3.12.1 (a) Let $u(x, t)$ satisfy the equation $u_{tt} = c^2 u_{xx}$, $c = $ const, in some region of the (x, t) plane. Show that the quantity $(u_t - cu_x)$ is constant along each straight line defined by $x - ct = $ const, and that $(u_t + cu_x)$ is constant along each straight line of the form $x + ct = $ const. These straight lines are called *characteristics*; we will refer to typical members of the two families as C_+ and C_- curves, respectively; thus $(x - ct = $ const) is a C_+ curve.

(b) Let $u(x, 0)$ and $u_t(x, 0)$ be prescribed for all values of x in $-\infty < x < \infty$, and let (x_0, t_0) be some point in the (x, t) plane, with $t_0 > 0$. Draw the C_+ and C_- curves through (x_0, t_0) and let them intersect the x-axis at points A, B (Fig. 3.6). Use the properties of C_+ and C_- derived in part (a) to determine $u_t(x_0, t_0)$ in terms of initial data at points A and B. Using a similar technique to obtain u_t at each point (x_0, τ) with $0 < \tau < t_0$, deter-

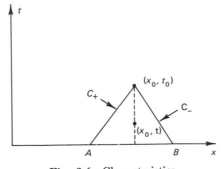

Fig. 3.6   Characteristics.

mine $u(x_0, t_0)$ by integration with respect to τ, and compare with Eq. (3.7). Observe that this "method of characteristics" again shows that $u(x_0, t_0)$ depends only on that part of the initial data between points A and B.

3.12.2   Use the method of characteristics to solve the problems outlined in Problem 3.4.3. [Note that in part (a), for example, a C_- characteristic joining some point on the positive x-axis to a point K on the t-axis will determine $u_t + cu_x$ at K; since $u_t = 0$ on the t-axis, we can now obtain u_x at K, and thus also determine $u_t - cu_x$ along a C_+ characteristic emanating from K.]

3.12.3   For a semi-infinite string in which $u_{tt} = c^2 u_{xx}$, $x > 0$, $t > 0$, with $c = \text{const}$, let $u(x, 0) = f(x)$, $u_t(x, 0) = 0$. Let the end support point move with constant velocity V, so that the end condition becomes $u(Vt, t) = 0$. Find $u(x, t)$ by (a) the method of characteristics, (b) a Laplace transform in t, (c) a preliminary transformation of (x, t) variables, which results in a fixed endpoint. Are all the methods feasible? [*Note:* consider $V < 0$, $V > 0$, $|V| < c$, $|V| > c$.]

3.12.4   An infinite string has a mass m attached at the origin. At $t = 0$, a sinusoidal wave, proceeding at speed c toward the right, occupies the region $x < 0$; the region $x > 0$ is at rest. Analyze the resulting wave motion. Show that the final steady-state solution, consisting of the initial wave, a reflected wave, and a transmitted wave, can alternatively be obtained by assuming the existence of appropriate waves and simply matching displacements and forces at $x = 0$.

4

THE POTENTIAL EQUATION

4.1 LAPLACE'S AND POISSON'S EQUATIONS

In this chapter we will be largely concerned with functions $\phi(x, y)$ that satisfy *Poisson's equation*

$$\phi_{xx} + \phi_{yy} = f(x, y) \tag{4.1}$$

in some region of the (x, y) plane, where $f(x, y)$ is a prescribed function. In the special case in which $f \equiv 0$, Eq. (4.1) is called *Laplace's equation*.

Our experience to date indicates that Eq. (4.1) alone is not sufficient to completely determine ϕ; some accessory or boundary conditions must also be prescribed. Before proceeding to consider various kinds of boundary conditions, and the corresponding solutions of Eq. (4.1), it is useful to provide some guidance (for example, as to the kinds of boundary conditions to be reasonably expected) by discussing a simple physical problem in which Eq. (4.1) arises. Although we choose a problem in thermal diffusion, it need hardly be remarked that many problems in the diffusion of chemical or biological species may lead to similar equations.

Let a thin plate of thickness h be bounded by a closed curve C lying in the (x, y) plane. We are interested in the steady-state temperature distribution resulting from the application of certain time-independent thermal boundary conditions to the edges of the plate. The top and bottom surfaces are insulated.

The physical configuration implies that the temperature ϕ is a function

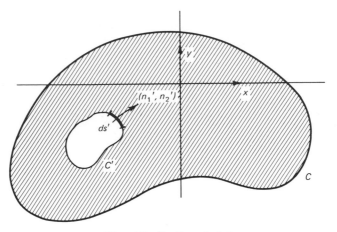

Fig. 4.1 Portion of plate.

of x and y only.† To derive the governing differential equation, consider that part of the plate contained within some chosen curve C' (see Fig. 4.1). Since the (constant) plate thickness is h, the surface area element associated with a portion ds' of C' is $h\,ds'$; the rate at which heat flows across this surface element is

$$k\,\frac{\partial \phi}{\partial n}\,h\,ds'$$

where k is the thermal conductivity and where $\partial \phi / \partial n$ denotes the rate of change of ϕ in the direction of the outward normal to C'. If the outward normal unit vector to C' has components n_1' and n_2' in the neighborhood of our area element $h\,ds'$, then this heat flow rate may be written (by use of the usual formula for a directional derivative)

$$k(\phi_x n_1' + \phi_y n_2')h\,ds'$$

At thermal equilibrium the net rate of heat flow across C' into our plate portion must be zero. Thus

$$\int_{C'} k(\phi_x n_1' + \phi_y n_2')h\,ds' = 0$$

The two-dimensional form of the divergence theorem enables us to rewrite

---

† A similar situation would hold for the case of a very long cylinder with its generators perpendicular to the (x, y) plane and with boundary conditions for temperature prescribed in such a way that they do not depend on the z coordinate.

this result as

$$\int_{A'} [(k\phi_x)_x + (k\phi_y)_y] h \, dA' = 0$$

where A' is the area enclosed by C'. But this result must hold for any such curve C' and area A'; by reasoning analogous to that used in Section 1.1, we therefore conclude that

$$(k\phi_x)_x + (k\phi_y)_y = 0 \tag{4.2}$$

If k is constant throughout the plate, Eq. (4.2) reduces to Laplace's equation

$$\phi_{xx} + \phi_{yy} = 0 \tag{4.3}$$

More generally, if there had been a distribution of heat sources within the plate of intensity $g(x, y)$ per unit volume, the reader should show that Eq. (4.2) is replaced by

$$(k\phi_x)_x + (k\phi_y)_y + g = 0 \tag{4.4}$$

and Eq. (4.3) by Poisson's equation (4.1), where $f = -g/k$.

We now ask what kind of accessory or boundary condition is appropriate. In this thermal equilibrium problem, it is physically reasonable to expect the temperature $\phi(x, y)$ to be completely determined by a specification of its value everywhere on C. The corresponding mathematical problem— find $\phi(x, y)$ such that $\phi_{xx} + \phi_{yy} = 0$ (say) inside C, and such that ϕ is given at each point of C—is frequently termed a *Dirichlet problem*. If, on the other hand, the normal gradient of temperature $\partial\phi/\partial n$ is specified on the boundary, rather than ϕ itself, the problem is called a *Neumann problem*. We note at once that if the heat transfer rate ($k \, \partial\phi/\partial n$ per unit area) is specified at each point on the boundary, then the net rate of heat flow into the plate can be computed as

$$\int_C hk \frac{\partial\phi}{\partial n} \, ds$$

(where ds is the element of arc length of C). If there is no heat generation or absorption within the plate, then unless this integral vanishes, the temperature distribution in the plate cannot possibly attain a steady state. For the special case of Eq. (4.1) it follows that a consistency condition, which cannot be violated if a Neumann problem is to have a solution, is that

$$\int_C \frac{\partial\phi}{\partial n} \, ds = 0 \tag{4.5}$$

Finally, a *mixed boundary value problem* would result if we specified at each boundary point the value of some linear combination $\alpha\phi + \beta(\partial\phi/\partial n)$, where α and β may be functions of position along C. Physically, such a condition can be thought of as a radiation or cooling condition—the local thermal gradient (cooling rate) depends on the local edge temperature.

The expression $\partial\phi/\partial n$ will occur frequently. As in the above derivation, we will always interpret it as the rate of change of ϕ in the direction of the *outward-pointing* unit normal. Thus on C, if n_1 and n_2 are the x and y components, respectively, of the outward unit normal vector, we have

$$\partial\phi/\partial n = \phi_x n_1 + \phi_y n_2$$

Equations of form (4.1)–(4.4) occur frequently in applications. For historical reasons, we will generally describe such equations as being of *potential* type (cf. Problem 4.2.1).

4.2 PROBLEMS

4.2.1 The electrostatic potential at a distance r cm from a point charge of q esu is equal to the work required to bring a unit test charge from infinity to that position; it is easily computed to be equal to (q/r) ergs. Show that, for the electrostatic field resulting from a number of such fixed charges, the potential function $V(x, y, z)$ satisfies the three-dimensional Laplace equation

$$V_{xx} + V_{yy} + V_{zz} = 0$$

in any region of space not occupied by charges. Obtain a similar result for the gravitational potential of a collection of point masses. In either case, how could a two-dimensional problem (at least approximately) be devised?

4.2.2 Let a thin membrane be stretched across a wire loop lying in the (x, y) plane. Let the membrane undergo a small vertical displacement $w(x, y)$ as the result of either a vertical motion of a portion of its boundary loop or the application of a pressure force distribution (of intensity $f(x, y)$ per unit area) to its surface. Using the result of Section 3.7 to the effect that the tension T per unit length of line element is nearly constant, derive the equation governing w. [Make approximations similar to those used in Section 3.7.] This so-called "membrane problem" is frequently invoked as a mental device to help visualize solutions of potential-type equations; using this idea, sketch the expected contour lines of the solution to the problem

$$\phi_{xx} + \phi_{yy} = 0 \quad \text{in} \quad R$$

where R is the square region $0 < x < 1$, $0 < y < 1$, and where $\phi = 0$ everywhere on the boundary except on the segment $0 < x < 1$, $y = 0$, where $\phi = x(\frac{1}{2} - x)(1 - x)$.

4.2.3   Use the coordinate transformation $x = r \cos \theta$, $y = r \sin \theta$, $\psi(r, \theta) = \phi(r \cos \theta, r \sin \theta)$, to show that if ϕ satisfies Eq. (4.3) then ψ satisfies the equation

$$\psi_{rr} + (1/r)\psi_r + (1/r^2)\psi_{\theta\theta} = 0$$

(Note the dimensional compatibility of each term.) Can you find a solution function that depends only on r? Only on θ?

Obtain a similar result for elliptical coordinates, defined by $x = \cosh \alpha \cos \beta$, $y = \sinh \alpha \sin \beta$.

4.3 SIMPLE PROPERTIES OF HARMONIC FUNCTIONS

A function $\phi(x, y)$ satisfying Laplace's equation $\phi_{xx} + \phi_{yy} = 0$ in a region R of the (x, y) plane is said to be *harmonic* in R. Physically, we can visualize ϕ as representing some steady-state temperature distribution throughout a plate (Section 4.1), and this fact suggests some properties that ϕ can be expected to possess. For one thing, we would not expect to encounter a maximum of ϕ anywhere inside R, for in the corresponding plate problem there would be continual heat flow in all directions away from this maximum point, and so thermal equilibrium would be impossible (unless, of course, the temperature everywhere had this same value— i.e., unless ϕ were constant). Arguing similarly, we would expect that the value of ϕ at any point inside R must equal some kind of average of the temperatures at surrounding points in order that the heat flow rates toward and away from the point in question balance. Also, as a result of our visualization of ϕ as a temperature distribution, we expect from Eq. (4.5) that

$$\int_C \frac{\partial \phi}{\partial n} \, ds = 0 \tag{4.6}$$

for any (sufficiently smooth) closed curve C drawn in R. (As a special case, C could be the boundary of R).

Guided by these expectations, we now proceed more formally; we begin with a derivation of Eq. (4.6). Since $\phi_{xx} + \phi_{yy} = 0$ in R, we have

$$\int_A \left[(\phi_x)_x + (\phi_y)_y \right] dA = 0$$

where A is the area enclosed by C. Transforming to a contour integral around C, we obtain Eq. (4.6) at once; we have in fact merely reversed our original procedure leading to the derivation of Eq. (4.3).

Next, consider any point P with coordinates (x_0, y_0) inside the region R and let (r, θ) be a polar coordinate system with origin P. Define $\psi(r, \theta)$ to be the value of $\phi(x, y)$ at the corresponding point—i.e., $\psi(r, \theta) = \phi(x_0 + r \cos \theta, y_0 + r \sin \theta)$. Construct a circle C_ρ with center P and radius ρ. The average value of ψ on this circle is defined by

$$\Psi(\rho) = \frac{1}{2\pi} \int_0^{2\pi} \psi(\rho, \theta) \, d\theta \tag{4.7}$$

Differentiating with respect to ρ and using $\psi_\rho = \partial\psi/\partial n$, we obtain

$$\Psi_\rho(\rho) = \frac{1}{2\pi} \int_0^{2\pi} \psi_\rho(\rho, \theta) \, d\theta = \frac{1}{2\pi\rho} \int_0^{2\pi} \frac{\partial\psi}{\partial n} \, ds$$

and by Eq. (4.6) this last integral must equal zero. Thus $\Psi(\rho)$ is independent of ρ, and since $\Psi(0) = \psi(0, \theta)$, we have shown that *the value of a harmonic function at a point P is equal to the average of its values on the circumference of any circle centered on P.* (The whole circle must, of course, lie within R.) An easy corollary is that *the value of a harmonic function at a point P is equal to the average†of its values over the area of any circle centered on P.* These two essentially equivalent results are referred to collectively as the *mean value theorem* for harmonic functions; this theorem is a mathematical statement of the average-value property expected physically as a result of the discussion in the first paragraph of this section.

We can now use the mean value theorem to show that if ϕ is harmonic inside a region R, then the maximum of ϕ always occurs on the boundary of R. More precisely, we will show that ϕ cannot attain a local maximum value at a point P inside R unless it has that same value everywhere in R. We prove this result by the time-honored method of assuming the opposite and being led to a contradiction. Suppose ϕ is indeed a maximum at P, and suppose that there is some point Q in the immediate neighborhood of P

† This average would be defined by

$$\frac{1}{\pi\rho^2} \int_0^\rho \int_0^{2\pi} \psi(r, \theta) r \, dr \, d\theta$$

such that ϕ at Q is actually less than ϕ at P. Draw a circle of radius λ centered on P and passing through Q. The average value of ϕ on this circle is obtained by multiplying ϕ by $d\theta$, integrating around the circle, and dividing by 2π. Since the value of ϕ at each point on the circle is necessarily less than or equal to the value of ϕ at P by the hypothesis, and since ϕ at the point Q on the circle is actually less than ϕ at P, it follows that this average value of ϕ on the circle must be less than the value of ϕ at P. But this contradicts the mean value theorem, and so our assertion is proved.

If ϕ is harmonic, so is $(-\phi)$; since a maximum for $(-\phi)$ corresponds to a minimum for ϕ, we can also conclude that a harmonic function cannot attain a local minimum at any point interior to R unless it has that same value everywhere.

An immediate consequence of the maximum property of harmonic functions is that the solution to the following problem is unique: find $\phi(x, y)$ satisfying $\phi_{xx} + \phi_{yy} = f(x, y)$ in a region R bounded by a curve Γ, where ϕ is specified on Γ. Suppose that $\phi_1(x, y)$ and $\phi_2(x, y)$ are two solutions to this problem. Then the difference function $(\phi_1 - \phi_2)$ satisfies Laplace's equation and so is harmonic; it vanishes on the boundary, and since its value in R cannot be greater than the greatest value attained on the boundary nor less than the smallest such value, it follows that the difference function must vanish everywhere in R.

Another simple consequence of the maximum property is that if u and v are harmonic in R, with $v \geq u$ on the boundary Γ, then $v \geq u$ everywhere in R—for otherwise the maximum of $u - v$ would not be attained on the boundary.

To conclude this section, we observe that Eq. (4.6) is easily generalized to hold for regions with "holes"—i.e., for multiply connected regions.† For example, let C_0 be a curve drawn inside R, and let C_1 and C_2 be two curves lying within C_0 (Fig. 4.2). Then again

$$\int_A (\phi_{xx} + \phi_{yy}) \, dA = 0$$

(where A is the region inside C_0 but outside C_1 and C_2) and again we can use a transformation to a contour integral. For a multiply connected region, however, the contour integral must include all boundaries (one way of showing this is to introduce new artificial boundaries, shown dashed in the figure, so as to yield a simply connected region; the dashed-

† Formally, a region is said to be simply connected if any simple closed curve lying in the region can be shrunk down to a point without ever leaving the region. Otherwise, the region is said to be multiply connected.

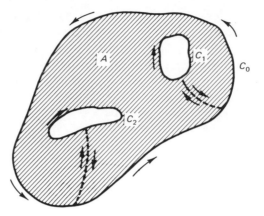

Fig. 4.2   Multiply connected region.

line portions are traversed twice, and with opposite signs for $\partial\phi/\partial n$, so that their contributions to the contour integral vanish). Thus

$$\int_{C_0} \frac{\partial\phi}{\partial n}\, ds + \int_{C_1} \frac{\partial\phi}{\partial n}\, ds + \int_{C_2} \frac{\partial\phi}{\partial n}\, ds = 0$$

Note that all unit normal vectors point *out* of the shaded region A.

4.4   SOME SPECIAL SOLUTIONS—SERIES

Separation of variables is again a useful technique. If in $\phi_{xx} + \phi_{yy} = 0$ we set $\phi = X(x)\cdot Y(y)$, we obtain $X''/X = -Y''/Y = \lambda^2$; with λ real, this leads to solutions typified by $\phi = e^{\lambda x}\sin\lambda y$, and with $\lambda = \alpha + i\beta$ (α, β real), we find solutions like $\phi = e^{\alpha x - \beta y}\cos(\alpha y + \beta x)$. The possibility $\lambda = 0$ should not be forgotten; here we get $\phi = a + bx + cy + dxy$ as a solution, where a, b, c, d are constants. Similarly, using polar coordinates so that $r^2\psi_{rr} + r\psi_r + \psi_{\theta\theta} = 0$, we find that expressions like $\psi = r^\lambda\sin\lambda\theta$ or $\psi = e^{\lambda\theta}\cos(\lambda\ln r)$ or $\psi = (a + b\theta)(c + d\ln r)$ (where λ, a, b, c, d are constants) are solutions.

Let (x_0, y_0) be any point in the plane. The function

$$g(x, y) = \ln[(x - x_0)^2 + (y - y_0)^2]^{1/2} \tag{4.8}$$

satisfies Laplace's equation for all points (x, y) different from (x_0, y_0). It is easily checked to be (within an additive and multiplicative constant) the only solution of Laplace's equation that depends only on the distance from the chosen point (x_0, y_0), and for that reason it is frequently termed a *fundamental solution* of Laplace's equation. It is also referred to as a *logarithmic potential* (cf. Problem 4.2.1).

We can also construct polynomials in x and y that satisfy Laplace's equation. Consider for example

$$\phi = A + Bx + Cy + Dx^2 + Exy + Fy^2$$

Substitution into $\phi_{xx} + \phi_{yy} = 0$ shows that we must have $D + F = 0$; apart from this condition, the constants A, B, C, D, E, F may be chosen arbitrarily. For the cubic

$$\phi = A + Bx + Cy + Dx^2 + Exy + Fy^2 + Gx^3 + Hx^2y + Ixy^2 + Jy^3$$

we must similarly have $D + F = 3G + I = H + 3J = 0$, and for an nth-degree polynomial we would similarly find (cf. Problem 4.5.1) that only $(2n + 1)$ of the coefficients may be arbitrarily chosen.

Series solutions for problems with suitable geometries can be constructed much as in Sections 1.5 and 3.5. Problems 4.5.2 and 4.5.3 will provide some practice. We will, however, consider one case that leads to an interesting general formula.

Using polar coordinates (r, θ), let it be required that a function $\psi(r, \theta)$ be found such that ψ is harmonic inside a circle of radius R centered on the origin and has prescribed boundary values on the circumference of that circle. Thus

$$\psi_{rr} + (1/r)\psi_r + (1/r^2)\psi_{\theta\theta} = 0 \qquad \text{for} \quad r < R$$

$$\psi(R, \theta) = \text{prescribed function of } \theta \tag{4.9}$$

We want to expand $\psi(r, \theta)$ in a series of θ-dependent eigenfunctions; a suitable set [use separation of variables in Eq. (4.9), plus periodic end conditions] is $\{\cos n\theta\}$ or $\{\sin n\theta\}$, where n is an integer. We are thus led to a conventional Fourier series:

$$\psi(r, \theta) = \tfrac{1}{2}a_0(r) + \sum_{n=1}^{\infty} [a_n(r) \cos n\theta + b_n(r) \sin n\theta] \tag{4.10}$$

where

$$a_n(r) = \frac{1}{\pi} \int_0^{2\pi} \psi(r, \theta) \cos n\theta \, d\theta, \qquad b_n(r) = \frac{1}{\pi} \int_0^{2\pi} \psi(r, \theta) \sin n\theta \, d\theta$$

Multiplying Eq. (4.9) by $\cos n\theta$ or $\sin n\theta$ and integrating with respect to θ from 0 to 2π, we obtain after an integration by parts,

$$a_n'' + (1/r)a_n' - (n^2/r^2)a_n = 0, \qquad b_n'' + (1/r)b_n' - (n^2/r^2)b_n = 0$$

respectively. The solutions of these equations, subject to the conditions that a_n and b_n be finite at the origin and that they agree with the Fourier coefficients of $\psi(R, \theta)$ when $r = R$, are

$$a_n = \frac{1}{\pi}\left(\frac{r}{R}\right)^n \int_0^{2\pi} \psi(R, \alpha) \cos n\alpha \, d\alpha, \qquad b_n = \frac{1}{\pi}\left(\frac{r}{R}\right)^n \int_0^{2\pi} \psi(R, \alpha) \sin n\alpha \, d\alpha$$

Inserting these results in Eq. (4.10), we obtain

$$\psi(r, \theta) = \frac{1}{2\pi}\int_0^{2\pi} \psi(R, \alpha) \, d\alpha$$

$$+ \frac{1}{\pi} \sum_{n=1}^{\infty} \left[\left(\frac{r}{R}\right)^n \int_0^{2\pi} \psi(R, \alpha) \cos n(\alpha - \theta) \, d\alpha\right] \qquad (4.11)$$

If $\psi(R, \theta)$ is reasonably well behaved (e.g., piecewise continuous), the orders of integration and summation may be interchanged. Also, the series

$$\sum_{n=1}^{\infty} \left(\frac{r}{R}\right)^n \cos n(\alpha - \theta)$$

is easily summed, since this series is simply the real part of $\sum z^n$ where $z = (r/R) \exp[i(\alpha - \theta)]$; since $|z| < 1$, we have in fact $\sum z^n = z/(1 - z)$. Taking the real part of this expression and inserting the result in Eq. (4.11), we obtain the *Poisson integral formula*

$$\psi(r, \theta) = \frac{1}{2\pi}\int_0^{2\pi} \psi(R, \alpha) \frac{R^2 - r^2}{R^2 + r^2 - 2Rr \cos(\alpha - \theta)} \, d\alpha \qquad (4.12)$$

Equation (4.12) provides an explicit formula for a harmonic function $\psi(r, \theta)$, at a point (r, θ) interior to a circle of radius R, in terms of its boundary values on the circle. As a special case, the choice $r = 0$ leads again to the mean value theorem.

4.5 PROBLEMS

4.5.1 Show that if a harmonic function has the form of an nth-degree polynomial in x and y, then only $(2n + 1)$ of the coefficients can be arbitrary. [*Hint:* use polar coordinates.]

4.5.2 Let $\phi(x, y)$ satisfy Laplace's equation in the rectangle $0 < x < 1$, $0 < y < 2$, with $\phi(x, 2) = x(1 - x)$ and with $\phi = 0$ on the other three sides. Use a series solution to determine ϕ inside the rectangle. How many terms are required to give $\phi(\frac{1}{2}, 1)$ with about 1% (also 0.1%) accuracy; how about $\phi_x(\frac{1}{2}, 1)$?

4.5.3 (a)   In Problem 4.5.2, modify (only) the boundary condition on $y = 0$ to read $\phi(x, 0) = x^2(1 - x)$ and solve the resulting problem. Explain what a convenient series solution approach would be if the boundary conditions were such that ϕ vanished on the sides $y = 0$ and $y = 2$ but not on $x = 0$ or $x = 1$. What if ϕ vanishes on no sides?
  (b)   In the same rectangle, let

$$\phi_{xx} + \phi_{yy} = \sin \pi x \cdot \cos \pi y$$

with $\phi(0, y) = \phi(1, y) = 0$ for $0 < y < 2$, and with $\phi(x, 0) = \phi_y(x, 2) = 0$ for $0 < x < 1$. Determine ϕ via each of two expansions—one in terms of x-dependent eigenfunctions and one in terms of y-dependent eigenfunctions.

4.5.4 (a)   Use a series expansion to find $\psi(r, \theta)$ satisfying the equation

$$\psi_{rr} + (1/r)\psi_r + (1/r^2)\psi_{\theta\theta} = -\frac{\cos \theta}{r^2}$$

in the sector $1 < r < 3, 0 < \theta < \pi/2$, where the boundary values on the sector are given by

$$\psi(1, \theta) = 0, \quad \psi(3, \theta) = \tfrac{2}{3} \cos \theta, \quad \psi(r, 0) = (r - 1)/r, \quad \psi(r, \pi/2) = 0$$

  (b)   Formulate a Dirichlet problem for the sector $0 < r < R, 0 < \theta < \Theta$ (where R, Θ are given), involving Poisson's equation (4.1), and obtain a general series solution.

4.5.5 Set $\psi(R, \alpha) = 1$ in Poisson's integral (4.12) and deduce, from the fact that $\psi(r, \theta)$ must then also equal one, the integral identity

$$\int_0^{2\pi} \frac{d\alpha}{A + B \cos \alpha} = \frac{2\pi}{(A^2 - B^2)^{1/2}}$$

for constants A and B satisfying $A > 0, |B| < A$.
  Obtain more general results of this nature by making other choices for $\psi(R, \alpha)$ corresponding to known harmonic functions (e.g., $r^n \cos n\theta$, etc.).

4.5.6 Obtain, to whatever extent you can, a formula analogous to the Poisson integral formula to solve each of
  (a)   the Neumann problem for the circular region $0 \le r < R$,
  (b)   the Dirichlet problem for the annular region $R_0 < r < R_1$,
  (c)   the "exterior Dirichlet problem" for the region $r > R$ with $\psi(r, \theta) \to A$ as $r \to \infty$, where A is the average of values of ψ on $r = R$.

[*Partial answer:* for (a)

$$\psi(r, \theta) = -\frac{R}{2\pi} \int_0^{2\pi} \psi_r(R, \alpha) \ln\left[1 - 2\frac{r}{R}\cos(\theta - \alpha) + \frac{r^2}{R^2}\right] d\alpha$$

within an arbitrary additive constant.]

4.5.7   (a)   Show that a function $v(x, y)$ that satisfies the mean value theorem—i.e., the value at any point is equal to the average of its values on the circumference of any circle centered on that point—is necessarily harmonic. [*Hint:* we can deduce that the maximum and minimum of v must be attained on the boundary of a region; consider $v - u$, where u is a harmonic function having the same boundary values.]

(b)   Let $\phi(x, y)$ be harmonic in a region R of the upper half plane, one of whose boundaries is the portion of the x-axis defined by $a < x < b$. Let $\phi(x, y) \to 0$ as $y \to 0$, for x in (a, b). Let R' denote the "reflection" of the region across the x-axis—i.e., if a point (x, y) is in R, then $(x, -y)$ is in R'. Define a function $\psi(x, y)$ for (x, y) in R' by the formula $\psi(x, y) = -\phi(x, -y)$. Show that the composite function

$$w(x, y) = \begin{cases} \phi(x, y) & \text{for} \quad (x, y) \quad \text{in} \quad R \\ \psi(x, y) & \text{for} \quad (x, y) \quad \text{in} \quad R' \\ 0 & \text{for} \quad x \quad \quad \text{in} \quad (a, b), \ y = 0 \end{cases}$$

is harmonic in the overall region obtained by combining R, R', and the interval (a, b). [This result is sometimes termed the *Schwarz reflection principle.*]

4.6   DISCONTINUOUS BOUNDARY DATA

The real world rarely has discontinuities—apparent sharp corners or sudden transitions from, say, a conductor at one potential to a conductor at another potential become blurred when examined microscopically. However, it is often convenient to simplify the exact boundary conditions, in which there are narrow transition regions, by permitting discontinuities in boundary data. We now investigate one such problem.

Let it be required that we find a function $\phi(x, y)$, harmonic and bounded†
in the region $0 < x < \infty$, $0 < y < 1$, with $\phi(x, 0) = \phi(x, 1) = 0$ for
$0 < x < \infty$ and with $\phi(0, y) = 1 - y$ for $0 < y < 1$. There is a discon-

---

† The condition of boundedness ensures uniqueness. See Problem 4.8.8.

tinuity in the boundary data at the point $(0, 0)$; our immediate purpose is to analyze the behavior of $\phi(x, y)$ near that point. We do this by constructing a series expansion for ϕ and then summing this series so as to provide a simple closed-form solution for the problem.

An appropriate series expansion is given by

$$\phi(x, y) = \sum_{1}^{\infty} a_n e^{-n\pi x} \sin n\pi y$$

and the condition $\phi(0, y) = 1 - y$ requires that $a_n = 2/(n\pi)$. Thus

$$\phi(x, y) = (2/\pi) \operatorname{Im}\{z + \tfrac{1}{2}z^2 + \tfrac{1}{3}z^3 + \tfrac{1}{4}z^4 + \cdots\}$$

where Im is "the imaginary part of" and $z = \exp[-\pi x + i\pi y]$. This z-series is simply the power series expansion of $-\ln(1 - z)$, so that

$$\phi(x, y) = -(2/\pi) \operatorname{Im}\{\ln(1 - e^{-\pi x + i\pi y})\}$$

$$= -(2/\pi) \operatorname{Im}\{\ln[(1 - e^{-\pi x} \cos \pi y) - ie^{-\pi x} \sin \pi y]\}$$

$$= (2/\pi) \tan^{-1}\left(\frac{e^{-\pi x} \sin \pi y}{1 - e^{-\pi x} \cos \pi y}\right)$$

(since if $\zeta = Re^{i\alpha}$ is any complex number, $\ln \zeta = \ln R + i\alpha$, and therefore $\operatorname{Im}(\ln \zeta) = \alpha$). Our final result is therefore

$$\phi(x, y) = \frac{2}{\pi} \tan^{-1}\left(\frac{\sin \pi y}{e^{\pi x} - \cos \pi y}\right) \tag{4.13}$$

For a point (x, y) within our region and close to the origin, we have $\sin \pi y \cong \pi y$ and $e^{\pi x} - \cos \pi y \cong \pi x$; thus

$$\phi(x, y) = (2/\pi) \tan^{-1}(y/x)$$

This means that if we approach the origin along a ray making an angle θ with the x-axis, $\phi(x, y) \to (2/\pi)\theta$. Thus, depending on the direction of approach that we choose, ϕ can approach any value between zero and one as we approach the discontinuity point. As required, ϕ is certainly bounded, although its derivatives are not. We note in passing that our use of a series expansion carried with it the implicit assumption that the behavior of ϕ was not so bad as to make the series nonexistent; in effect, we "filtered out" any possible behavior of ϕ that the series could not manage.

An alternative approach to the problem of determining the behavior of ϕ near a point of discontinuity will be found in Problem 4.8.9; this alternative approach does not require us to find an exact solution of the given problem and so is more generally applicable. It can also be used for problems in which one is willing to accept a somewhat stronger singularity in the behavior of the solution (the permissible strength of the singularity of the solution is usually constrained by some such physical condition as finiteness of total energy or total force).

4.7 COMPLEX VARIABLES AND CONFORMAL MAPPING

Let $z = x + iy$ and $w = u + iv$ be complex variables, with x, y, u, v real. Let there be a functional relationship between z and w, so that $w = f(z)$; moreover, let this function be differentiable at each point z within some region of the complex z plane, in the sense that

$$f'(z) = \frac{df}{dz} = \lim_{\Delta z \to 0} \frac{f(z + \Delta z) - f(z)}{\Delta z} \tag{4.14}$$

exists for any mode of approach to zero of Δz. We then say that w is an analytic function of z within that region. We may think of each of u and v as conventional real functions of x and y, and the reader will recall that by considering two special modes of approach of Δz to zero—viz., parallel to the real or imaginary axis, respectively—one obtains the Cauchy–Riemann conditions

$$u_x = v_y, \qquad u_y = -v_x \tag{4.15}$$

Conversely, if two functions u and v satisfy these two conditions, it follows easily that $w = u + iv$ is a differentiable function of z.

Our first remark is that, from Eqs. (4.15), $u_{xx} + u_{yy} = v_{xx} + v_{yy} = 0$, so that *the real and imaginary parts of any analytic function are harmonic.* This fact provides us with a simple mechanism for generating harmonic functions. For example, each of the functions z^n, e^z, $\sin z = (1/2i)(e^{iz} - e^{-iz})$, $\sinh z = \frac{1}{2}(e^z - e^{-z})$, $\ln z$ (for $z \neq 0$) is easily seen to be analytic [just by application of the test (4.14)], so that each of

$$\mathrm{Re}\{(x + iy)^n\} = x^n - \tfrac{1}{2}n(n - 1)x^{n-2}y^2 + \cdots$$

$$\mathrm{Im}\{\sin(x + iy)\} = \cos x \sinh y$$

$$\mathrm{Re}\{e^{\sin z}\} = e^{\sin x \cosh y} \cos(\cos x \sinh y)$$

and so on, are necessarily harmonic. We note that polar coordinates are often convenient; thus we could write $\mathrm{Re}(z^{-3}) = \mathrm{Re}\{(re^{i\theta})^{-3}\} = r^{-3} \cos 3\theta$ and so deduce at once that $(\cos 3\theta)/r^3$ was harmonic (except, of course, at $r = 0$).

Our second remark is that, conversely, *any harmonic function is the real (or imaginary—whichever we prefer) part of some analytic $f(z)$.* Suppose for example that $u(x, y)$ is harmonic in a region R.† Define a function $v(x, y)$ by means of

$$v(x, y) = \int_{(x_0, y_0)}^{(x, y)} [(-u_y)\, dx + (u_x)\, dy]$$

[where (x_0, y_0) is an arbitrary fixed point, and where the path of integration lies in R] and observe that the standard condition that the result of this integration be independent of the path of integration is met, since $(-u_y)_y = (u_x)_x$. Thus $v(x, y)$ is a well-defined function, within an arbitrary constant. Moreover, partial differentiation with respect to each of x and y of the defining equation shows that $v_x = -u_y$, $v_y = u_x$, so that the Cauchy–Riemann conditions are met, and we conclude that the function $(u + iv)$ is an analytic function of $(x + iy)$. We say that v is *conjugate* to u.

These two results show that the theory of harmonic functions in two dimensions is essentially equivalent to the theory of analytic functions of a complex variable.

We turn now to a technique for solving certain kinds of potential theory problems, based on *conformal mapping*. We will content ourselves with an explanation of the basic idea and with one simple example; for a more extensive treatment, see Carrier *et al.* (1966, Chap. 4).

Let $w = f(z)$ be analytic, with $z = x + iy$ and $w = u + iv$. We can visualize this relationship as a mapping‡ from the z plane into the w plane; in particular, a region R with boundary Γ in the z plane will be carried into a region R' with boundary Γ' in the w plane (see Fig. 4.3). We restrict ourselves to the case in which there is only one point in R that maps into any chosen point in R'. Suppose now that $\psi(u, v)$ is a harmonic function of the two real variables u and v. Define a function $\phi(x, y)$ to be the "image" of ψ under the mapping $w = f(z)$—i.e., the value of ϕ at any point (x, y) in R is to be equal to the value of ψ at that point (u, v) into which (x, y) is carried by the mapping. Then

$$\phi_x = \psi_u u_x + \psi_v v_x$$

$$\phi_{xx} = \psi_{uu}(u_x)^2 + \psi_u u_{xx} + 2\psi_{uv}u_x v_x + \psi_{vv}(v_x)^2 + \psi_v v_{xx}$$

and similarly

$$\phi_{yy} = \psi_{uu}(u_y)^2 + \psi_u u_{yy} + 2\psi_{uv}u_y v_y + \psi_{vv}(v_y)^2 + \psi_v v_{yy}$$

† For simplicity, we take R to be simply connected; the extension to multiply connected regions is easy, although it may result in multiple-valued analytic functions.

‡ Such a mapping defined by an analytic function is termed *conformal* because of its property of preservation of angles between directions.

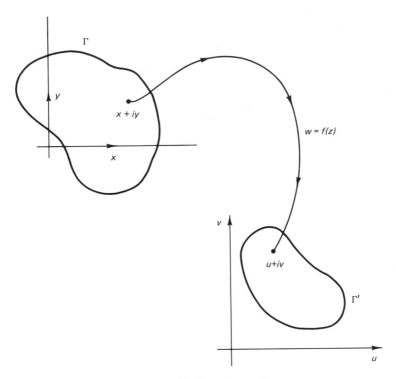

Fig. 4.3   Conformal mapping.

Consequently, using $u_x = v_y$, $u_y = -v_x$, $u_{xx} + u_{yy} = v_{xx} + v_{yy} = 0$, we find

$$\phi_{xx} + \phi_{yy} = (\psi_{uu} + \psi_{vv})(u_x{}^2 + u_y{}^2) = 0$$

Thus the image function $\phi(x, y)$ is a harmonic function in the region R.

An example will show how this result can be used. Suppose we want to determine a function ϕ, harmonic in the annular region R lying between two nonconcentric circles defined by $|z| = 1$ and $|z - 1| = \frac{5}{2}$ in the z plane (Fig. 4.4), and satisfying $\phi = 0$ on the inner circle and $\phi = 1$ on the outer circle. The mapping

$$w = (z + \tfrac{1}{4})/(z + 4) \tag{4.16}$$

carries R into a region R', in the w plane, contained between the two *concentric* circles $|w| = \frac{1}{4}$ and $|w| = \frac{1}{2}$ (verify this result). In terms of the R' region, our problem has become that of determining a function ψ that is harmonic in R' and that satisfies $\psi = 0$ on $|w| = \frac{1}{4}$ and $\psi = 1$ on

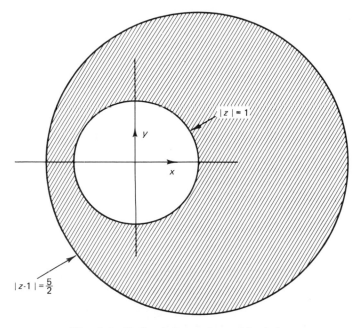

Fig. 4.4 Region between concentric circles.

$|w| = \frac{1}{2}$. But this is easy; the solution is obviously

$$\psi(r, \theta) = \ln(4r)/\ln 2$$

where (r, θ) are polar coordinates in the w plane (i.e., $w = re^{i\theta}$). The value of ϕ at any point $z = \rho e^{i\alpha}$ in R must equal the value of ψ at the point w that corresponds to z via Eq. (4.16); thus [in terms of polar coordinates (ρ, α)]

$$\phi(\rho, \alpha) = \frac{1}{\ln 2} \ln\left(4\left|\frac{z + \frac{1}{4}}{z + 4}\right|\right) = \frac{1}{2\ln 2} \ln\left\{\frac{16\rho^2 + 1 + 8\rho \cos \alpha}{\rho^2 + 16 + 8\rho \cos \alpha}\right\}$$

For the conformal mapping technique to be successful, we must find a function $f(z)$ that maps the given region into one for which the Dirichlet problem can be solved easily—e.g., into a circle. There are standard techniques that will do this for regions of polygonal shape, regions bounded by arcs of circles, regions involving "slits," etc.; many such situations will be found listed in tables of conformal mappings.† For regions that are "almost" of such a standard shape, approximation techniques are available; again, we refer to books on complex-variable techniques for further discussion.

† See, for example, Kober (1952).

4.8 PROBLEMS

4.8.1   (a)   Show that the solution to the problem of Section 4.6, when the boundary condition for $\phi(0,y)$ is modified to read $\phi(0, y) = 1$ for $0 < y < 1$, is given by

$$\phi(x, y) = \frac{2}{\pi} \tan^{-1}\left(\frac{\sin \pi y}{\sinh \pi x}\right)$$

(b)   More generally, let $\phi(0, y) = f(y)$ for $0 < y < 1$, where $f(y)$ is some given function. Obtain as compact a form for the solution as you can.

4.8.2   In the problem of Section (4.6), replace the boundary condition for $\phi(0, y)$ by $\phi_x(0, y) = 1 - y$ for $0 < y < 1$, and obtain a compact expression for the solution. [One form of the solution, valid also for the boundary line $x = 0$, is

$$\phi(x, y) = \left(\frac{\ln 2}{\pi} - x\right)y + \frac{1}{\pi}\int_0^y \ln(\cosh \pi x - \cos \pi \eta)\, d\eta]$$

4.8.3   Consider the complex z plane with a "barrier" along that part of the real axis satisfying $-1 \le x \le 1$—i.e., we are not allowed to cross over this line segment. Discuss the properties of the imaginary part of the function $\ln[(z + 1)/(z - 1)]$, and in particular compute its different limiting values as we approach points on the top, or on the bottom, of the barrier. What boundary value problem does this function solve?

4.8.4   Discuss the properties of the mapping function

$$w = \tfrac{1}{2}(z + 1/z)$$

What happens to circles and straight-line rays in the z plane? Can this mapping function, together with the result of Problem 4.5.6c, be used to solve the exterior Dirichlet problem for a region outside an ellipse? If so, provide a recipe.

4.8.5   With $w = u + iv$, $z = x + iy$, let $w = f(z)$ be analytic in a region of the z plane. Show that the (x, y)-plane curves $u = \text{const}$, $v = \text{const}$ intersect orthogonally. Show also that the directional derivative of u in any direction is equal to the directional derivative of v in the orthogonal direction, and use this fact to explain how a Neumann problem for a finite region R with boundary Γ (having a continuously turning tangent) can be transformed into a Dirichlet problem for the conjugate function.

4.8.6   In terms of polar coordinates, let $\phi(r, \theta)$ be harmonic inside some circle of radius R. In the region $r > R$, define a function

$$\psi(r, \theta) = \phi(R^2/r, \theta)$$

and show that ψ is harmonic in this region. Does this relationship correspond to a conformal mapping?

4.8.7 Let ϕ be harmonic and positive in a region A of the (x, y) plane. Choose a point P inside A as the origin of a polar coordinate system, and consider a circle of radius R, center P, lying wholly within A. In the Poisson integral formula, Eq. (4.12), observe that $-1 \leq \cos(\alpha - \theta) \leq 1$, and hence obtain two bounds on $\phi(r, \theta)$ (for $r < R$) by replacing the cosine term in the integrand by either -1 or $+1$. This result is termed *Harnack's inequality*:

$$\phi_P \frac{R - r}{R + r} \leq \phi(r, \theta) \leq \phi_P \frac{R + r}{R - r}$$

(where ϕ_P denotes the value of ϕ at the point P). Sketch the two bounds as functions of r. What happens if $R = \infty$?

4.8.8 (a) Consider a region R bounded by a curve Γ (Fig. 4.5). The uniqueness theorem of Section 4.3, stated somewhat more carefully, requires that the solution to the following problem be unique. Let continuous boundary data be prescribed on Γ. Find a function ϕ, harmonic inside R, such that as we approach any boundary point on Γ, the value of ϕ approaches the given boundary data at that point.

We now modify these conditions. Let the given boundary data be continuous everywhere on Γ except at the point P, where there is a discontinuity. Show that the solution to the following problem is unique: "Find ϕ, harmonic *and bounded* inside R, such that as we approach any boundary point on Γ other than P, the value of ϕ approaches the given

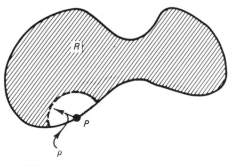

Fig. 4.5 Data discontinuity point.

boundary data at that point." [*Hint:* consider the function $u = \varepsilon \ln(D/r)$, where $0 < \varepsilon \ll 1$, r is the distance from P, and D is the greatest diameter of R. Then u is positive in R. Delete from R that portion cut out by a circle of radius ρ centered on P, where ρ is so small that $u > A$ (where A is some chosen constant) on that part of the boundary of the circle within R. Next, let ϕ_1 and ϕ_2 be two solutions to the original problem, and define $\psi = \phi_1 - \phi_2$. Then choose A such that $u \geq |\psi|$ on the boundary of the deleted region, and hence at any interior point, etc.] Note that the extension to several points of discontinuity is easily made.

(b)   Show that if boundedness is not imposed, the solution need not be unique. [*Hint:* consider the semicircle $0 < r < 1$, $0 < \theta < \pi$, and functions like $(r - 1/r) \sin \theta$, $(r^2 - 1/r^2) \sin 2\theta$, etc.]

4.8.9   Carry out the following alternative approach to investigating the nature of the singularity of the function ϕ of Section 4.6, near the point $(0, 0)$. From physical expectation (e.g., via the heat conduction interpretation of Section 4.1), we expect the nature of ϕ near $(0, 0)$ to depend only on the boundary values near that point. Hence for the polar coordinate problem: find $\phi(r, \theta)$, harmonic and bounded in the region $0 < r < \infty$, $0 < \theta < \pi/2$, with $\phi(r, 0) = 0$ and $\phi(r, \pi/2) = 1$ for $r > 0$, the solution should exhibit the same behavior near $(0, 0)$. But solutions to this problem, via separation of variables, look like $(A + B\theta)(C + D \ln r)$, $(Ar^\lambda + Br^{-\lambda})(C \sin \lambda\theta + D \cos \lambda\theta)$, etc.

4.8.10   In the quarter-plane region $x > 0$, $y < 0$, let $\phi(x, y)$ satisfy $\phi_{xx} + \phi_{yy} = 0$. The boundary conditions are that $\phi_x(0, y) = 0$, for $y < 0$, and $\phi_y(x, 0) = \phi(x, 0)$ for $x > 0$. Moreover, $\phi \sim e^{y-ix}$ as $x \to \infty$, for any value of y, and $\phi/\ln(x^2 + y^2)$ is to be bounded as $x^2 + y^2 \to 0$. Determine ϕ. [*Hint:* use a new unknown function $\psi = \phi_y - \phi$, and experiment with functions like $\psi = Ar^n \sin n\theta$ for appropriate (negative) values of n. In manipulating the formula for ϕ_x, integration by parts and use of a new variable of integration $(x\tau)$ may be useful.]

5

CLASSIFICATION OF SECOND-ORDER EQUATIONS

We turn now to the general linear second-order equation in two independent variables:

$$A\phi_{xx} + 2B\phi_{xy} + C\phi_{yy} + D\phi_x + E\phi_y + F\phi + G = 0 \qquad (5.1)$$

where A, B, C, D, E, F, and G are given functions of x and y. (The factor 2 in front of B is for future convenience.) In previous chapters we have discussed three special cases—the diffusion equation, the wave equation, and the potential equation. In each of these, the kinds of initial and/or boundary conditions that are natural to impose, and the properties of the solutions, are rather different. It might appear that still other special cases of interest could be extracted from Eq. (5.1); we will now show, however, that (at least locally) a change in variables can always be found such that Eq. (5.1) takes the form of one of the three special equations we have already studied.

5.1 CAUCHY DATA ON y-AXIS

To motivate our classification procedure, suppose that we try to set up a standard kind of initial or boundary value problem in terms of which solutions to Eq. (5.1) can be discussed.

An instructive experiment along these lines is to prescribe ϕ along some part of the y-axis; let $\phi(0, y) = f(y)$ for some range of values of y. Then

$\phi_y(0, y)$, $\phi_{yy}(0, y)$, $\phi_{yyy}(0, y)$, ... are obtained by repeated differentiation of $f(y)$. However, we can say nothing about $\phi_x(0, y)$, $\phi_{xx}(0, y)$, $\phi_{xxx}(0, y)$— apart of course from the fact that Eq. (5.1) imposes a certain relation between $\phi_x(0, y)$ and $\phi_{xx}(0, y)$. Let us therefore also prescribe $\phi_x(0, y) = g(y)$, say, for the same range of y values. Then we obtain $\phi_{xy}(0, y)$, $\phi_{xyy}(0, y)$, ... by repeated differentiation of $g(y)$. Moreover, we can now obtain $\phi_{xx}(0, y)$ using Eq. (5.1) (since all other terms are known, for $x = 0$; we assume for the present that $A \neq 0$) and differentiation gives all y derivatives of $\phi_{xx}(0, y)$. Differentiation of Eq. (5.1) with respect to x permits the calculation of $\phi_{xxx}(0, y)$, and proceeding similarly we can now compute all partial derivatives of ϕ on that part of the y-axis on which ϕ and ϕ_x are prescribed. We have assumed here, of course, that f, g, and all coefficient functions in Eq. (5.1) are adequately differentiable.

We could now use the calculated values of the x derivatives of ϕ on the y-axis to find, via a Taylor series, values of ϕ off that axis:

$$\phi(x, y) = \phi(0, y) + x\phi_x(0, y) + \frac{x^2}{2!}\phi_{xx}(0, y) + \frac{x^3}{3!}\phi_{xxx}(0, y) + \cdots$$

This procedure would break down if one of the above derivatives did not exist, or if the series did not converge. There is, however, a formal theorem (the Cauchy–Kowalewski Theorem†) that states that if $f(y)$, $g(y)$, and all of the functions B/A, C/A, D/A, ... are expansible in power series in x and y (i.e., are analytic functions) in the neighborhood of some point $(0, y_0)$, then this procedure will indeed generate a unique solution $\phi(x, y)$ analytic in some neighborhood of $(0, y_0)$.

In any event, we see that a prescription of ϕ and ϕ_x along a portion of the y-axis is sufficient to characterize rather thoroughly the solution function ϕ of Eq. (5.1). Note that ϕ_x is the derivative of ϕ in the direction normal to the y-axis. If we now consider, more generally, *any* curve Γ in the (x, y) plane, it is reasonable to formulate the following problem: given ϕ and $\partial\phi/\partial n$ (the normal derivative) on a curve Γ, determine the solution ϕ of Eq. (5.1) in some region adjacent to Γ. For brevity, we will call such boundary data *Cauchy data*. It will turn out that, for some equations of the form (5.1), this problem can be solved no matter how Γ is chosen, whereas for other equations this is not always the case (consider for example the possiblity $A \equiv 0$ in the example just discussed, in which Γ is a part of the y-axis). This dichotomy of possibilities will suggest a classification procedure for Eq. (5.1).

---

† For a proof of this theorem, see Garabedian (1964, p. 16) or Courant and Hilbert (1962, p. 39).

5.2 CAUCHY DATA ON ARBITRARY CURVE

Consider a curve Γ, defined by $\xi(x, y) = 0$, say, on which Cauchy data is prescribed. We will try to generalize the procedure of Section 5.1 so as to determine all derivatives of ϕ on Γ. For convenience, consider adjacent curves of the form $\xi(x, y) = \text{const}$, where various values of the constant are chosen. Introduce also a family of curves $\eta(x, y) = \text{const}$, in such a way that we can use the values of ξ and η as a local coordinate system (Fig. 5.1). In particular, we require that the Jacobian

$$J\left(\frac{\xi, \eta}{x, y}\right) = \xi_x \eta_y - \xi_y \eta_x \tag{5.2}$$

be nonzero.† Note that the ξ and η families of curves need not be mutually orthogonal.

We now think of ϕ as a function of ξ and η. At any point on Γ, we know from the prescribed data what the rate of change of ϕ is in each of the tangential and normal directions, and so in any direction. Thus we can

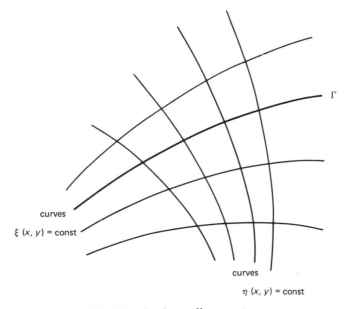

curves
$\xi(x, y) = \text{const}$

curves
$\eta(x, y) = \text{const}$

Fig. 5.1 (ξ, η) coordinate system.

† The condition that two curves $\xi(x, y) = \text{const}$ and $\eta(x, y) = \text{const}$ be tangent to one another at a point is easily seen to be equivalent to the condition that $\xi_x \eta_y - \xi_y \eta_x = 0$ at that point.

calculate ϕ_ξ and ϕ_η at each point of Γ (see Problem 5.3.2). We now proceed just as in Section 5.1 to compute higher derivatives. Since ϕ and ϕ_ξ are known functions of η on Γ, we can differentiate with respect to η to obtain values of ϕ_η, $\phi_{\eta\eta}$, ..., $\phi_{\xi\eta}$, $\phi_{\xi\eta\eta}$, ... on Γ. To determine $\phi_{\xi\xi}$ on Γ we must, however, use Eq. (5.1). But

$$\phi_x = \phi_\xi \xi_x + \phi_\eta \eta_x$$

$$\phi_{xx} = \phi_{\xi\xi}\xi_x^2 + 2\phi_{\xi\eta}\xi_x\eta_x + \phi_{\eta\eta}\eta_x^2 + \phi_\xi \xi_{xx} + \phi_\eta \eta_{xx}$$

and so on, so that Eq. (5.1) becomes

$$\phi_{\xi\xi}(A\xi_x^2 + 2B\xi_x\xi_y + C\xi_y^2) + 2\phi_{\xi\eta}(A\xi_x\eta_x + B[\xi_x\eta_y + \eta_x\xi_y] + C\xi_y\eta_y)$$

$$+ \phi_{\eta\eta}(A\eta_x^2 + 2B\eta_x\eta_y + C\eta_y^2) + \phi_\xi(A\xi_{xx} + 2B\xi_{xy} + C\xi_{yy} + D\xi_x + E\xi_y)$$

$$+ \phi_\eta(A\eta_{xx} + 2B\eta_{xy} + C\eta_{yy} + D\eta_x + E\eta_y) + F\phi + G = 0 \qquad (5.3)$$

Our ability to determine $\phi_{\xi\xi}$ on Γ then hinges on whether or not

$$A\xi_x^2 + 2B\xi_x\xi_y + C\xi_y^2 = 0 \qquad (5.4)$$

on Γ. Suppose first that there is *no* real function $\xi(x, y)$ for which Eq. (5.4) is satisfied (that this situation can occur is demonstrated by the simple example $A = C = 1$, $B = 0$). Then no matter what curve Γ is chosen, the coefficient of $\phi_{\xi\xi}$ in Eq. (5.3) cannot vanish; so $\phi_{\xi\xi}$ can be determined along Γ. All higher derivatives of ϕ can now also be determined on Γ by repeated differentiation, just as in Section 5.1. Suppose second, however, that a function $\xi(x, y)$ for which Eq. (5.4) is satisfied along the curve $\xi = 0$ *does* exist. Then the curve Γ defined by $\xi(x, y) = 0$ is one for which the above procedure fails, since the coefficient of $\phi_{\xi\xi}$ in Eq. (5.3) now vanishes along Γ.

Thus there are two possibilities associated with the rather reasonable Cauchy data problem we have chosen as a standard problem in terms of which to discuss Eq. (5.1). Either the problem is always solvable, or alternatively, there exist special curves for which the problem is not solvable. In examining the possibility of finding solutions for Eq. (5.4), we note that the left-hand side of this equation is a quadratic form in either (ξ_x/ξ_y) or (ξ_y/ξ_x); the existence or nonexistence of real roots of this quadratic form depends on the sign of the discriminant $B^2 - AC$. There are three possible cases: $B^2 - AC > 0$, $B^2 - AC = 0$, and $B^2 - AC < 0$; we consider these three cases in Sections 5.4, 5.5, and 5.6, respectively.

5.3 PROBLEMS

5.3.1 (a) Let $u(x, y)$ satisfy $u_{xx} + u_{yy} = 0$, and let $u(0, y) = \sin y$, $u_x(0, y) = y$, for all values of y. Use the procedure of Section 5.1 to deter-

mine the Taylor series expansion of $u(x, y)$, and sum the series to show
that $u = \sin y \cdot \cosh x + xy$.

(b) Given that $\phi_t = \phi_{xx}$, with $\phi(0, t) = 1$ for $t > 0$ and $\phi_x(0, t) = t$
for $t > 0$, determine all derivatives of ϕ on the line $x = 0, t > 0$. What is ϕ?

(c) Let $\phi_t = \phi_{xx}$, with $\phi(x, 0) = 1$ for $x > 0$ and $\phi_t(x, 0) = x$ for
$x > 0$. Can you determine ϕ? Why not?

5.3.2 Let $\xi(x, y)$ and $\eta(x, y)$ be two given functions of x and y for which
the Jacobian of Eq. (5.2) is nonzero over some region of interest. Let the
curve Γ be defined by the condition $\xi(x, y) = 0$; equivalently, if λ is arc
length along this curve, let Γ be defined by $x = X(\lambda), y = Y(\lambda)$. Finally,
let a function ϕ and its normal derivative ϕ_n be prescribed along Γ, say
$\phi = f(\lambda)$ and $\phi_n = g(\lambda)$ on Γ. Show that

$$\phi_\xi = \frac{\dot{X}[\dot{f}\eta_y - g\eta_x] - \dot{Y}[\dot{f}\eta_x + g\eta_y]}{\xi_x\eta_y - \xi_y\eta_x}$$

on Γ, and obtain a similar formula for ϕ_η. (In traversing Γ in the direction
of increasing λ, the positive direction for the unit normal vector is here
taken to the left.)

5.3.3 For the diffusion equation $u_{xx} - u_t = 0$ satisfied by a function
$u(x, t)$, Cauchy data on the line $x = 0, t > 0$, would correspond to a specification of $u(0, t) = f(t)$ and $u_x(0, t) = g(t)$. Use a transform method to
solve this problem for the region $x > 0, t > 0$. [*Hint:* if $u(x, 0)$ were known,
a conventional Laplace transform method, using the given value of $u(0, t)$,
would work. Carry this through formally, and then use the fact that
$u_x(0, t) = g(t)$ in order to obtain $u(x, 0)$. Show in particular that

$$\int_0^\infty e^{-\alpha x} u(x, 0) \, dx = \alpha F(\alpha^2) + G(\alpha^2)$$

so that $u(x, 0)$ may be obtained by inverting a Laplace transform with
respect to x.]

5.4 CASE I: $B^2 - AC > 0$

Let $B^2 - AC > 0$ throughout some region R of the (x, y) plane. Assuming temporarily that $A \neq 0$, write Eq. (5.4) in the form

$$\xi_x/\xi_y = [-B \pm (B^2 - AC)^{1/2}]/A \tag{5.5}$$

But the slope of the curve $\xi(x, y) = \text{const}$ is given by $dy/dx = -\xi_x/\xi_y$,
so we conclude that Cauchy data given along any curve $y(x)$ for which

$$dy/dx = [B + (B^2 - AC)^{1/2}]/A \tag{5.6}$$

or

$$dy/dx = [B - (B^2 - AC)^{1/2}]/A \qquad (5.7)$$

will be insufficient. The right-hand sides are real functions of x and y, so curves satisfying these equations may be found by conventional means (e.g., numerical procedures). Such curves are called *characteristics*.

To be explicit, we choose Eq. (5.6) as that defining the family $\xi(x, y) =$ const.† The curves $\eta(x, y) =$ const can be any other set of curves nowhere tangent to the curves $\xi(x, y) =$ const; let us in particular choose that family defined by Eq. (5.7), noting that the condition $B^2 - AC > 0$ ensures that the slopes (5.6) and (5.7) are nowhere the same. In this special (ξ, η) coordinate system, Eq. (5.3) takes a remarkably simple form. The curves $\xi(x, y) =$ const satisfy Eq. (5.6) and hence Eq. (5.4); similarly, as a consequence of Eq. (5.7), we have

$$A\eta_x^2 + 2B\eta_x\eta_y + C\eta_y^2 = 0.$$

Thus the coefficients of $\phi_{\xi\xi}$ and $\phi_{\eta\eta}$ in Eq. (5.3) vanish, and after dividing through by the coefficient of $\phi_{\xi\eta}$, we obtain the *canonical form*

$$\phi_{\xi\eta} + \alpha\phi_\xi + \beta\phi_\eta + \gamma\phi + \delta = 0 \qquad (5.8)$$

where α, β, γ, δ are functions of ξ and η.

In the above discussion we assumed $A \neq 0$. If $A = 0$, but $C \neq 0$, then Eq. (5.4) can be solved for ξ_y/ξ_x and the characteristics can be defined in terms of dx/dy; we again obtain Eq. (5.8). If both A and C vanish, then the characteristics are straight lines parallel to the x and y axes, and division of Eq. (5.1) by the quantity $2B$ (which is nonzero since $B^2 - AC > 0$) gives Eq. (5.8) at once.

If we start with Eq. (5.1) and change to any other coordinate system instead of (x, y), then the existence or nonexistence of real characteristics cannot be affected. Thus, if the condition $B^2 - AC > 0$ is fulfilled in the (x, y) system, the analogous condition must be fulfilled in any other system. We can verify this directly. Let (ξ, η) be any other system (not necessarily characteristic); then Eq. (5.1) becomes Eq. (5.3). Denote the coefficients of $\phi_{\xi\xi}$, $\phi_{\xi\eta}$, and $\phi_{\eta\eta}$ by a, $2b$, c, respectively. Direct calculation now shows that

$$b^2 - ac = (B^2 - AC)(\xi_x\eta_y - \xi_y\eta_x)^2 \qquad (5.9)$$

so that, as a result of the requirement that the Jacobian J of Eq. (5.2) be nonzero, we see that the sign of $B^2 - AC$ is the same as the sign of $b^2 - ac$. As a special case, we notice that if ξ and η are chosen as characteristic co-

---

† Equation (5.5) is a linear first-order PDE for ξ. A systematic discussion of such equations begins in Chapter 6.

ordinates so that $a = c = 0$, then the condition $b^2 > 0$ means $b \neq 0$, so that our previous division by $2b$ [in obtaining Eq. (5.8)] was legitimate.

If $B^2 - AC > 0$ in a region, Eq. (5.1) is said to be *hyperbolic*† in that region. Thus, if Eq. (5.1) is hyperbolic, we conclude that there exist two real families of characteristic curves with the property that along any curve of either family Cauchy data is inadequate. Moreover, by using these characteristics as new coordinate curves, Eq. (5.1) reduces to the canonical form (5.8). One example of a hyperbolic equation is the wave equation of Chapter 3.

5.5 CASE II: $B^2 - AC = 0$

If $B^2 - AC = 0$ in a region R, it follows that both A and C cannot vanish throughout R, for then B would also vanish, and we would no longer have a second-order equation. We postulate therefore—restricting ourselves to a subregion of R if necessary—that $A \neq 0$ in our region of interest. (If $A = 0$ and $C \neq 0$, the discussion in terms of dx/dy rather than dy/dx is analogous.) Equation (5.4) now has only the single solution

$$\xi_x/\xi_y = -B/A$$

so that there is only a single family of characteristics defined by

$$dy/dx = B/A$$

Thus in the so-called *parabolic case* there is only one family of characteristics (curves along which Cauchy data are insufficient), in place of the two independent families we encountered in the hyperbolic case.

Again, we anticipate some simplification if we alter the (x, y) coordinate system to a (ξ, η) system in which the $\xi = $ const curves are characteristics. From Eq. (5.4), the coefficient of $\phi_{\xi\xi}$ must then vanish—but as a result of Eq. (5.9), the coefficient of $\phi_{\xi\eta}$ must then also vanish. Dividing by the coefficient of $\phi_{\eta\eta}$ (necessarily nonzero, for otherwise the η and ξ families would coincide), we obtain the canonical form for the parabolic case:

$$\phi_{\eta\eta} + \alpha\phi_\xi + \beta\phi_\eta + \gamma\phi + \delta = 0 \tag{5.10}$$

where α, β, γ, δ are functions of ξ and η. The diffusion equation provides one example of a parabolic equation.

† One reason for this name is that the condition for a quadratic form

$$\alpha x^2 + 2\beta xy + \gamma y^2 + \delta x + \varepsilon y = \text{const}$$

to represent a hyperbola is that $\beta^2 - \alpha\gamma > 0$.

5.6   CASE III: $B^2 - AC < 0$

If $B^2 < AC$ in a region R, Eq. (5.1) is said to be *elliptic* in that region. Laplace's equation is one example. The condition $B^2 < AC$ means that Eqs. (5.6) and (5.7) have no real solutions, and thus that there are *no* curves along which Cauchy data is insufficient.

We now look for new variables (ξ, η) such that the form of Eq. (5.3) is simplified. To start with, we try to make the coefficient of $\phi_{\xi\eta}$ equal to zero. This requires

$$\xi_x(A\eta_x + B\eta_y) + \xi_y(B\eta_x + C\eta_y) = 0 \qquad (5.11)$$

If we choose any reasonable function $\eta(x, y)$, Eq. (5.11) tells us that the slope of a curve along which ξ is constant must be given by

$$\left.\frac{dy}{dx}\right|_{\xi=\text{const}} = -\frac{\xi_x}{\xi_y} = \frac{B\eta_x + C\eta_y}{A\eta_x + B\eta_y} \qquad (5.12)$$

and this gives us a differential equation that may be solved (at least numerically) to find those curves along which $\xi = $ const.† We can therefore find a new coordinate system in which the coefficient of the second mixed-derivative term vanishes, at least in some subregion of R. Let us rename as x and y the variables in terms of which this has been done; thus $B = 0$ in Eq. (5.1). We now look for new variables, renamed ξ and η, in terms of which Eq. (5.3) (with B now equal to zero) can be further simplified.

Actually, what we shall show is that we can find (ξ, η) variables such that the coefficients of $\phi_{\xi\xi}$ and $\phi_{\eta\eta}$ are equal and that of $\phi_{\xi\eta}$ is zero. This requires

$$A\xi_x^2 + C\xi_y^2 = A\eta_x^2 + C\eta_y^2, \qquad A\xi_x\eta_x + C\xi_y\eta_y = 0 \qquad (5.13)$$

where we have made use of $B = 0$. Since $AC > B^2$, A and C must have the same sign, which can be taken as positive, and neither can vanish. One way‡ in which to solve Eq. (5.13) is to require

$$\sqrt{A}\,\xi_x = \sqrt{C}\,\eta_y, \qquad \sqrt{C}\,\xi_y = -\sqrt{A}\,\eta_x \qquad (5.14)$$

---

† We remark that the families $\xi = $ const and $\eta = $ const are necessarily independent— for otherwise $\xi_x/\xi_y = \eta_x/\eta_y = \rho$, say, and Eq. (5.12) would then read

$$-\rho = (B\rho + C)/(A\rho + B)$$

but this quadratic equation for ρ has no real roots.

‡ Equations (5.13) are easily combined to give $A(\xi_x + i\eta_x)^2 = -C(\xi_y + i\eta_y)^2$; taking the square root and equating real and imaginary parts leads either to Eqs. (5.14) or to these equations with signs reversed.

Now the question is whether we can indeed find two functions, $\xi(x, y)$ and $\eta(x, y)$, such that Eqs. (5.14) are simultaneously satisfied. If η were known, we could determine ξ via

$$\xi(x, y) = \text{const} + \int_{(x_0, y_0)}^{(x,y)} [(C/A)^{1/2}\eta_y\, dx - (A/C)^{1/2}\eta_x\, dy] \quad (5.15)$$

where (x_0, y_0) is any chosen fixed point; Eq. (5.14) would then necessarily be satisfied. But Eq. (5.15) makes sense only if the integral is independent of the path of integration, and the standard condition for this is that

$$[(C/A)^{1/2}\eta_y]_y = -[(A/C)^{1/2}\eta_x]_x \quad (5.16)$$

This potential-type equation is analogous to that governing two-dimensional thermal equilibrium (with anisotropic conductivity), so that we can confidently expect such functions η to exist.

Choosing η to satisfy Eq. (5.16) and ξ to satisfy Eq. (5.15), the final result is that, in the elliptic case, Eq. (5.1) can be transformed into the canonical form

$$\phi_{\xi\xi} + \phi_{\eta\eta} + \alpha\phi_\xi + \beta\phi_\eta + \gamma\phi + \delta = 0 \quad (5.17)$$

where $\alpha, \beta, \gamma, \delta$ are functions of ξ and η.

5.7 PROBLEMS

5.7.1 Let $u(x, y)$ satisfy the equation

$$u_{xx} - 2u_{xy} + u_{yy} + 3u_x - u + 1 = 0$$

in a region of the (x, y) plane. Classify the equation and find its characteristics. Transform it into canonical form.

Construct a solution, if possible, for each of the two following Cauchy data problems:

(a) $u = 2, u_y = 0$ on the line $y = 0$
(b) $u = 2, \partial u/\partial n = 0$ on the line $x + y = 0$

Are the data in case (b) consistent with the governing equation in canonical form?

5.7.2 Classify as hyperbolic, parabolic, or elliptic, in R, each of the equations for $u(x, t)$:

(a) $u_t = (pu_x)_x$
(b) $u_{tt} = c^2 u_{xx} - \gamma u$
(c) $(qu_x)_x + (qu_t)_t = 0$

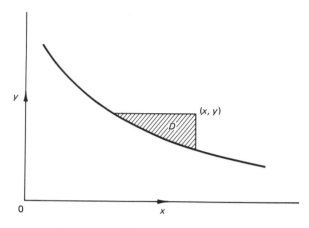

Fig. 5.2 Integration over a region D.

where $p(x)$, $c(x, t)$, $q(x, t)$, and $\gamma(x)$ are given functions that take on only positive values in a region R of the (x, t) plane.

5.7.3 Transform each of the equations for $\phi(x, y)$:

(a) $\phi_{xx} - y^2\phi_{yy} + \phi_x - \phi + x^2 = 0$
(b) $\phi_{xx} + y\phi_{yy} - x\phi_y + y = 0$
(c) $\phi_{xx} + x\phi_{yy} = 0$ [Tricomi equation]
(d) $\phi_{xy} + y\phi_{yy} + \sin(x + y) = 0$

into canonical form, in appropriate regions. Use the canonical form in part (d) to obtain the general solution

$$\phi = e^x \int_0^x e^{-\alpha} \cos(\alpha + ye^{\alpha-x}) \, d\alpha + e^x F(ye^{-x}) + G(x)$$

where F and G are arbitrary functions of their arguments.

5.7.4 (a) Show that an alternative canonical form to that of Eq. (5.8) is

$$\phi_{\xi\xi} - \phi_{\eta\eta} + \alpha\phi_\xi + \beta\phi_\eta + \gamma\phi + \delta = 0$$

[*Hint:* start with Eq. (5.8) and define $\xi_1 = \xi + \eta$, $\eta_1 = \xi - \eta$.]
(b) Let $\phi_{xy} = f(x, y)$, where f is a given function. Let Γ be a curve in the (x, y) plane (Fig. 5.2) that is nowhere tangent to either coordinate axis. From a point (x, y), draw lines parallel to the axes; denote by D that region (shaded in the figure) bounded by those two lines and by Γ. Show

that

$$\phi(x, y) = A(x) + B(y) + \int_D f(\alpha, \beta) \, d\alpha \, d\beta$$

where A and B are arbitrary functions. If u and $\partial u/\partial n$ are prescribed along Γ, can A and B be determined? [*Hint:* use the above formula to compute ϕ_x and ϕ_y, and let the point (x, y) approach Γ.]

5.7.5 The classification procedure of this chapter tacitly assumes that the important terms are those involving second-order derivatives; various examples considered in previous chapters indicate that this is indeed the usual situation insofar as the basic properties (e.g., signal speed, nature of appropriate boundary conditions) are concerned. Consider however (a) the two equations $u_{xx} = 0$, $u_{xx} - u_y = 0$ and (b) the two equations $u_{xy} = 0$, $u_{xy} + u_x = 0$, and draw whatever conclusions seem appropriate. In the general case of Eq. (5.1), *with constant coefficients*, to what extent can the various canonical forms be further simplified by a transformation of the form $\phi(\xi, \eta) = e^{\alpha\xi+\beta\eta}\psi(\xi, \eta)$, where α and β are constants?

5.7.6 Let a function $u(\xi, \eta)$ satisfy the equation

$$u_{\xi\eta} + \alpha u_\xi + \beta u_\eta + \gamma u + \delta = 0$$

where α, β, γ, δ are functions of ξ and η. In the (ξ, η) plane, the characteristics are lines parallel to the coordinate axes. We have shown that Cauchy data on a characteristic is not adequate for computation of the solution elsewhere; show moreover that Cauchy data cannot be arbitrarily prescribed on a characteristic in any event because of a compatibility condition imposed by the equation itself. What does this result imply about Cauchy data on a (x, y) plane characteristic of Eq. (5.1) in the hyperbolic case? In the parabolic case? Is there any similar restriction in the elliptic case?

5.7.7 Let the curve Γ defined by $\xi(x, y) = 0$ be a characteristic curve, so that the coefficient of $\phi_{\xi\xi}$ in Eq. (5.3) vanishes along Γ. A prescription of ϕ and its normal derivative $\partial\phi/\partial n$ along Γ is equivalent to a prescription of ϕ and ϕ_ξ along Γ (cf. Problem 5.3.2); obtain the relation between ϕ and ϕ_ξ that must be satisfied along Γ as a result of Eq. (5.3). Show that if this consistency condition is satisfied, many solutions of Eq. (5.3) compatible with the given Cauchy data can be found. How many? [*Hint:* a study of simple special cases may be useful.]

5.8   DISCONTINUITIES; SIGNAL PROPAGATION

In our discussion of the wave equation $u_{tt} = c^2 u_{xx}$ in Chapter 3, we found that disturbances (or signals) traveled with velocity c. In the (x, t) plane the trajectory of a disturbance front therefore is a curve Γ satisfying the condition $dx/dt = c$ (or the condition $dx/dt = -c$). This curve Γ divides the (x, t) plane into two regions; in one the disturbance has been experienced and in the other it has not. The functional form of u must therefore be different in the two regions, so that in some sense there is a discontinuity of u across Γ. We observe next that the curves defined by $dx/dt = \pm c$ are characteristics of the wave equation, and this raises now the more general question of whether the characteristics of Eq. (5.1) are also curves across which ϕ can exhibit some kind of discontinuity.

It turns out (cf. Problem 5.9.1) that the possibility of a discontinuity in either ϕ itself, or in its first partial derivative, is not particularly interesting; consequently we consider the following situation. Let Γ defined by $\xi(x, y) = 0$ be a curve in the (x, y) plane—we pause to introduce a (ξ, η) coordinate system as in Section 5.2—and let $\phi(\xi, \eta)$ be a function possessing continuous first and second derivatives throughout a region containing Γ, with the single exception that $\phi_{\xi\xi}$ may be discontinuous across Γ. Moreover, ϕ is to satisfy Eq. (5.3) on each side of Γ.

We will now show that this situation is possible only if the chosen curve Γ is a characteristic.

Let us approach a point on Γ from each of the two sides; denote values on one side of Γ by a superscript $+$ and values on the other side by a superscript $-$. Now Eq. (5.3) holds on each side of Γ, so that as we approach the point on Γ, we can write

$$\phi_{\xi\xi}^{+}(A\xi_x^2 + 2B\xi_x\xi_y + C\xi_y^2) + (\text{other terms})^{+} = 0$$

$$\phi_{\xi\xi}^{-}(A\xi_x^2 + 2B\xi_x\xi_y + C\xi_y^2) + (\text{other terms})^{-} = 0$$

where the "other terms" are all continuous across Γ. Subtracting, we obtain in the limit of approach

$$(\phi_{\xi\xi}^{+} - \phi_{\xi\xi}^{-})(A\xi_x^2 + 2B\xi_x\xi_y + C\xi_y^2) = 0$$

so that the only way in which $\phi_{\xi\xi}^{+} \neq \phi_{\xi\xi}^{-}$, i.e., in which $\phi_{\xi\xi}$ can be discontinuous across Γ, is if

$$A\xi_x^2 + 2B\xi_x\xi_y + C\xi_y^2 = 0 \qquad (5.18)$$

But this is again the equation defining a characteristic.

A similar result would have been obtained had we asked for a curve Γ across which $\phi_{\xi\xi\xi}$ (or a higher derivative) was discontinuous, all other derivatives of interest being continuous everywhere. For differentiation of Eq. (5.3) with respect to ξ gives

$$\phi_{\xi\xi\xi}(A\xi_x^2 + B\xi_x\xi_y + C\xi_x^2) + \text{other terms} = 0$$

and it is clear that the condition $\phi_{\xi\xi\xi}^+ \neq \phi_{\xi\xi\xi}^-$ would lead again to Eq. (5.18).

We therefore have the general result that solutions of Eq. (5.1), well behaved everywhere save that some second or higher order ξ derivative is discontinuous across a curve Γ, can be found only in the real characteristic case—i.e., only if Eq. (5.1) is hyperbolic or parabolic. We can reasonably conclude that if an equation of the form (5.1) is to possess a signal propagation property, the equation must be hyperbolic or parabolic—and only the former case will generally lead to a finite signal speed.

The above discussion may be framed in terms of a geometric visualization. A solution function $\phi(x, y)$ of Eq. (5.1) can be represented as a surface in (x, y, ϕ) space. A prescription of the value of ϕ at each point of some curve Γ in the (x, y) plane then defines a space curve through which the solution surface must pass; if $\partial\phi/\partial n$ is also prescribed at each point of Γ, the orientation of a local tangent plane to the solution surface, at each point of the space curve, is then also determined. Thus Cauchy data along Γ determine an "initial strip" in (x, y, ϕ) space through which the solution surface must pass, and tangentially so. Let us next define a "branch strip" as a strip of the above kind along which two different solution surfaces touch. [That is, let ϕ_1 and ϕ_2 be two solutions of Eq. (5.1), defined on the two sides of Γ. Along Γ, these two functions have the same values and the same slopes, so they contain the same initial strip. However, on Γ, ϕ_1 and ϕ_2 are to differ in some higher derivative.]

The analysis of this section then implies that the only possible branch strips are those for which the projections into the (x, y) plane are characteristics. Moreover, if Γ is a characteristic, we emphasize (cf. Problems 5.7.7 and 5.9.2) that Cauchy data along Γ cannot be arbitrarily prescribed; it must satisfy a consistency condition.

5.9 PROBLEMS

5.9.1 Consider any chosen straight line $ax + by + c = 0$ in the (x, y) plane, where a, b, c are constants. For any point (x, y) in the plane, define $\phi(x, y) = |ax + by + c|$. Show that ϕ satisfies Laplace's equation $\phi_{xx} + \phi_{yy} = 0$ on each side of the above line, and that ϕ is continuous across the line but that its normal derivative is not. Thus any straight line can serve as a carrier of a first-derivative discontinuity for Laplace's equation (which possesses no characteristics), and this simple example suggests that one could hardly expect to be led to characteristics on the basis of first-order derivative discontinuities.

More generally, let Γ be any curve across which a solution ϕ of Eq. (5.1) is to be continuous, with $\partial\phi/\partial n$ discontinuous. In effect, different Cauchy data are then prescribed on each side of Γ; explain why we can generally expect to be able to determine such a function ϕ for almost any choice of Γ. Consider similarly the possibility of a discontinuity in ϕ itself.

5.9.2 Let a curve Γ in the (x, y) plane be defined by $x = X(\lambda), y = Y(\lambda)$, where λ is arc length. If $\phi(x, y)$ is a function of x and y, we will use the notation $p = \phi_x$, $q = \phi_y$, $r = \phi_{xx}$, $s = \phi_{xy}$, $t = \phi_{yy}$ for its first and second partial derivatives. If Cauchy data are prescribed along Γ, then (cf. Problem 5.3.2) we can equivalently say that ϕ, p, and q are prescribed along Γ in such a way as to satisfy the "strip relation"

$$\dot\phi = p\dot X + q\dot Y$$

If we now require ϕ to be a solution of Eq. (5.1), the preceding sections lead us to ask whether or not we can now also determine r, s, and t along Γ. [We remark that if r, s, and t can be uniquely determined, then the initial strip determined by Γ, ϕ, p, q cannot be a branch strip in the sense of Section 5.8. Why not?]

Derive each of the following relations along Γ:

$$r\dot X + s\dot Y = \dot p, \qquad s\dot X + t\dot Y = \dot q$$
$$Ar + 2Bs + Ct = -Dp - Eq - F\phi - G$$

Since p, q, ϕ are known functions of λ along Γ, these three linear equations can be solved so as to determine r, s, and t along Γ, provided the coefficient determinant is nonzero. Show that this determinant will vanish only if Γ is a characteristic. If Γ is indeed a characteristic, the three equations can be consistent only if the right-hand sides satisfy a certain condition; determine this consistency condition.

5.9.3 Let the curve Γ, defined by $\xi(x, y) = 0$ as in Section 5.2 (we use the notation of that section), be a characteristic curve for Eq. (5.1). Let ϕ satisfy Eq. (5.1) on each side of Γ, and let $\phi_{\xi\xi}$ be discontinuous across Γ, with all lower derivatives continuous across Γ as in Section 5.8. Denote the difference in values of $\phi_{\xi\xi}$ on the two sides of Γ by $[\phi_{\xi\xi}]$. If Eq. (5.3) is written as

$$2b\phi_{\xi\eta} + c\phi_{\eta\eta} + d\phi_\xi + e\phi_\eta + f\phi + g = 0$$

show that $[\phi_{\xi\xi}]$ satisfies, along Γ, the *ordinary* differential equation

$$2b[\phi_{\xi\xi}]_\eta + d[\phi_{\xi\xi}] = 0$$

A consequence of this equation is that $[\phi_{\xi\xi}]$ vanishes either nowhere or everywhere, along Γ.

5.9.4 An equation such as (5.1) can be written in terms of a set of coupled equations involving derivatives of lower order. For example, let

$$\phi_{xx} - \phi_{yy} + \phi_\tau + y = 0$$

Then defining $\phi_x = u$, $\phi_y = v$, we have

$$u_y - v_x = 0, \qquad u_x - v_y + u + y = 0$$

Now let u and v (Cauchy data) be specified along some given curve Γ, defined as before by $\xi(x, y) = 0$. Transform this pair of coupled equations into a (ξ, η) system, and show that derivatives of u and v (specifically, u_ξ and v_ξ) can be uniquely determined only if Γ is not a characteristic. Generalize to the case of Eq. (5.1).

5.9.5 Let $\phi(x, y)$ satisfy the equation

$$A\phi_{xx} + 2B\phi_{xy} + C\phi_{yy} + D = 0 \tag{5.19}$$

where A, B, C, and D are functions of x, y, ϕ, ϕ_x, and ϕ_y. Note that only the second-order derivatives of ϕ occur linearly in this equation; the equation is said to be *quasi-linear*. Much of the discussion in this chapter can be extended so as to apply to Eq. (5.19); the main difference is that specified values of ϕ, ϕ_x, and ϕ_y along a curve Γ will have to be taken into account in deciding whether or not that curve is a characteristic. Carry out such an extension. Which parts of the theory *cannot* be carried over?

5.10 SOME REMARKS

We have now encountered the idea of a characteristic of Eq. (5.1) in several different way. It is a curve along which a prescription of Cauchy data is inadequate to uniquely determine the solution in an adjoining region; it is a curve along which Cauchy data cannot be freely specified but must satisfy a consistency condition; it is a curve along which different solutions may be joined smoothly together; or it is a curve that can represent a signal trajectory. We have seen that these various definitions are equivalent to one another, at least for the case of the linear equation (5.1).

Moreover, we have observed that, depending on whether there are one, two, or no families of real characteristics, Eq. (5.1) can be rewritten by means of a coordinate transformation in essentially the form of a diffusion equation, a wave equation, or a potential equation, respectively. The nature of the anticipated initial or boundary conditions to be specified, and the properties of the solution obtained, must be compatible with the particular category of equation under consideration. We have obtained useful experience with this matter in Chapters 1–4; probably the safest general guidance one can provide is that if a problem is physically reasonable it will also be mathematically reasonable.

An example involving Laplace's equation, due to Hadamard, is instructive. In dealing with an elliptic equation (such as Laplace's), it is typical that a linear combination of ϕ and $\partial\phi/\partial n$ will be specified on a closed boundary surrounding the region of interest. One almost never encounters a realistic problem for which Cauchy data are specified along some curve, despite the Cauchy–Kowalewski theorem of Section 5.1, which assures us that a unique solution of such an initial value problem could be constructed, at least if everything is analytic. Consider however the equation $\phi_{xx} + \phi_{yy} = 0$, with $\phi(0, y) = 0$, $\phi_x(0, y) = (1/N) \sin Ny$, where N is a positive integer. The solution is $(1/N^2) \sin Ny \sinh Nx$. As N becomes large, the Cauchy data approach zero uniformly along the y-axis. However, for (almost) any chosen point (x, y), the solution oscillates with increasing amplitude as N becomes large, so that the solution does not approach the value (zero) that is appropriate for the limiting case of zero Cauchy data. In other words, the solution is not continuously dependent on the initial data. Such a situation would not be reasonable in a practical problem, for one would normally expect that a slight alteration in the initial data would have a correspondingly slight effect on the solution. As a result of examples like this, we say that a Cauchy data problem for an elliptic equation is not "well posed."

From examples such as that of Problem 5.9.1 we can conclude that a discontinuity in ϕ or in its first derivative is of little interest, since it can be "carried" by almost any curve Γ. In the case of nonlinear equations, such lower order discontinuities ("shocks") may, however, be of much greater interest. We will encounter this situation in Chapter 13, where we will find that the use of certain accessory conditions permits shock curves to be delineated. In general, these curves will not coincide with characteristics.

6

FIRST-ORDER EQUATIONS

So far we have directed most of our attention toward second-order equations, primarily because simple physical problems often lead to such equations. However, first-order equations are by no means uncommon, and we now take up the subject of linear and quasi-linear first-order equations. Fully nonlinear first-order equations will be discussed in Chapter 12.

6.1 LINEAR EQUATION EXAMPLES

Let it be required that a function $u(x, y)$ be found satisfying the equation

$$u_x + 2xu_y = y \tag{6.1}$$

subject to the condition

$$u(0, y) = 1 + y^2 \quad \text{for} \quad 1 < y < 2 \tag{6.2}$$

The left-hand side of Eq. (6.1) is exactly the expression we would use to calculate $(d/dx)u(x, y(x))$, where $y(x)$ satisfies the condition $dy/dx = 2x$ (i.e., $y = x^2 + c_1$, where c_1 is a constant). Thus we can interpret Eq. (6.1) as the statement that, along any curve $y = x^2 + c_1$,

$$(d/dx)u(x, y(x)) = y = x^2 + c_1$$

so that, along such a curve,

$$u = \tfrac{1}{3}x^3 + c_1 x + c_2 \tag{6.3}$$

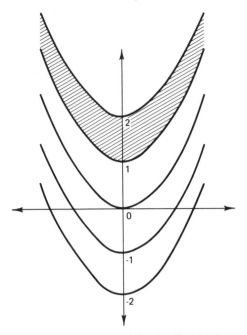

Fig. 6.1 Characteristics for Eq. (6.1).

where c_2 is a constant. The parabolas $y = x^2 + c_1$ are sketched in Fig. 6.1.
Along any one of these curves (the choice of a particular curve involves
a specification of c_1), Eq. (6.3) is satisfied; if we know u at some point on
that curve, we can use Eq. (6.3) to determine c_2, and thus u will be known
at all points of that curve. In the present case, Eq. (6.2) specifies $u(0, y)$
for $1 < y < 2$, so that we will be able to determine $u(x, y)$ at any point
(x, y) that can be linked to the segment (1, 2) of the y-axis by one of our
parabolas. This can be done for any point in the shaded region of Fig. 6.1.

In fact, let (x_0, y_0) be such a point. The parabola through (x_0, y_0) has
the equation

$$y = x^2 + (y_0 - x_0^2) \tag{6.4}$$

and this parabola cuts the y-axis at the point $y = y_0 - x_0^2$, at which point
Eq. (6.2) requires that

$$u = 1 + (y_0 - x_0^2)^2$$

The value of u on any other point of the parabola (6.4) is now given via
Eq. (6.3):

$$u = \tfrac{1}{3}x^3 + (y_0 - x_0^2)x + \{1 + (y_0 - x_0^2)^2\}$$

In particular, since the parabola passes through the point (x_0, y_0), we have

$$u(x_0, y_0) = \tfrac{1}{3}x_0^3 + (y_0 - x_0^2)x_0 + \{1 + (y_0 - x_0^2)^2\}$$

But (x_0, y_0) was any point in the shaded region; thus, replacing x_0 by x and y_0 by y, we have the result that at any point (x, y) in this region

$$u(x, y) = x^4 - \tfrac{2}{3}x^3 - 2x^2y + xy + y^2 + 1 \qquad (6.5)$$

and it may be verified by direct substitution that Eqs. (6.1) and (6.2) are satisfied.

This procedure has *uniquely* determined u throughout the shaded region, but it tells us nothing about u outside that region. In fact, we could assign u in any arbitrary (but say continuously differentiable) manner along that part of the y-axis outside the segment $(1, 2)$ and then use the above procedure to determine u throughout the rest of the plane. No matter how we chose these additional values of u along the y-axis, all conditions of the problem would be satisfied; it follows that the given data (6.2) are sufficient to determine u uniquely *only* within the shaded region.

The example

$$xu_x + yu_y = 1 + y^2 \qquad (6.6)$$

$$u(x, 1) = x + 1 \qquad (6.7)$$

suggests a slight modification of the above technique. We could divide the equation through by x, and this would lead us to consider the curves defined by $dy/dx = y/x$, along which $du/dx = (1/x) + (y^2/x)$. However, this formulation would introduce a singularity at $x = 0$ that does not seem inherent in Eqs. (6.6) and (6.7). Also, in general the process of dividing through by the coefficient of u_x (or of u_y) might lead to un-necessarily awkward differential equations. We avoid these inconveniences by utilizing a parametric representation.

Consider an (x, y) plane curve defined by

$$dx/dt = x, \qquad dy/dt = y \qquad (6.8)$$

where t is a parameter varying along the curve. Then along such a curve,

$$du/dt = u_x(dx/dt) + u_y(dy/dt) = xu_x + yu_y$$

so that, using Eq. (6.6),

$$du/dt = 1 + y^2 \qquad (6.9)$$

Equations (6.8) and (6.9) provide a set of three differential equations, which may be solved to give

$$x = c_1 e^t, \qquad y = c_2 e^t, \qquad u = t + \tfrac{1}{2} c_2^2 e^{2t} + c_3 \qquad (6.10)$$

where c_1, c_2, c_3 are constants.

We conclude that if $u(x, y)$ satisfies Eq. (6.6), then along any curve defined by the first two of Eqs. (6.10), u must satisfy the third of Eqs. (6.10). We are given initial data on the line $y = 1$, and we therefore want to consider curves $x = c_1 e^t$, $y = c_2 e^t$ emanating from this line. [For convenience, we can set $t = 0$ at the point of emanation for any one of these curves, since a different choice for t at this point simply corresponds to a different choice for the c_i.] The completion of this example is left for an exercise.

In these two examples we have obtained special curves in (x, y, u) space that lie on the solution surface. These curves are called *characteristics*; their projections on the (x, y) plane are called *characteristic traces* (or sometimes just characteristics—usage is not consistent). We will shortly see that these special curves have analogous properties to the characteristics of Chapter 5, so that the terminology is appropriate.

6.2 PROBLEMS

6.2.1 Complete the second example above to show that

$$u = \ln y + \tfrac{1}{2} y^2 + \tfrac{1}{2} + x/y \qquad \text{for} \quad y > 0$$

and explain, with the aid of a diagram, why we cannot uniquely determine $u(x, y)$ for $y \le 0$.

6.2.2 Use the method of characteristics to solve anew Eq. (I.6)—i.e., $\phi_x + \phi_y = 0$—and compare results.

6.2.3 Solve Eq. (6.1) subject to the condition $u(0, y) = f(y)$, where $f(y)$ is reasonably arbitrary. Do we obtain a solution with continuous values of u_x and u_y if $f'(y)$ is continuous?

6.2.4 Solve the problem

$$xu_x + (x + y)u_y = 1$$

$$u(1, y) = y \qquad \text{for} \quad 0 < y < 1$$

and describe the region over which the solution is uniquely determined.

6.2.5 Discuss the problem $u_x + 2xu_y = y$, with the initial conditions

(a) $u(x, x^2) = 1$, for $-1 < x < 1$, and (b) $u(x, x^2) = \frac{1}{3}x^3 + \pi$, for $-1 < x < 1$. How many solutions exist in each case? Note that the initial data are here prescribed along a characteristic trace.

6.2.6 The equation

$$u_x + 2xu_y = y + xu$$

differs from Eq. (6.1) in that a term involving u has been added to the right-hand side. It is, however, still linear. Show that the method of characteristics continues to be appropriate and leads to the requirement that

$$du/dx = y + xu$$

along any curve for which $dy/dx = 2x$. Find $u(x, y)$ if $u(0, y) = 1 + y^2$ for $1 < y < 2$ (in what region?), and verify the solution.

6.2.7 Let $u(x, y)$ be *homogeneous of degree n* in x and y, in the sense that $u(\lambda x, \lambda y) = \lambda^n u(x, y)$ for any constant λ, over the range of independent variables for which u is defined. Obtain a partial differential equation satisfied by u, and conversely, solve that equation. [*Hint:* differentiate the defining equation with respect to λ, and set $\lambda = 1$.]

6.3 QUASI-LINEAR CASE

We now discuss more formally the method of Section 6.1, and at the same time extend it slightly. The equation

$$f(x, y, u)u_x + g(x, y, u)u_y = h(x, y, u) \tag{6.11}$$

(where f, g, h are given functions of x, y, and u) is linear in the derivatives u_x and u_y, but not necessarily so in u. It is said to be *quasi-linear*. That special case in which f and g are functions of x and y only, and in which $h = r(x, y) \cdot u + s(x, y)$ (where r and s are functions of x and y) is of course termed *linear*; it turns out that the quasi-linear case is in principle no harder to treat than the linear case.

The method of characteristics is again useful. We consider the *characteristic curves* in (x, y, u) space defined by

$$dx/dt = f(x, y, u), \qquad dy/dt = g(x, y, u), \qquad du/dt = h(x, y, u)$$

$$\tag{6.12}$$

If f, g, and h are reasonably well-behaved functions, then a curve satisfying Eqs. (6.12) will be determined uniquely by the specification of any point

(x_0, y_0, u_0) through which it passes. The relationship between the curves (6.12) and solutions to Eqs. (6.11) is a consequence of the following two results:

(1)   If $u = \phi(x, y)$ is a solution to Eq. (6.11) representable as a surface in (x, y, u) space, and if (x_0, y_0, u_0) is a point on this surface, then the characteristic through this point lies in the surface. For the curve defined by

$$dx/dt = f(x, y, \phi(x, y)), \qquad dy/dt = g(x, y, \phi(x, y)), \qquad u = \phi(x, y)$$

certainly lies in the solution surface, and on this curve

$$du/dt = \phi_x\, dx/dt + \phi_y\, dy/dt = \phi_x f + \phi_y g = h \qquad [\text{from Eq. (6.11)}]$$

so that this curve satisfies the same three differential equations as does a characteristic; since the two curves have the point (x_0, y_0, u_0) in common, they must coincide.

(2)   Conversely, suppose that we start with some (noncharacteristic) curve defined parametrically by $x = x(\alpha)$, $y = y(\alpha)$, $u = u(\alpha)$, and construct the family of characteristics emanating from this curve (Fig. 6.2). Then the surface so generated must represent a solution of Eq. (6.11), for Eq. (6.11) is equivalent to the condition that $du/dt = h$ along any curve defined by $dx/dt = f$, $dy/dt = g$, and this condition is exactly that used to construct the surface.

Thus we can generally use the method of characteristics to construct a solution surface passing through a given initial curve $x = x(\alpha), y = y(\alpha)$,

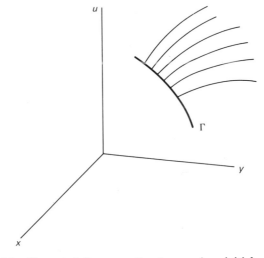

Fig. 6.2   Characteristics emanating from a given initial curve Γ.

$u = u(\alpha)$; moreover, this solution will be unique. We have implicitly assumed that the initial curve, and the functions f, g, and h, are adequately well behaved. We also assume that the characteristic traces do not intersect one another in the region of interest. Finally, we must exclude the case in which the initial curve is such that, at some point along it, $x'(\alpha) : y'(\alpha) :: f[x(\alpha), y(\alpha), u(\alpha)] : g[x(\alpha), y(\alpha), u(\alpha)]$, for as we will see in the next section, this could lead to an inconsistency. We turn now to an example.

Let

$$x^2 u_x + u u_y = 1 \qquad (6.13)$$

with $u = 0$ on the line $x + y = 1$, $x > 0$. From Eqs. (6.12), the characteristics satisfy

$$dx/dt = x^2, \qquad dy/dt = u, \qquad du/dt = 1$$

whence

$$x = 1/(c_1 - t), \qquad y = \tfrac{1}{2}t^2 + c_3 t + c_2, \qquad u = t + c_3$$

where the c_i are constants. That characteristic emanating from the initial curve at the point $x = s$, $y = 1 - s$ (we set $t = 0$ at this point) is given by

$$x = s/(1 - st), \qquad y = \tfrac{1}{2}t^2 + 1 - s, \qquad u = t \qquad (6.14)$$

Thus each of x, y, and u is a function of the two parameters s and t. Eliminating these parameters, we obtain

$$x/(1 + ux) = \tfrac{1}{2}u^2 + 1 - y \qquad (6.15)$$

which defines $u(x, y)$ implicitly.

6.4 PROBLEMS

6.4.1 (a) Sketch several typical characteristics as given by Eqs. (6.14) and make appropriate comments.

(b) Since Eq. (6.15) is a cubic in u, there presumably are three values of u corresponding to any choice for x and y; how could we choose the right value of u?

(c) By neglecting small quantities, simplify Eq. (6.15) so as to obtain an explicit formula for $u(x, y)$ valid near the initial curve. How does u_x behave as we approach the point $(0, 1)$?

6.4.2 Use the method of characteristics to show that the solution of

$$u u_x + u_y = 0, \qquad u(x, 0) = f(x)$$

is given implicitly by

$$u = f(x - uy)$$

and verify the result by direct differentiation. In what region is this result valid?

6.4.3 (a) Verify by differentiation that Eq. (6.15) is indeed a solution of Eq. (6.13).

(b) The form of Eq. (6.15) suggests that it might have been easier to consider x and u as independent variables, with $y = y(x, u)$. Show that $u_y = 1/y_u$, $u_x = -y_x/y_u$, and substitute these results into Eq. (6.13) so as to obtain an equation for $y(x, u)$. Solve it.

6.4.4 In a rotating-fluid problem† it is required that a function $v(r, \tau)$ be found in the region $0 < r < a$, $\tau > 0$, satisfying

$$v_\tau - (1 - v/r)(rv)_r = 0$$

$$v(r, 0) = 0, \qquad v(a, \tau) = a$$

Use the method of characteristics to obtain the solution

$$v = \frac{re^{2\tau} - (a^2/r)}{e^{2\tau} - 1}, \qquad r \geq ae^{-\tau}$$

$$v = 0, \qquad\qquad r \leq ae^{-\tau}$$

Sketch the two families of characteristics and discuss the nature of the solution near the point $r = a$, $\tau = 0$.

6.4.5 (a) Show that if two functions $u(x, y)$ and $w(x, y)$ are such that their Jacobian $u_x w_y - u_y w_x$ vanishes throughout a region, then their contour curves coincide, and therefore each of u and w is a function of the other. Use this fact to obtain the general solution of

$$u_x + 2xu_y = 0$$

noting that a suitable w function is given by $w = y - x^2$. Is this method as powerful as the method of characteristics?

(b) More generally, show that if $w = w(x, y, u)$, then the general solution of $u_x w_y - u_y w_x = 0$ is still given implicitly by $u = f(w)$.

6.4.6 Let

$$I(x, y) = \int_0^\infty \frac{t \exp[-y(1 + t^2)^{1/2}] \sin xt}{(1 + t^2)^{1/2}} \, dt$$

† Greenspan (1968, p. 165).

Use integration by parts to show that

$$I = x/y \int_0^\infty \exp[-y(1 + t^2)^{1/2}] \cos xt \, dt$$

and use differentiation with respect to each of x and y to obtain the result

$$I_y = [(y/x)I]_x$$

Solve this equation to obtain the result that $I = xf(x^2 + y^2)$, and find a simple form for the function f.

6.4.7 A typical glacier moves downhill as a result of a combination of (a) internal plastic deformation and (b) sliding over its bed.† For a thick cold glacier (a) is dominant, whereas for a thin warm glacier (b) is dominant. Consider a glacier of the latter kind and take the flow velocity u as constant over the thickness h. Assume its bed to be a plane surface of (small, constant) slope α, and as a first approximation take the shear stress τ between the glacier and its bed to be the slope component of the glacier weight, so that $\tau = \rho g h \alpha$, where ρ is the density of the ice and g the acceleration of gravity. Assume also that the relationship between τ and u has the form of a power law: $u = B\tau^m$, where B and m are constants.

Let the thickness h and velocity u be functions of position x and time t. Derive the continuity law $(hu)_x + h_t = 0$ (for the case of no accretion), and obtain an equation for h. In one set of experiments for a glacier of this kind, it was found that a zone of constant thickness moved at an average velocity of 0.2 km/yr, whereas the actual ice velocity was about 0.07 km/yr. Determine the appropriate value of m.

6.4.8 A continuum model of automobile traffic moving along a highway can be obtained by denoting the local density by ρ (number of automobiles per unit of length) and velocity by u, with each of ρ and u some function of position x and time t. A conservation-of-automobiles condition leads to $(\rho u)_x + \rho_t = 0$; let u be some reasonable function of ρ, and investigate solutions of this equation (e.g., consider the effect of a traffic light).

6.5 FURTHER PROPERTIES OF CHARACTERISTICS

In discussing Eq. (6.1) [or Eq. (6.11)] we were led to the idea of special curves in (x, y, u) space—characteristics—by noticing the similarity of the left-hand side of the equation to the formula for the derivative along

† See Embleton and King (1968, Chap. 4).

a curve of a function $u(x, y)$. We indicated that the name "characteristic" would be shown to be appropriate, in that there exists an alternative approach very similar to that used in our discussion of characteristics for second-order equations in Chapter 5. We now follow such an approach.

We have solved several cases of Eq. (6.11) in which u is specified along some curve Γ in the (x, y) plane—i.e., in which a solution surface is to pass through some curve in (x, y, u) space, whose projection on the (x, y) plane is Γ. For a first-order equation, this kind of initial data would seem to correspond to the Cauchy data of Section 5.1, and we are therefore led to the question: Does there exist a curve Γ, and a set of associated u values along this curve, such that we are not able to uniquely determine u in some region adjacent to Γ?

If there exists such a curve, let it be defined by $\xi(x, y) = 0$. For convenience, utilize other curves of the family $\xi(x, y) = $ const and of an intersecting family $\eta(x, y) = $ const, so as to provide a new local coordinate system (ξ, η) to replace (x, y). Then Eq. (6.11) becomes

$$u_\xi(f\xi_x + g\xi_y) + u_\eta(f\eta_x + g\eta_y) = h \qquad (6.16)$$

Now u is specified on Γ; since ξ is constant on Γ but η varies, we can differentiate the given data to determine u_η on Γ. Equation (6.16) would then normally determine u_ξ on Γ (note that the calculation of f and g on Γ would generally use the given u values on Γ). The exceptional case of present interest is that in which the coefficient $(f\xi_x + g\xi_y)$ vanishes along Γ. Since ξ is constant along Γ, the vanishing of $f\xi_x + g\xi_y$ along Γ requires the slope of Γ to be given by

$$dy/dx = -\xi_x/\xi_y = g/f$$

(if $f = 0$, use dx/dy instead of dy/dx).

We can now state that if values of u are specified along a curve Γ such that the slope of Γ satisfies the condition $dy/dx = g/f$ at each point of Γ, then an attempt to determine u in a region adjoining Γ fails at once, since we cannot even find u_ξ on Γ. The relationship to the first two of Eqs. (6.12) is clear, and the appropriateness of the word "characteristic" is thereby justified.

Let us now consider this kind of initial condition in more detail. Let u be specified along some curve Γ, and let $dy/dx = g/f$ at each point of Γ. Then along Γ, we must have

$$du/dx = u_x + u_y \, dy/dx = f^{-1}[fu_x + gu_y]$$

Two cases now arise. If the u values on Γ are such that

$$du/dx = h(x, y, u)/f(x, y, u) \qquad (6.17)$$

then Γ is a characteristic trace, and the (x, y, u) curve representing the given initial values is a characteristic. By using the method of characteristics, we can then construct a solution surface containing this particular characteristic and passing through any other space curve that intersects it; thus the initial value problem has an infinity of solutions. If, however, the condition (6.17) is violated, it is clear that no solution of Eq. (6.11) satisfying the initial condition can exist.

In Chapter 5, we found that we could also approach characteristics via the idea of finding curves along which certain discontinuities could exist. The same path can be followed here also, as is explained in Problem 6.6.4.

6.6 PROBLEMS

6.6.1 Construct an actual quasi-linear equation of the form (6.11) (in which f and g do involve u) and also a curve Γ along which u values are specified, so as to illustrate the two cases discussed in Section 6.5.

6.6.2 To what extent can the discussion of Section 6.5 be simplified if f and g do not involve u?

6.6.3 Let u be specified along a curve Γ in the (x, y) plane. Let s be a parameter along Γ and let t be a parameter along each characteristic emanating from the (x, y, u) curve representing the initial data. Then, just as in the example of Eq. (6.13), we can in principle determine each of x, y, and u, in some neighborhood of Γ, as functions of s and t. If the Jacobian $x_s y_t - x_t y_s$ does not vanish, we can solve for s and t as functions of x and y, and so obtain $u(x, y)$—at least in principle. If, however, this Jacobian vanishes at some point on Γ, we run into trouble; interpret such a possibility geometrically and relate it to the discussion of Section 6.5.

6.6.4 Consider an (x, y) plane curve Γ, defined by $\xi(x, y) = 0$, and an associated set of coordinate curves $\xi(x, y) = $ const, $\eta(x, y) = $ const. Let a function u be such that u_x and u_y are continuous on each side of Γ, with u but not u_ξ continuous across Γ. Moreover, let Eq. (6.11) be satisfied on each side of Γ. Explore the consequences of this supposition. Could we equally well have required u_ξ to be continuous across Γ, with $u_{\xi\xi}$ discontinuous?

6.7 MORE VARIABLES

The extension of the method of characteristics to a first-order quasi-linear equation involving more variables is so straightforward that we might as well make it at this point. We illustrate the general situation by considering the case of three independent variables. Let $u(x, y, z)$ satisfy

$$f(x, y, z, u)u_x + g(x, y, z, u)u_y + h(x, y, z, u)u_z = m(x, y, z, u) \quad (6.18)$$

where f, g, h, and m are given functions of x, y, z, and u. Also, let u be prescribed on a two-dimensional surface S in (x, y, z) space.

The characteristics are now the four-dimensional curves in (x, y, z, u) space defined by

$$dx/dt = f(x, y, z, u), \qquad dy/dt = g(x, y, z, u)$$

$$(6.19)$$

$$dz/dt = h(x, y, z, u), \qquad du/dt = m(x, y, z, u)$$

Let (x_0, y_0, z_0) be a typical point on S, and let the associated value of u be u_0. One curve of the family (6.19) will emanate from the point (x_0, y_0, z_0, u_0); denote it, say, by $x = x(t), y = y(t), z = z(t), u = u(t)$. The first three of these functions describe a curve in (x, y, z) space emanating from the point (x_0, y_0, z_0) on S; along this curve we now know u. As we consider similarly all other characteristics, we will in principle be able to determine u along curves in (x, y, z) space emanating from all of the various points of S—i.e., we will determine u in some neighborhood of S.

As a simple example, the reader may use the method of characteristics to show that the solution of

$$xu_x + yu_y + uu_z = 0$$

$$u(x, y, 0) = xy \qquad \text{for} \quad x > 0, \quad y > 0$$

is given implicitly by $u = xy \exp(-2z/u)$ in an appropriate region.

Returning to the general problem, we can also ask if there is a surface S, and associated values of u, such that we cannot uniquely determine u in an adjoining region. We leave it to the reader to explore, in a manner analogous to that of Section 6.5, the possibility of such a situation.

7

EXTENSIONS

In the preceding chapters we have simplified our discussion of basic features of partial differential equations by restricting ourselves to the case of one dependent and two independent variables. We now illustrate, by a series of examples and exercises, that many of the general features of this preceding discussion carry over to the case of a larger number of variables.

We begin with two instructive examples. The first one again deals with the diffusion equation, but now in three space dimensions; the power of the divergence theorem becomes apparent. Second, we examine the three-dimensional time-dependent flow of a fluid; not only are the resulting equations applicable to problems in an extensive array of disciplines, but they are of great mathematical interest in themselves. Few sets of equations have been so much studied.

Series expansions in spherical coordinates frequently lead to the use of Legendre functions; we take the opportunity of inserting some basic results concerning these functions (see Section 7.5).

7.1 MORE VARIABLES

Consider first the diffusion of heat in a three-dimensional body; let (x, y, z) be position coordinates and let t be time. Denote the temperature above some chosen reference level by $\phi(x, y, z, t)$. The thermal conductivity is denoted by k and is permitted to depend on position but not direction.

Thus for any area element dS inside the body, oriented via the three components (n_1, n_2, n_3) of its unit normal vector, the rate of heat transfer across dS is equal to

$$k(\phi_x n_1 + \phi_y n_2 + \phi_z n_3)$$

since the expression in parentheses is simply the gradient of ϕ in the direction of the normal vector.

We now single out for examination some interior portion of the body, the surface and volume of which we denote by S and V, respectively. The rate N at which heat is entering across S is clearly given by

$$N = \int_S [k\phi_x n_1 + k\phi_y n_2 + k\phi_z n_3]\, dS$$

which may be rewritten by use of the divergence theorem as

$$N = \int_V [(k\phi_x)_x + (k\phi_y)_y + (k\phi_z)_z]\, dV$$

The rate R at which thermal energy is accumulating within V is given by

$$R = \int_V c\rho\phi_t\, dV$$

where c is the specific heat and ρ the density; c and ρ may be position dependent. If heat is being generated (electrically or chemically, say) inside the body at a given rate $g(x, y, z, t)$ per unit volume, energy conservation requires that

$$R = N + \int_V g\, dV$$

Inserting the expressions for R and N, we have

$$\int_V \{c\rho\phi_t - [(k\phi_x)_x + (k\phi_y)_y + (k\phi_z)_z] - g\}\, dV = 0$$

But this result must hold for any chosen subvolume V of the original body; we therefore deduce—as in Section 1.1—that

$$c\rho\phi_t - [(k\phi_x)_x + (k\phi_y)_y + (k\phi_z)_z] = g \tag{7.1}$$

We would expect the physical problem, and hence the mathematical problem, to be completely defined if we adjoin to Eq. (7.1) appropriate initial and boundary conditions—e.g., we might specify the initial temperature $\phi(x, y, z, 0)$ and also the surface temperature (or its normal gradient, or some combination of the two) for all $t > 0$.

The form of Eq. (7.1) is similar to that of Eq. (1.1), and it is reasonable to expect the general nature of solutions to be similar also. Of course if the initial and boundary conditions associated with Eq. (7.1) are such that the solution ϕ does not depend on y or z, we recover the one-dimensional situation as a special case.

Another interesting special case is that of thermal equilibrium, in which $\phi_t = 0$; Eq. (7.1) then becomes

$$(k\phi_x)_x + (k\phi_y)_y + (k\phi_z)_z = -g \qquad (7.2)$$

which is a generalization of our previous two-dimensional Poisson's equation (4.4). Setting $g = 0$, $k = $ const, we obtain the three-dimensional form of Laplace's equation

$$\Delta\phi = \phi_{xx} + \phi_{yy} + \phi_{zz} = 0 \qquad (7.3)$$

For either Eq. (7.2) or (7.3), reasonable boundary conditions would involve the specification of some combination of ϕ and its normal derivative on the surface of the body.

A similar increase in the number of variables may be encountered in the wave equation. The reader can easily show, for example, that if the thin membrane considered in Section 3.7 or in Problem 4.2.2 is endowed with mass ρ per unit area, and if the displacement w is permitted to be time dependent, then the governing equation becomes

$$w_{tt} = (T/\rho)(w_{xx} + w_{yy}) \qquad (7.4)$$

Again, appropriate initial and boundary conditions would be of the same general kind as those considered for the one-dimensional wave equation in Chapter 3.

One frequently encounters, rather than a single equation, sets of partial differential equations involving several dependent variables. As an example, consider the motion of an ideal gas possessing neither viscosity nor heat conductivity. From thermodynamics, we can write the relationship between pressure p and density ρ for such a gas as

$$p = A\rho^\gamma \qquad (7.5)$$

where the constant γ is the ratio of specific heats and where we take the constant A as the same for all portions of the gas. Let the three components of velocity be (u_1, u_2, u_3).

Begin with an arbitrary closed surface S enclosing a volume V, drawn in fixed position within the fluid. The fluid moves through this surface, but since mass is conserved we require

$$\int_S \rho(u_1 n_1 + u_2 n_2 + u_3 n_3) \, dS = -\int_V \frac{\partial \rho}{\partial t} \, dV$$

where (n_1, n_2, n_3) are the components of the outward-pointing unit normal vector at the surface element dS. Using the divergence theorem to change the surface integral into a volume integral, and writing the equation as

$$\int_V \left[\frac{\partial \rho}{\partial t} + \frac{\partial}{\partial x} (\rho u_1) + \frac{\partial}{\partial y} (\rho u_2) + \frac{\partial}{\partial z} (\rho u_3) \right] dV = 0$$

we conclude from the arbitrariness of V that the integrand must vanish. This yields the *equation of continuity:*

$$\frac{\partial \rho}{\partial t} + \frac{\partial}{\partial x} (\rho u_1) + \frac{\partial}{\partial y} (\rho u_2) + \frac{\partial}{\partial z} (\rho u_3) = 0 \qquad (7.6)$$

Take again a closed surface S (enclosing a volume V) drawn inside the fluid, but this time permit the surface to move (and deform) with the fluid, so as always to contain the same fluid particles.† At any instant, the total mass acceleration in the ith direction ($i = 1$, 2, or 3, corresponding to parallelism to the x, y, or z axes) must equal the net force in this direction, so that

$$\int_V \frac{du_i}{dt} (\rho \, dV) = -\int_S p n_i \, dS$$

where du_i/dt denotes the acceleration of a fluid particle of mass $(\rho \, dV)$. We have neglected body forces, such as that due to gravity. Rewriting this result as a volume integral (we contemplate some instantaneous position of S), and using the fact that S is arbitrary, we deduce

$$\rho \frac{du_1}{dt} = -\frac{\partial p}{\partial x}, \qquad \rho \frac{du_2}{dt} = -\frac{\partial p}{\partial y}, \qquad \rho \frac{du_3}{dt} = -\frac{\partial p}{\partial z}$$

Finally, the acceleration term (the rate of change of velocity for a chosen

---

† A "fluid particle," although minute from the viewpoint of continuum mechanics, actually consists of a large number of molecules. It moves with the average velocity of these molecules, and increases or decreases in volume so as always to contain the same total mass, although individual molecules may of course migrate in to or out of the "particle."

fluid particle) may be rewritten via the chain rule as

$$\frac{du_i}{dt} = \frac{\partial u_i}{\partial t} + \frac{\partial u_i}{\partial x}\frac{dx}{dt} + \frac{\partial u_i}{\partial y}\frac{dy}{dt} + \frac{\partial u_i}{\partial z}\frac{dz}{dt}$$

$$= \frac{\partial u_i}{\partial t} + \frac{\partial u_i}{\partial x}u_1 + \frac{\partial u_i}{\partial y}u_2 + \frac{\partial u_i}{\partial z}u_3$$

so that

$$\rho\left(\frac{\partial u_1}{\partial t} + \frac{\partial u_1}{\partial x}u_1 + \frac{\partial u_1}{\partial y}u_2 + \frac{\partial u_1}{\partial z}u_3\right) = -\frac{\partial p}{\partial x}$$

$$\rho\left(\frac{\partial u_2}{\partial t} + \frac{\partial u_2}{\partial x}u_1 + \frac{\partial u_2}{\partial y}u_2 + \frac{\partial u_2}{\partial z}u_3\right) = -\frac{\partial p}{\partial y} \qquad (7.7)$$

$$\rho\left(\frac{\partial u_3}{\partial t} + \frac{\partial u_3}{\partial x}u_1 + \frac{\partial u_3}{\partial y}u_2 + \frac{\partial u_3}{\partial z}u_3\right) = -\frac{\partial p}{\partial z}$$

Equations (7.5)–(7.7) constitute a set of five equations for the five dependent variables p, ρ, u_1, u_2, u_3; each of these is a function of (x, y, z, t). Again, physically reasonable boundary and initial conditions must be imposed, and we will give some examples of this in the sequel. However, we note at this point the crucial fact that these equations involve a more serious complication than that arising simply from an increase in the number of variables; they are inherently nonlinear, so that conventional solution techniques (summing series, taking transforms, etc.) are much less useful.

If we are concerned with an incompressible fluid, rather than with a gas, Eqs. (7.7) are still applicable, but Eq. (7.5) is replaced by $\rho = $ const, and Eq. (7.6) by

$$\frac{\partial u_1}{\partial x} + \frac{\partial u_2}{\partial y} + \frac{\partial u_3}{\partial z} = 0$$

so that we have a total of four equations in the four unknowns u_1, u_2, u_3, p.

7.2 PROBLEMS

7.2.1 Obtain the form of Eq. (7.1) appropriate for a spherical polar coordinate system (r, θ, α), where $x = r\sin\theta\cos\alpha$, $y = r\sin\theta\sin\alpha$, $z = r\cos\theta$, and also for a cylindrical coordinate system (r, θ, z) in which $x =$

$r \cos \theta$, $y = r \sin \theta$. [*Note:* It is often easier to derive an equation directly in a desired coordinate system than to transform the equation from one system into another. The method of considering a small element of volume, with faces as defined by adjoining coordinate surfaces, works well here.]

7.2.2 (a) Find a solution for Eq. (7.1) (with c, ρ, k all constant and $g \equiv 0$) of the form

$$\phi = f(t) \cdot \sin \alpha x \sin \beta y \sin \gamma z$$

where α, β, and γ are constant. Invent a physical problem for which this would be an appropriate solution and comment on the nature of the solution and on the possibility of generalization.

(b) Prove uniqueness for the solution to the problem you invented in part (a). Will the maximum of ϕ always occur on the boundary of the space–time region?

7.2.3 Let each of $\psi_1(x, t)$, $\psi_2(x, t)$ $\psi_3(x, t)$ be a solution of $\psi_t = a^2 \psi_{xx}$, where $a^2 = $ const. Show that

$$\phi(x, y, z, t) = \psi_1(x, t) \cdot \psi_2(y, t) \cdot \psi_3(z, t)$$

is a solution of

$$\phi_t = a^2 (\phi_{xx} + \phi_{yy} + \phi_{zz})$$

and use this idea, together with Eq. (2.16), to obtain the solution

$$\phi = \frac{1}{(4\pi a^2 t)^{3/2}} \exp \left[\frac{-r^2}{4a^2 t} \right]$$

corresponding to the release of a concentrated amount of heat, at time zero, at a point inside an infinite body; here r is the distance from the release point.

7.2.4 (a) Let $\phi(x, y, z)$ be harmonic—i.e., let it satisfy Laplace's equation in three dimensions. Show that its value at any point is equal to the average of its values on the surface (or through the interior) of any sphere centered on that point, and that the maximum or minimum value is always attained on the boundary of a region. Find a "fundamental solution" of Laplace's equation in three dimensions—i.e., a nontrivial solution depending only on the radial distance from the origin.

(b) A Dirichlet problem in three dimensions is one in which ϕ satisfies Poisson's equation in a region and is specified on the boundary of that region. If $\partial\phi/\partial n$ is specified on the boundary instead of ϕ, the problem is a Neumann one. Show in general that (1) the solution to a Dirichlet problem is unique, and (2) the solution to a Neumann problem does not exist unless the boundary data satisfy a certain consistency condition, but that if this is the case, any solution is unique within an arbitrary additive constant. Examine also the possibility of singularities and nonuniqueness near boundary points at which the given boundary data are discontinuous.

7.2.5 Let $\phi(r, \theta, \alpha)$ be harmonic inside a sphere of radius R centered on the origin, where (r, θ, α) are spherical polar coordinates (as in Problem 7.2.1). For $r > R$, define a function

$$\psi(r, \theta, \alpha) = \frac{1}{r}\, \phi\left(\frac{R^2}{r}, \theta, \alpha\right)$$

Show that ψ is harmonic in the region exterior to the sphere. What would the corresponding result in two dimensions be? This result is termed *Kelvin inversion* with respect to a sphere; is similar inversion with respect to a plane possible?

7.2.6 In the wave equation

$$\phi_{tt} = c^2(\phi_{xx} + \phi_{yy})$$

let c be a function of x and y. In (x, y, t) space, Cauchy conditions would—in analogy with the discussion of Section 5.1—consist in the specification of ϕ and its normal derivarive $\partial\phi/\partial n$ on some surface $\zeta(x, y, t) = $ const. [Note the special case in which this surface is simply the plane $t = 0$; here such Cauchy data are certainly reasonable physically, since they correspond to specifying ϕ and ϕ_t—the initial displacement and velocity of a membrane, for example.] Introduce additional functions $\xi(x, y, t)$ and $\eta(x, y, t)$, so that (ξ, η, ζ) represent a new coordinate system, and carry out a discussion similar to that of Chapter 5 so as to answer the questions:

(a) Are there surfaces $\zeta = $ const (characteristics) on which the specification of Cauchy data does not permit us to determine ϕ in an adjoining region?

(b) Are there surfaces $\zeta = $ const across which there may be a derivative discontinuity in a solution?

(c) What is the signal speed? [*Hint:* consider the intersection of the $(\zeta = $ const) surface with planes $t = t_1, t = t_1 + \Delta t$.]

Sketch a similar discussion for the three-dimensional wave equation

$$\phi_{tt} = c^2(\phi_{xx} + \phi_{yy} + \phi_{zz})$$

7.2.7 Let ϕ be a function of the n variables $(x_1, x_2, \ldots, x_n)$, satisfying in some region the linear PDE of second order

$$\sum_{i,j=1}^{n} a_{ij}\frac{\partial^2\phi}{\partial x_i\, \partial x_j} + \sum_{i=1}^{n} b_i\frac{\partial\phi}{\partial x_i} + c\phi + d = 0$$

where the a_{ij}, b_i, c, and d are constants, with $a_{ij} = a_{ji}$ (this latter symmetry requirement clearly involves no loss in generality).

Use the theory of quadratic forms to show that there exists a transformation

$$\xi_i = \sum_{j=1}^{n} \gamma_{ij} x_j$$

with each $\gamma_{ij} = $ const, and with $\det | \gamma_{ij} | \neq 0$, such that in terms of ξ_i coordinates the equation becomes

$$\sum_{i=1}^{n} r_i \frac{\partial^2 \phi}{\partial \xi_i^2} + \cdots = 0$$

where the r_i are constants—i.e., the second-order terms have become "diagonalized." Show that (within an overall sign change) the number of positive, negative, and zero r_i values is the same for all such diagonalizing transformations, and relate these numbers to the eigenvalues of the matrix (a_{ij}). Explain the reasonableness of the following classification scheme:

(a)   If all r_i are nonzero and have the same sign, the original PDE is termed *elliptic*.

(b)   If all r_i are nonzero and all except one have the same sign, the original PDE is termed *hyperbolic*.

(c)   If some of the r_i are zero, the PDE is termed *parabolic*.

(d)   Finally, if the a_{ij} are functions of position rather than constants, the nature of the PDE at a point is (by convention) obtained from the above classification by assigning to the a_{ij} their values at that point. Give an example to show that it is not in general possible, for the case of more than two independent variables, to "diagonalize" the PDE over a whole region.

Would it be reasonable to base the classification of the nonlinear PDE

$$F(x_1, \ldots, x_n, \phi, \phi_{x_1}, \ldots, \phi_{x_n}, \ldots, \phi_{x_i x_j}, \ldots) = 0$$

on the coefficients

$$a_{ij} = \partial F / \partial (\phi_{x_i x_i})?$$

7.2.8   (a)   The transformation $p = \phi_x$, $q = \phi_y$, permits any second-order equation

$$F(x, y, \phi, \phi_x, \phi_y, \phi_{xx}, \phi_{xy}, \phi_{yy}) = 0$$

to be written as a coupled set of three first-order PDEs. Show that the converse of this is not true.

(b)   Consider the system of equations for n unknowns $\phi_1, \phi_2, \ldots, \phi_n$

dependent on x and t:

$$\begin{pmatrix} \partial\phi_1/\partial t \\ \cdot \\ \cdot \\ \partial\phi_n/\partial t \end{pmatrix} + (A) \begin{pmatrix} \partial\phi_1/\partial x \\ \cdot \\ \cdot \\ \partial\phi_n/\partial x \end{pmatrix} = (B)$$

where A is an $n \times n$ matrix whose elements are functions of x and t, and B is a column matrix of such functions. Determine conditions on A under which there are curves $\xi(x, t) = $ const such that Cauchy data prescribed on ξ are inadequate. Is it reasonable to say that this system is hyperbolic if A is symmetric and $\det|\,A\,| \neq 0$?

7.2.9   Using Eq. (7.5), it follows that Eq. (7.7) can be written in vector form as

$$\frac{\partial\mathbf{u}}{\partial t} + \mathbf{u}\cdot(\nabla\mathbf{u}) = -\frac{A\gamma}{\gamma-1}\,\nabla(\rho^{\gamma-1})$$

so that the curl of the left-hand side must vanish. Define $\boldsymbol{\xi} = \nabla \times \mathbf{u}$, and deduce that $d\boldsymbol{\xi}/dt$ (the time rate of change of $\boldsymbol{\xi}$ as seen by a moving particle) must be a linear combination of the three components of $\boldsymbol{\xi}$, so that if $\boldsymbol{\xi} = 0$ at time zero, then $\boldsymbol{\xi} = 0$ for that particle for all time. Conclude that, if the fluid emanates from a state of rest or uniform motion, a velocity potential $\phi(x, y, z, t)$ must exist, with the property that $\mathbf{u} = \nabla\phi$.

Next, take the scalar product of each side of the above equation with $d\mathbf{r}$ [having components (dx, dy, dz), and in the direction of $\mathbf{u}$] to show that if the motion is steady—so that nothing depends on t—then along any streamline of the motion we must have

$$\tfrac{1}{2}(\gamma - 1)q^2 + \gamma(p/\rho) = K, \text{ constant}$$

where $q^2 = \mathbf{u}\cdot\mathbf{u} = \nabla\phi\cdot\nabla\phi$.

Show, finally, that in the steady-motion potential flow case, with K the same for all streamlines, $\phi(x, y, z)$ satisfies the equation

$$[K - \tfrac{1}{2}(\gamma - 1)q^2]\,\Delta\phi = \tfrac{1}{2}\nabla\phi\cdot\nabla q^2$$

where Δ denotes the Laplacian operator.

7.2.10   Consider a large number of minute particles moving through some portion of space (e.g., the molecules of a gas or even the stars in a galaxy, if an appropriate scale is used). In terms of a Cartesian coordinate system (x, y, z) and velocity components (u, v, w), let the number of particles that lie within a volume element $dx\,dy\,dz$ at time t, and whose velocities lie in the range u to $u + du$, v to $v + dv$, w to $w + dw$, be given by $\rho(x, y, z,$

u, v, w, t) $dx\,dy\,dz\,du\,dv\,dw$. We term ρ the distribution function. Derive an equation of continuity for ρ. [*Hint:* for the corresponding one-dimensional problem, the number of particles in the range x to $x + dx$, u to $u + du$ is given by $\rho(x, u)\,dx\,du$. A particle that at time t_0 is at position x_0 and has velocity u_0 will move, in a time interval δt, to a position x given by $x = x_0 + u_0\,\delta t + \frac{1}{2}a_0\,\delta t^2 + \cdots$, and will have a velocity given by $u = u_0 + a_0\,\delta t + \cdots$, where a_0 is the acceleration at time t_0. Thus the Jacobian $\partial(x, u)/\partial(x_0, u_0)$ will have the form $J = 1 + (\partial a_0/\partial u_0)\,\delta t + \cdots$. A moving element of phase volume will always contain the same number of particles, so that $d/dt(\rho J) = 0$, and this leads to $\rho_t + \rho_x u + \rho_u a + \rho a_u = 0$.]

7.2.11 The wave equation $\phi_{tt} = a^2\phi_{xx}$ (with $a = $ const) studied in Chapter 3 possesses a traveling harmonic wave solution of the form $\phi = A \sin k(x - ct)$; the velocity c with which this wave moves equals a and so is independent of the choice of wave number k. Consider, however, each of the following equations, and show that the propagation velocity c of any such solution is now a function of k:

(a) $\phi_{tt} = a^2\phi_{xx} - \lambda\phi$ (a, λ constant)
(b) $\phi_t + a\phi_x + \lambda\phi_{xxx} = 0$ (a, λ constant)
(c) $u_{tt} + a^2 u_{xxxx} = 0$ (a constant)

A situation in which c depends on k is termed *dispersive* (the nomenclature is borrowed from optics). Consider next a more general wave pattern in a dispersive situation. This pattern can be thought of as the superposition of a large number of Fourier components whose wave numbers are very close together if the interval of interest is large. Let k_1 and k_2 be two adjoining wave numbers (i.e., $k_2 = k_1 + dk$), the corresponding component waves being $\phi_1 = A_1 \sin[k_1(x - c_1 t) + \theta_1]$, $\phi_2 = A_2 \sin[k_2(x - c_2 t) + \theta_2]$. Since $k_1 \cong k_2$, we expect $A_1 \cong A_2$. The superposition of these two waves produces a "beat" pattern; show that this beat pattern moves with a velocity g given by $g = d(ck)/dk$. This velocity g is called the *group velocity* (as contrasted to the *phase velocity* c) It represents the velocity with which that part of the Fourier-decomposed pattern associated with the wave number k moves. It may or may not represent the velocity with which energy or information is transferred; see Carrier *et al.* (1966, Sec. 6.5).

7.3 SERIES AND TRANSFORMS

To illustrate the way in which such methods may be used for higher-dimensionality situations, consider the problem of wave motion in a finite circular cylinder, at all of whose surfaces a fixity condition is imposed. Let

$u(x, r, \theta, t)$ satisfy

$$u_{tt} = c^2[u_{xx} + u_{rr} + (1/r)u_r + (1/r^2)u_{\theta\theta}] + h(x, r, \theta, t) \qquad (7.8)$$

in $0 < x < l, 0 < r < a, 0 < \theta < 2\pi, 0 < t$, with

$$u(x, r, \theta, 0) = f(x, r, \theta), \qquad u_t(x, r, \theta, 0) = 0,$$

and with

$$u(0, r, \theta, t) = u(l, r, \theta, t) = u(x, a, \theta, t) = 0.$$

Here c, l, and a are constants, and f, h are prescribed functions. For simplicity, we have chosen the boundary conditions to be homogeneous; were they not so, we would make a preliminary change of dependent variables— just as in Section 1.7—so as to strengthen the convergence of whatever series expansion we contemplate.

Now any function $u(x, r, \theta, t)$ can be expanded in terms of a complete set of x-dependent eigenfunctions $p_n(x)$ as

$$u = \sum_n \gamma_n(r, \theta, t) \cdot p_n(x)$$

where the γ_n functions are the expansion coefficients. In turn, if we have a complete set of θ-dependent eigenfunctions $\rho_m(\theta)$, we can expand each γ_n function in terms of them via

$$\gamma_n = \sum_m \mu_{nm}(r, t)\rho_m(\theta)$$

so that u becomes

$$u = \sum_n \sum_m \mu_{nm}(r, t)\rho_m(\theta)p_n(x)$$

Finally, if the μ_{nm} functions can be expanded similarly, we obtain

$$u = \sum_n \sum_m \sum_j \eta_{nmj}(t) \cdot \xi_j(r)\rho_m(\theta)p_n(x) \qquad (7.9)$$

To find suitable expansion functions ξ_j, ρ_m, and p_n, we use separation of variables in the differential operator associated with Eq. (7.8), together with appropriate homogeneous boundary conditions. Consider therefore the substitution $w = T(t) \cdot R(r)\Theta(\theta)X(x)$ in

$$w_{tt} - c^2[w_{xx} + w_{rr} + (1/r)w_r + (1/r^2)w_{\theta\theta}] = 0$$

We are led to

$$X'' + \lambda^2 X = 0, \qquad X(0) = X(l) = 0$$

$$\Theta'' + m^2\Theta = 0, \qquad \Theta \text{ periodic, with period } 2\pi$$

$$R'' + (1/r)R' - (m^2/r^2)R = -k^2R, \qquad R(a) = 0, \qquad R(0) \text{ finite}$$

Here λ, m, k are constants. Solving these eigenvalue problems, and using the results in Eq. (7.9), we obtain

$$u = \sum_{n=1}^{\infty} \sum_{m=0}^{\infty} \sum_{j=1}^{\infty} [a_{nmj}(t) \cos m\theta + b_{nmj}(t) \sin m\theta] \cdot \sin(n\pi x/l) J_m(\zeta_{mj} r/a)$$

(7.10)

where ζ_{mj} is the jth zero of the Bessel function J_m.

We must now substitute Eq. (7.10) into Eq. (7.8); also, let

$$h = \sum_n \sum_m \sum_j (h_{nmj}^{(1)}(t) \cos m\theta$$

$$+ h_{nmj}^{(2)}(t) \sin m\theta) \cdot \sin(n\pi x/l) J_m(\zeta_{mj} r/a)$$

(7.11)

where the $h^{(1)}$ and $h^{(2)}$ coefficients can be obtained by the usual integration process applied to products of h with these eigenfunctions (including weighting factors as required). We obtain the equations

$$a_{nmj}'' = c^2[-(n^2\pi^2/l^2) - (\zeta_{mj}^2/a^2)]a_{nmj} + h_{nmj}^{(1)}(t)$$

$$b_{nmj}'' = c^2[-(n^2\pi^2/l^2) - (\zeta_{mj}^2/a^2)]b_{nmj} + h_{nmj}^{(2)}(t)$$

(7.12)

The initial conditions require $a_{nmj}(0)$ and $b_{nmj}(0)$ to equal the expansion coefficients [as in Eq. (7.11)] of the given function f, and also that $a_{nmj}'(0) = b_{nmj}'(0) = 0$. Thus, Eqs. (7.12) can be solved for a_{nmj} and b_{nmj}, and our solution is complete.

Because of the triple summations, the final result for u looks laborious, to say the least. However, the functions f and h may be such that only a few terms of the series are important; even if this is not the case, the availability of a computer may enable one to sum numerically (with care!) a large number of terms. One great advantage possessed by the series solution is that it is exact, so that if numerical summation is feasible, the answer is reliable. Moreover, competitive techniques have their own disadvantage—for example, the accurate numerical solution of a difference equation approximating Eq. (7.8) becomes very time-consuming in three space dimensions.

In this series solution we used eigenfunction expansions in r, θ, and x, but not in t. The reason is that the t domain is semi-infinite, so that we cannot easily construct a Sturm–Liouville-type problem so as to generate suitable t eigenfunctions. We can, however, make use of this semi-infinite domain by taking a Laplace transform of Eq. (7.8) with respect to t. We obtain

$$-sf(x, r, \theta) + s^2 U = c^2[U_{xx} + U_{rr} + (1/r)U_r + (1/r^2)U_{\theta\theta}]$$

$$+ H(x, r, \theta, s)$$

(7.13)

so that, at the cost of a future Laplace inversion, the dimensionality of the PDE has been reduced. We can now write

$$U = \sum_n \sum_m \sum_j (\alpha_{nmj} \cos m\theta + \beta_{nmj} \sin m\theta) \sin \frac{n\pi x}{l} J_m\left(\frac{\zeta_{mj}}{a} r\right)$$

where the coefficients $\alpha_{nmj}(s)$ and $\beta_{nmj}(s)$ are determined by substitution into Eq. (7.13); this will lead to exactly the same final result as before. Alternatively, we can expand only with respect to r and θ, say

$$U = \sum_m \sum_j [A_{mj}(x, s) \cos m\theta + B_{mj}(x, s) \sin m\theta] J_m\left(\frac{\zeta_{mj}}{a} r\right)$$

and substitute into Eq. (7.13) so as to obtain ordinary differential equations for A_{mj} and B_{mj}—this process may, as in earlier chapters, lead to a form of the solution that is particularly useful for small values of t.

We remark finally that if spatial regions are infinite or semi-infinite in extent, then *Fourier transforms* may be useful. We do not use Fourier transform methods in this book, since the full power of such methods depends on more knowledge of complex variable theory than we assume on the part of the reader. Such methods are discussed in detail in Carrier *et al.* (1966, Chap. 7). One can, however, sometimes carry out the equivalent of a Fourier transform process by replacing the infinite or semi-infinite interval by a finite one, using a conventional Fourier series expansion, and then letting the ends (or end) of the interval go to infinity; on occasion, we shall do this.

7.4 PROBLEMS

7.4.1 Let $u(x, y, z)$ satisfy Laplace's equation in the region $0 < x < a$, $0 < y < b$, $0 < z < c$, with $u = 0$ on all surfaces except for the surface $x = 0$, where $u = 1$. Show that

$$u = \frac{4}{\pi^2} \sum \frac{1}{mn} [(-1)^m - 1][(-1)^n - 1]$$

$$\times \frac{\sinh \pi[(m/b)^2 + (n/c)^2]^{1/2}(a - x)}{\sinh \pi[(m/b)^2 + (n/c)^2]^{1/2}a} \sin \frac{m\pi y}{b} \sin \frac{n\pi z}{c}$$

and obtain a similar result for the case $a = \infty$, with u bounded. Can you obtain any information concerning the nature of the singularity in u near the y axis? Obtain also an alternative series for the solution, involving products of three sine functions (cf. Section 1.7 and Problem 1.8.3).

7.4.2   Modify Problem 7.4.1 by replacing the condition $u(x, 0, z) = u(x, b, z) = 0$ by the new condition $u_y(x, 0, z) = u(x, b, z) + u_y(x, b, z) = 0$, and solve the new problem. Does your series converge?

7.4.3   A semi-infinite prism occupies the region $0 < x < a$, $0 < y < b$, $0 < z$. It has a constant thermal diffusivity a^2. At time $t = 0$, the temperature u is everywhere equal to 1; for $t > 0$, $u = 0$ on all surfaces. Find $u(x, y, z, t)$ in the form of a series. Having obtained the solution, discuss it in the light of Problem 7.2.3.

7.4.4   Repeat Problem 7.4.3 for a semi-infinite right cylinder whose cross section is half of a disk; i.e., find $u(r, \theta, z, t)$ for the region $0 < r < a$, $0 < \theta < \pi$, $0 < z$, for the same initial and boundary conditions as in Problem 7.4.3.

7.4.5   Let $u(x, y, z, t)$ satisfy the wave equation (with $c = $ const)

$$u_{tt} = c^2(u_{xx} + u_{yy} + u_{zz})$$

in the prismatical region $0 < x < a$, $0 < y < b$, $0 < z$. The medium is initially at rest, so that $u(x, y, z, 0) = u_t(x, y, z, 0) = 0$. On the lateral sides we require $\partial u/\partial n = 0$; i.e., $u_x(0, y, z, t) = u_x(a, y, z, t) = u_y(x, 0, z, t) = u_y(x, b, z, t) = 0$. On the surface $z = 0$, we apply an oscillatory forcing function corresponding to

$$u(x, y, 0, t) = \cos(A\pi x/a)\cos(B\pi y/b)\sin \omega t$$

where A, B, ω are positive constants, with A and B integral. Find $u(x, y, z, t)$ and discuss the surprising way in which the normal modes (obtained as $t \to \infty$) depend in character on the value of ω. Consider next the case of a more general forcing function, $u(x, y, 0, t) = f(x, y)\sin \omega t$, and in light of the previous question, discuss the solution for large z and large t.

7.4.6   Solve for $u(r, x, t)$ if

$$u_t + u_x = u_{rr} + (1/r)u_r + u_{xx}$$

for the region $0 < r < 1$, $x > 0$, where $u(r, x, 0) = 1$, $u(r, 0, t) = 1$, $u(1, x, t) = 0$.

7.5 LEGENDRE FUNCTIONS

We will shortly find that the use of separation of variables for Laplace's equation in spherical coordinates leads to the ordinary differential equation (Legendre's equation)

$$(1 - x^2)y'' - 2xy' + \nu(\nu + 1)y = 0 \tag{7.14}$$

for $-1 < x < 1$, where ν is a constant. In this section, we collect some facts concerning solutions of Eq. (7.14) and of a generalization of this equation.

We begin with the Legendre polynomials $P_n(x)$ defined† as the coefficients in the expansion in powers of h of

$$(1 - 2hx + h^2)^{-1/2} = P_0(x) + P_1(x) \cdot h + P_2(x) \cdot h^2 + \cdots$$
$$+ P_n(x) \cdot h^n + \cdots \qquad (7.15)$$

[If h is small enough that $| (-2hx + h^2) | < 1$, this power series expansion certainly exists, for the binomial theorem is then applicable.] It is clear that $P_n(x)$ is a polynomial of degree n. Direct computation gives the first few of the $P_n(x)$ as

$$P_0(x) = 1, \qquad P_1(x) = x, \qquad P_2(x) = \tfrac{3}{2}x^2 - \tfrac{1}{2}$$
$$P_3(x) = \tfrac{5}{2}x^3 - \tfrac{3}{2}x, \qquad P_4(x) = \tfrac{1}{8}(35x^4 - 30x^2 + 3)$$
$$P_5(x) = \tfrac{1}{8}(63x^5 - 70x^3 + 15x)$$

If we replace x by $(-x)$ and h by $(-h)$ in Eq. (7.15), we see that $P_n(x)$ is an even or odd function of x according as n is even or odd. Setting $x = -1$ or $+1$, we find that $P_n(-1) = (-1)^n$, $P_n(1) = 1$. Differentiation of both sides of Eq. (7.15) with respect to h leads to

$$(n + 1)P_{n+1} = (2n + 1)xP_n - nP_{n-1}, \qquad n = 1, 2, \ldots \quad (7.16)$$

(which permits us to extend the above P_n table very easily), whereas differentiation with respect to x leads to

$$P_n = P'_{n+1} - 2xP_n' + P'_{n-1}, \qquad n = 1, 2, \ldots \quad (7.17)$$

Manipulation of these two equations (notice the possibility of replacing n by $n - 1$ or $n + 1$, as one device) gives various other recursion relations, typified by

$$nP_n = xP_n' - P'_{n-1}$$
$$(2n + 1)P_n = P'_{n+1} - P'_{n-1}$$
$$(1 - x^2)P_n' = nP_{n-1} - nxP_n \qquad (7.18)$$
$$P_n' = xP'_{n-1} + nP_{n-1}$$

If we differentiate the third of Eqs. (7.18) and use the first of Eqs. (7.18),

---

† From any origin in three-dimensional space, draw two rays subtending an angle θ between them. Measure off distances 1 and h along the two rays. The left side of Eq. (7.15), with $x = \cos \theta$, is then the reciprocal of the distance between these points and so is proportional to the potential at one point due to a point mass at the other point.

we find that $P_n(x)$ satisfies the equation

$$(1 - x^2)P_n'' - 2xP_n' + n(n+1)P_n = 0 \tag{7.19}$$

Thus, we have found a class of solutions of Eq. (7.14) for those cases in which ν is integral. [Note that if $\nu = -n$, with $n > 0$, then $\nu(\nu+1) = (n-1)n$, so that P_{n-1} is a solution.] The fact that the leading coefficient, $1 - x^2$, of Eq. (7.14) vanishes at $x = \pm 1$ would normally alert us to the possibility of a singularity at $x = \pm 1$ in a solution of the equation; we observe however that $P_n(x)$ is well behaved at these values of x (and in fact for all values of x). In a moment we will deduce the interesting fact that if ν is nonintegral then Eq. (7.14) has *no* solution that is well behaved at both $x = +1$ and $x = -1$.

Let us, however, first return to Eq. (7.14) and suppose that we have found well-behaved solutions for two essentially different† values of ν (not necessarily integral). Let these two values of ν be α and β, and let the corresponding solutions be denoted by y_α and y_β. Then if we write the two resulting equations satisfied by y_α and y_β we have

$$[(1 - x^2)y_\alpha']' + \alpha(\alpha+1)y_\alpha = 0$$
$$[(1 - x^2)y_\beta']' + \beta(\beta+1)y_\beta = 0$$

Multiplying the first of these by y_β, multiplying the second by y_α, subtracting the results, and integrating from -1 to 1, we find

$$\int_{-1}^{1} \{(1 - x^2)(y_\alpha'y_\beta - y_\beta'y_\alpha)\}' \, dx$$

$$+ [\alpha(\alpha+1) - \beta(\beta+1)] \int_{-1}^{1} y_\alpha y_\beta \, dx = 0$$

whence

$$\int_{-1}^{1} y_\alpha y_\beta \, dx = 0 \tag{7.20}$$

so that well-behaved solutions corresponding to different values of ν are orthogonal. Thus, for example, if m and n are positive integers, with $m \neq n$, $\int_{-1}^{1} P_m P_n \, dx = 0$, from which we deduce easily that P_n is orthogonal to all polynomials in x of degree less than n. To evaluate $\int_{-1}^{1} P_n^2 \, dx$, we multiply the first of Eqs. (7.18) by P_n and integrate by parts:

$$n \int_{-1}^{1} P_n^2 \, dx = \left\{ \left[x \cdot \tfrac{1}{2} P_n^2 \right]_{-1}^{1} - \int_{-1}^{1} (\tfrac{1}{2}P_n^2) \, dx \right\} - 0$$

---

† Since the replacement of ν by $-1 - \nu$ does not alter the value of the coefficient $\nu(1 + \nu)$, the statement that α and β are to be essentially different means that $\alpha \neq \beta$ and also $\alpha \neq (-1 - \beta)$.

whence

$$\int_{-1}^{1} P_n{}^2 \, dx = \frac{2}{2n + 1} \tag{7.21}$$

We can now use Eq. (7.20) to show that Eq. (7.14) does not possess solutions that are well behaved at both endpoints $x = -1$ and $x = +1$, if ν is nonintegral. For suppose the contrary; let F_ν be a well-behaved solution (not identically zero) corresponding to some nonintegral choice of ν. Then for each of our P_n, we must have $\int_{-1}^{1} F_\nu P_n \, dx = 0$, and consequently—since any polynomial in x can obviously be written as a finite linear combination of the P_n's—we deduce that $\int_{-1}^{1} F_\nu \phi \, dx = 0$, for any polynomial $\phi(x)$. But by Weierstrass's approximation theorem,† any function continuous in the closed interval $[-1, 1]$ can be approximated uniformly by a polynomial, so that no matter how small $\varepsilon > 0$ is, there is a polynomial J such that $F_\nu = J + \psi$ for $-1 \le x \le 1$, where $| \psi(x) | < \varepsilon$. Then choose $\phi = J$, and we find

$$0 = \int_{-1}^{1} F_\nu J \, dx = \int_{-1}^{1} F_\nu (F_\nu - \psi) \, dx$$

and since ε can be made as small as we please, we deduce that $\int_{-1}^{1} F_\nu{}^2 \, dx = 0$, so that $F_\nu = 0$, which contradicts our assumption.

Thus the only values of ν for which Eq. (7.14) has solutions well behaved at both (± 1) are the integral values. When $\nu = n$, a well-behaved solution does exist—it is $P_n(x)$; by use of the usual Wronksian process, the second solution in this case has the form

$$P_n(x) \cdot \int^x \frac{d\xi}{(1 - \xi^2) P_n{}^2(\xi)} \tag{7.22}$$

and so is certainly not well behaved at either of $x = \pm 1$. We now rephrase this discussion in terms of an eigenvalue problem and generalize it somewhat.

We have in fact shown that the $P_n(x)$ are eigenfunctions of the problem: find λ and $y(x)$, with $y(x)$ finite at $x = \pm 1$, such that

$$[(1 - x^2) y']' + \lambda y = 0 \tag{7.23}$$

The corresponding eigenvalues are $\lambda = n(n + 1)$, with n integral. Given any such eigenvalue problem, we can construct an associated one by dif-

† For a simple proof of Weierstrass's theorem, see Courant and Hilbert (1953, p. 65).

ferentiation. Define $z = y'$, and differentiate Eq. (7.23) to obtain

$$(1 - x^2)z'' - 4xz' + (\lambda - 2)z = 0 \qquad (7.24)$$

We can say at once that if $\lambda = n(n + 1)$ for some integer n, then $z = P_n'(x)$ is a solution of this equation that is finite at $x = \pm 1$. There are no other eigenvalues or eigenfunctions, for if there were we could integrate† to obtain an eigenfunction of Eq. (7.23) not contained in the $P_n(x)$ set. More generally, we can differentiate Eq. (7.23) m times and set $y^{(m)} = z$ to obtain

$$(1 - x^2)z'' - (2 + 2m)xz' + [\lambda - m(m + 1)]z = 0 \qquad (7.25)$$

and by a similar argument we can deduce that Eq. (7.25) has solutions finite at $x = \pm 1$ only if λ has the form $n(n + 1)$, n integral, in which case $z = P_n^{(m)}(x)$. Equation (7.25) is not self-adjoint but it is easily made so via the transformation $w = (1 - x^2)^{m/2}z$, which yields

$$[(1 - x^2)w']' - m^2w/(1 - x^2) + \lambda w = 0, \qquad -1 < x < 1 \quad (7.26)$$

known as *Legendre's associated equation*. The eigenvalue problem constructed by adjoining to Eq. (7.26) the condition of finiteness as $x \to \pm 1$ has only the solutions $\lambda = n(n + 1)$, n integral, $w = P_n^m(x)$ where these *associated Legendre functions* are defined‡ by

$$P_n^m(x) = (-1)^m(1 - x^2)^{m/2}P_n^{(m)}(x) \qquad (7.27)$$

We require of course that $m \leq n$ to avoid the trivial solution $w \equiv 0$. As in the case of the $P_n(x)$, it follows easily by use of the differential equation satisfied by $P_n^m(x)$ that

$$\int_{-1}^{1} P_n^m(x)P_q^m(x)\, dx = 0, \qquad n \neq q$$

Using Rodrigues' formula of Problem 7.6.2 and repeated integrations by parts, the reader may show that

$$\int_{-1}^{1} P_n^m(x)P_n^m(x)\, dx = \frac{2}{2n + 1}\frac{(n + m)!}{(n - m)!}$$

† Note that the arbitrary constant of integration in the eigenfunction can be adjusted to make the integrated equation homogeneous.

‡ We use a bracketed superscript to denote differentiation. The factor $(-1)^m$ is inserted for agreement with the "Handbook of Mathematical Functions" (Nat. Bur. Std., 1964).

7.6 PROBLEMS

7.6.1 If one starts with the set of functions $1, x, x^2, x^3, \ldots$, and constructs from this set a new set $\phi_0(x), \phi_1(x), \phi_2(x), \ldots$ via

$$\phi_0 = 1$$

$$\phi_1 = x - \frac{\phi_0 \int_{-1}^{1} x\phi_0 \, dx}{\int_{-1}^{1} \phi_0^2 \, dx}$$

$$\phi_2 = x^2 - \frac{\phi_1 \int_{-1}^{1} x^2\phi_1 \, dx}{\int_{-1}^{1} \phi_1^2 \, dx} - \frac{\phi_0 \int_{-1}^{1} x^2\phi_0 \, dx}{\int_{-1}^{1} \phi_0^2 \, dx}$$

then these ϕ_j are orthogonal to one another (weight function 1) over the interval $(-1, 1)$. Show that ϕ_j is simply a constant times P_j.

7.6.2 Prove that the $P_n(x)$ satisfy *Rodriques' formula*

$$P_n(x) = \frac{1}{2^n n!} \frac{d^n}{dx^n} \left[(x^2 - 1)^n \right]$$

[*Hint:* Use induction; start with

$$[(x^2 - 1)^n]^{(n+1)} = [2xn(x^2 - 1)^{n-1}]^{(n)}$$

$$= 2n[x\{(x^2 - 1)^{n-1}\}^{(n)} + n\{(x^2 - 1)^{n-1}\}^{(n-1)}]$$

and use the recursion relations.]

7.6.3 Show that $P_n(x)$ has n zeros in $(-1, 1)$ and that $|P_n(x)| \leq 1$ for x in $[-1, 1]$.

7.7 SPHERICAL HARMONICS

In spherical coordinates (r, θ, ϕ), the Laplacian of a function u is given by

$$\Delta u = \frac{1}{r^2} \left[(r^2 u_r)_r + \frac{1}{\sin \theta} (\sin \theta \cdot u_\theta)_\theta + \frac{1}{\sin^2 \theta} u_{\phi\phi} \right] \tag{7.28}$$

Let us try to find a solution of the form $u = R(r) \cdot \Theta(\theta) \cdot \Phi(\phi)$ suitable for use in the annular region $a < r < b$. Substitution into $\Delta u = 0$ gives

$$\sin^2 \theta \left[\frac{(r^2 R')'}{R} + \frac{1}{\sin \theta} \frac{(\sin \theta \cdot \Theta')'}{\Theta} \right] + \frac{\Phi''}{\Phi} = 0 \qquad (7.29)$$

whence $\Phi''/\Phi = \text{const} = -m^2$, where m is an integer (because our solution must be periodic in ϕ if it is to hold throughout the annulus). Now rewrite Eq. (7.29) as

$$\frac{(r^2 R')'}{R} + \frac{1}{\sin \theta} \frac{(\sin \theta \cdot \Theta')'}{\Theta} - \frac{m^2}{\sin^2 \theta} = 0$$

whence

$$\frac{1}{\sin \theta} \frac{(\sin \theta \cdot \Theta')'}{\Theta} - \frac{m^2}{\sin^2 \theta} = \text{const} = -\lambda, \quad \text{say}$$

Setting $\cos \theta = \xi$, this equation becomes

$$[(1 - \xi^2) \Theta_\xi]_\xi - \frac{m^2}{1 - \xi^2} \Theta + \lambda \Theta = 0$$

Because the solution must be finite throughout the annular region, and so in particular at the poles, we know from Section 7.5 that λ must have the form $n(n + 1)$, where n is an integer, which can be taken to be nonnegative without loss of generality; moreover, $n \geq m$. With this choice for λ, $\Theta = P_n^m(\xi) = P_n^m(\cos \theta)$. We are left with the equation

$$(r^2 R')'/R = \lambda = n(n + 1)$$

whose solution is $R = r^n$ or $R = r^{-n-1}$. The final result is then that any product function solution of $\Delta u = 0$, finite throughout an annular region, must have the form

$$u = (ar^n + br^{-n-1}) \cdot P_n^m(\cos \theta) \cdot (\alpha \cos m\phi + \beta \sin m\phi) \qquad (7.30)$$

where m and n are nonnegative integers with $n \geq m$, and where a, b, α, β are constants. Such functions are called (solid) *spherical harmonics*.

For a fixed value of r—i.e., on a spherical surface—the behavior of u in Eq. (7.30) is governed by the second two factors. We define the *surface spherical harmonics* S_{mn} and C_{mn} by

$$C_{mn} = P_n^m(\cos \theta) \cos m\phi, \qquad S_{mn} = P_n^m(\cos \theta) \sin m\phi \qquad (7.31)$$

For $m = 0$, $C_{0.n}$ is termed a *zonal* surface harmonic, since it vanishes on n parallels of latitude and so divides the spherical surface into zones of latitude. If $m = n \neq 0$, then C_{mn} and S_{mn} are termed *sectorial* surface harmonics, since they vanish on great circles through the poles, which divide the spherical surface into sectors (like sections of an orange). If $m \neq n$, and $m \neq 0$, then C_{mn} and S_{mn} are termed *tesseral* harmonics.

We have already observed in Section 7.5 that the set $\{P_n(\cos \theta)\}$ is complete over $(0, \pi)$. For similar reasons, for any fixed choice of m, the set $\{P_n{}^m(\cos \theta)\}$ is complete (in the usual mean square sense) over $(0, \pi)$. Moreover, the set $\{\cos m\phi, \sin m\phi\}$ is certainly complete over $(0, 2\pi)$. Consequently, any function $f(\theta, \phi)$ defined over the surface of a sphere can be expanded in terms of surface harmonics. Thus,

$$f(\theta, \phi) = \sum_{m=0}^{\infty} [a_m(\theta) \cos m\phi + b_m(\theta) \sin m\phi]$$

$$= \sum_{m=0}^{\infty} \sum_{n=m}^{\infty} [a_{mn} \cos m\phi + b_{mn} \sin m\phi] P_n{}^m(\cos \theta)$$

$$= \sum_{m=0}^{\infty} \sum_{n=m}^{\infty} [a_{mn} C_{mn}(\theta, \phi) + b_{mn} S_{mn}(\theta, \phi)] \qquad (7.32)$$

where the $a_m(\theta)$ and $b_m(\theta)$ are the Fourier coefficient functions for $f(\theta, \phi)$, and where a_{mn}, b_{mn} are the coefficients of the expansions of these two functions in terms of the $\{P_n{}^m(\cos \theta)\}$. For a given $f(\theta, \phi)$, these coefficients are easily determined by use of the orthogonality conditions for the $P_n{}^m$ and for the trigonometric functions.

Consider now any function $\psi(r, \theta, \phi)$ that is harmonic in an annular region $a < r < b$, where a and b are constants. Because of the completeness of the set of surface spherical harmonics, we can write

$$\psi = \sum_{m=0}^{\infty} \sum_{n=m}^{\infty} [\alpha_{mn}(r) C_{mn}(\theta, \phi) + \beta_{mn}(r) S_{mn}(\theta, \phi)] \qquad (7.33)$$

where the functions $\alpha_{mn}(r)$ and $\beta_{mn}(r)$ are the expansion coefficient functions for ψ. Substitution into the equation $\Delta u = 0$ now shows that each of α_{mn} and β_{mn} must be a linear combination of r^n and r^{-n-1}. Equation (7.33) therefore becomes, if $\alpha_{mn}^{(1)}$, $\alpha_{mn}^{(2)}$, $\beta_{mn}^{(1)}$, $\beta_{mn}^{(2)}$ are constants,

$$\psi = \sum_{m=0}^{\infty} \sum_{n=m}^{\infty} [(\alpha_{mn}^{(1)} r^n + \alpha_{mn}^{(2)} r^{-n-1}) C_{mn}(\theta, \phi)$$
$$+ (\beta_{mn}^{(1)} r^n + \beta_{mn}^{(2)} r^{-n-1}) S_{mn}(\theta, \phi)] \qquad (7.34)$$

Or course if the spherical annulus includes the origin, we must discard all terms involving negative powers of r. If on the other hand it includes arbitrarily large radii, and if ψ is to be bounded as $r \to \infty$, then we must discard all positive powers of r.

Equation (7.34), together with Eq. (7.32), enables us to write at once a formal series solution for such problems as: find ψ if $\Delta\psi = 0$ for $r < a$, and $\psi = f(\theta, \phi)$ on $r = a$ where f is a given function. Alternatively, we could replace $r < a$ by $r > a$ in this problem statement, provided we adjoin some condition at ∞. A third possibility is to let $a < r < b$ and to specify $\psi(a, \theta, \phi) = f_1(\theta, \phi)$, $\psi(b, \theta, \phi) = f_2(\theta, \phi)$.

An interesting alternative technique is based on the fact that (as the reader may prove) an axially symmetric harmonic function is completely defined by its values along the axis. As an example of the idea, consider a ring located at $r = c$, $\theta = \alpha$, where c and α are constants. Let this ring carry a uniformly distributed charge of total amount Q. The electrostatic potential (defined as the work required to bring a unit test charge from infinity; the potential of a test charge q at a point distant ρ from q is equal to q/ρ, in esu) at any point on the z axis—say at $r = z$, $\theta = 0$, is clearly given by

$$\psi(z, 0, \phi) = \frac{Q}{(z^2 + c^2 - 2zc \cos \alpha)^{1/2}}$$

$$= \begin{cases} \dfrac{Q}{c} \displaystyle\sum_{n=0}^{\infty} \left(\dfrac{z}{c}\right)^n P_n(\cos \alpha), & \left|\dfrac{z}{c}\right| < 1 \\[4mm] \dfrac{Q}{c} \displaystyle\sum_{n=0}^{\infty} \left(\dfrac{c}{z}\right)^{n+1} P_n(\cos \alpha), & \left|\dfrac{z}{c}\right| > 1 \end{cases} \qquad (7.35)$$

by use of Eq. (7.15). We now claim that simple modifications of these series give ψ at other points in space; viz.,

$$\psi(r, \theta, \phi) = \begin{cases} \dfrac{Q}{c} \displaystyle\sum_{n=0}^{\infty} \left(\dfrac{r}{c}\right)^n P_n(\cos \alpha) P_n(\cos \theta), & \dfrac{r}{c} < 1 \\[4mm] \dfrac{Q}{c} \displaystyle\sum_{n=0}^{\infty} \left(\dfrac{c}{r}\right)^{n+1} P_n(\cos \alpha) P_n(\cos \theta), & \dfrac{r}{c} > 1 \end{cases} \qquad (7.36)$$

(or either of these, if $r = c$, provided $\theta \neq \alpha$). The proof consists merely in noting the harmonicity of each convergent series, and the fact that for $\theta = 0$ we have agreement with Eq. (7.35).

7.8 PROBLEMS

7.8.1 Use the formal mechanism of Eqs. (7.32) and (7.34) to find ψ, harmonic in $0 \le r < 1$, with $\psi(1, \theta, \phi) = \theta(\pi - \theta) \cos \phi$. Discuss the rate at which the series converges.

7.8.2 Let $\psi(r, \theta, \phi)$ be harmonic in each of the two regions $r < a$, $r > a$, with $\psi \to 0$ as $r \to \infty$. Let $\psi(a+, \theta, \phi) = \psi(a-, \theta, \phi)$, with

$$\psi_r(a+, \theta, \phi) - \psi_r(a-, \theta, \phi) = P_n{}^m(\cos \theta) \sin m\phi$$

where m and $n \ge m$ are integers. Find ψ (by inspection) in each of $r < a$, $r > a$. [In electrostatics, this problem corresponds to that of determining the potential due to a surface-harmonic-type charge distribution over the surface of a sphere; of course, any charge distribution may be decomposed into a series of such distributions.]

7.8.3 Let $\Delta\psi = 0$ for $r > a$, with $\psi \to r \cos \theta$ as $r \to \infty$ and with $\psi_r(a, \theta, \phi) = 0$. Find ψ, or explain why no such ψ exists.

7.8.4 Discuss the relationship between spherical harmonics and solutions of Laplace's equation that are homogeneous polynomials in x, y, and z. How many are there of degree n?

7.8.5 Let ψ be harmonic and bounded in the region between two infinite cones, $\alpha < \theta < \beta$, with $0 < \alpha$ and $\beta < \pi$; let $\psi(r, \alpha, \phi) = 0$ and $\psi(r, \beta, \phi) = 1$. Find a simple expression for ψ, and discuss the nature of the singularity near $r = 0$.

7.8.6 Discuss the possible use of spherical harmonics to solve $\Delta u + k^2 u = 0$ in a spherical annulus. This equation would arise from the wave equation $\psi_{tt} = c^2 \Delta\psi$ for the stationary mode case, via $\psi = e^{i\omega t} u(r, \theta, \phi)$. Here k, c, ω are constants, with $k = c/\omega$. Note that the r-dependent functions now involve $r^{-1/2} C_{n+1/2}(kr)$, where $C_{n+1/2}$ is a Bessel function.

7.8.7 Let an electrostatic potential $u(r, \theta)$ satisfy the equation†

$$\frac{1}{r^2}\left[(r^2 u_r)_r + \frac{1}{\sin\theta}(\sin\theta \cdot u_\theta)_\theta\right] = \frac{2q}{a^2}\,\delta\left(\theta - \frac{\pi}{2}\right)\cdot\delta(r - a)$$

† The product of delta functions on the right-hand side corresponds to a total charge q esu distributed over a ring lying in the plane $\theta = \pi/2$, centered on the origin, and of radius a. In fact, if we multiply this expression by the element of volume $r^2 \sin\theta\, dr\, d\theta\, d\phi$ and integrate over any volume containing the ring, we get $4\pi q$, so that the original equation corresponds to the familiar electrostatic equation $\Delta u = 4\pi\rho$, where ρ is the charge density.

Thus, physically, we are solving the problem of finding the field strength at the inner surface of a grounded spherical shell if a central "collector" ring possesses a total charge q.

in the interior of a sphere of radius b, with $b > a$ (i.e., in the region $0 < r < b$, $0 < \theta < 2\pi$). Find the field strength $u_r(b, \theta)$ if $u(b, \theta) = 0$.

[*Hint:* Write $u = \sum_0^\infty f_n(r)P_n(\cos\theta)$ and show that f_n must satisfy the equation

$$(r^2 f_n')' - n(n+1)f_n = [(2n+1)/a]qr^2 P_n(0)\cdot\delta(r-a)$$

and then treat the regions $r > a$, $r < a$ separately. The final result is

$$u_r(b, \theta) = \sum_{n=0}^{\infty} q\left(\frac{a}{b}\right)^n \frac{P_n(0)}{a^2}(2n+1)P_n(\cos\theta)$$

where $P_n(0) = 0$ for odd n and is a known rational number for even n.]

8

PERTURBATIONS

One frequently encounters a problem whose solution is facilitated by the fact that the governing equation, the boundary condition, or the shape of the region, is not very different from that of a simpler problem. It may then be possible to modify, or "perturb," the solution of the simpler problem so as to be applicable to the actual problem. In this chapter we will consider some problems in which the basic idea of a perturbation process is directly fruitful. More subtle situations will be discussed in Chapter 15.

8.1 A NONLINEAR PROBLEM

Let (r, θ) be polar coordinates. In the interior of the unit disk $r < 1$, let $\psi(r, \theta)$ satisfy $\Delta\psi + \frac{1}{3}\psi_r\psi_\theta = 0$, i.e.,

$$\psi_{rr} + \frac{1}{r}\psi_r + \frac{1}{r^2}\psi_{\theta\theta} + \frac{1}{3}\psi_r\psi_\theta = 0 \qquad (8.1)$$

and let $\psi(1, \theta) = \cos\theta$. We want to find $\psi(r, \theta)$ in the interior of the disk, but because of the nonlinearity of the last term in Eq. (8.1), we anticipate that our preceding techniques will encounter difficulties. For example, a solution in the form of a series of simple product-type solutions would not be feasible, both because of the difficulty of determining such simple solutions and because superposition is no longer valid.

We might hope, however, that the nonlinear term is not too important. If so, we could seek an approximate solution $\psi^{(0)}$ satisfying

$$\Delta\psi^{(0)} = 0, \qquad \psi^{(0)}(1, \theta) = \cos\theta$$

127

from which $\psi^{(0)} = r \cos \theta$. If the true solution ψ does not differ too much from $\psi^{(0)}$, then the term $\frac{1}{3}\psi_r\psi_\theta$ can be approximated by $\frac{1}{3}\psi_r^{(0)}\psi_\theta^{(0)} = -\frac{1}{6}r \sin 2\theta$, so that a presumably improved approximation $\psi^{(1)}$ could be obtained by solving

$$\Delta\psi^{(1)} = \tfrac{1}{6}r \sin 2\theta, \qquad \psi^{(1)}(1, \theta) = \cos \theta$$

which leads to

$$\psi^{(1)} = r \cos \theta + (1/30)(r^3 - r^2) \sin 2\theta \qquad (8.2)$$

The fact that $\psi^{(1)}$ is not too different from $\psi^{(0)}$ is encouraging, and it is now natural to proceed iteratively and find $\psi^{(j+1)}$ from $\psi^{(j)}$ by solving

$$\Delta\psi^{(j+1)} = -\tfrac{1}{3}\psi_r^{(j)}\psi_\theta^{(j)}, \qquad \psi^{(j+1)}(1, \theta) = \cos \theta \qquad (8.3)$$

Thus for $\psi^{(2)}$ we obtain

$$\psi^{(2)} = r \cos \theta + \frac{1}{30}(r^3 - r^2) \sin 2\theta - \frac{1}{2700}\left(\frac{r^7}{11} - \frac{r^6}{4} + \frac{2r^5}{9} - \frac{25r^4}{396}\right) \sin 4\theta$$

$$+ \frac{1}{4320}(r^5 - r) \cos \theta + \left(\frac{r^4}{315} - \frac{r^5}{576} - \frac{87r^3}{60480}\right) \cos 3\theta$$

and it is clear that the encouragement continues, although the labor is starting to increase. (We could of course program a computer to do this algebra for us.)

The smallness of the correction terms indicates that the process is very probably convergent. Another check on the adequacy of the technique would be to examine the extent to which any of the $\psi^{(j)}$ fails to satisfy Eq. (8.1). For example, if we stop with $\psi^{(1)}$, we find that

$$\Delta\psi^{(1)} + \tfrac{1}{3}\psi_r^{(1)}\psi_\theta^{(1)} = \frac{r^3}{2700}(3r - 2)(r - 1) \sin 4\theta - \frac{r^3}{180} \cos \theta$$

$$+ \frac{r^2}{180}(5r - 4) \cos 3\theta$$

and the size of this "residue," for $r < 1$, is in general very small compared to the size of a typical term in the left-hand side of the equation. Of course, these kinds of a posteriori checks can be deceptive in some cases.

A more systematic equivalent to the iteration process (8.3) is to replace Eq. (8.1) by

$$\Delta\psi + \varepsilon\psi_r\psi_\theta = 0 \qquad (8.4)$$

where ε is a parameter. We anticipate that ψ is a well-behaved function of ε, so that a power series expansion seems reasonable:

$$\psi = \phi^{(0)}(r, \theta) + \varepsilon\phi^{(1)}(r, \theta) + \varepsilon^2\phi^{(2)}(r, \theta) + \cdots \qquad (8.5)$$

Substitution into Eq. (8.4) gives the sequence of equations

$$\Delta\phi^{(0)} = 0$$

$$\Delta\phi^{(1)} = -\phi_r^{(0)}\phi_\theta^{(0)}$$

$$\Delta\phi^{(2)} = -\phi_r^{(0)}\phi_\theta^{(1)} - \phi_r^{(1)}\phi_\theta^{(0)}$$

whereas the boundary condition becomes

$$\phi^{(0)}(1, \theta) = \cos\theta, \qquad \phi^{(j)}(1, \theta) = 0 \qquad \text{for} \quad j = 1, 2, \ldots$$

The reader should determine $\phi^{(0)}$, $\phi^{(1)}$, ..., and compare the results with those obtained above; note that in general $\psi^{(j)} \neq \phi^{(j)}$, although the two processes should of course converge to the same final result.

We describe $\phi^{(1)}$ (or $\psi^{(1)} - \psi^{(0)}$), as well as the higher-order corrections, as being *perturbations* on the solution $\phi^{(0)}$ (or $\psi^{(0)}$). Intuitively, it seems reasonable to hope for success if the basic equation is close, in some sense, to the equation satisfied by $\phi^{(0)}$. A commonly applied criterion for closeness is that the coefficient of the extra term—here $\frac{1}{3}$—be small; in a physical problem, however, the size of any such coefficient would depend on the units being used, and for that reason it is customary to nondimensionalize prior to the instigation of a perturbation process. In any event, however, there is no precise meaning attached to "smallness"; if one is fortunate, the perturbation process may be valid even for unexpectedly large values of the coefficient in question.

A note of caution should be inserted. The fact that a parameter ε is small does not in itself guarantee the efficacy of a perturbation approach. A simple example is provided by the problem of finding $u(x, y)$ such that $\varepsilon u_{xx} + u_{yy} = 0$ in the square region $0 < x < 1$, $0 < y < 1$, where u is specified on the boundary of the square. For $\varepsilon > 0$, this problem is, in principle, solvable. However, a first approximation to the solution can generally *not* be obtained by neglect of the term εu_{xx}, for the solution of the resulting ordinary differential equation has the form $f(x) + yg(x)$, and (except for very special situations) there is no hope of choosing the arbitrary functions $f(x)$ and $g(x)$ so that the boundary conditions on all four sides of the square are satisfied. Some of the exercises in this chapter will further illustrate the need for care in some perturbation problems; we will also return, more systematically, to this topic in Chapter 15.

8.2 PROBLEMS

8.2.1 In Eq. (8.1), replace the term $\frac{1}{3}\psi_r\psi_\theta$ by the term $\varepsilon\psi^2$, but otherwise leave the problem statement unaltered. Construct a perturbation series for the solution and discuss its apparent rate of convergence as a function of ε.

8.2.2 Let ϕ satisfy the equation

$$(1 + \varepsilon x)\phi_{xx} + \phi_{yy} = 0$$

in the rectangle $0 < x < \pi$, $0 < y < 1$. Here ε is a (small) constant. Let ϕ be required to vanish on all sides of this rectangle except for the side at $y = 1$, where $\phi = \sin x$. Determine ϕ in two ways: (a) by treating the term $\varepsilon x \phi_{xx}$ as a perturbation term, and (b) by a direct series expansion of the original problem. Compare results.

8.2.3 Solve each of the problems

(a) $\phi_t = [(1 + \varepsilon \sin x)\phi_x]_x$, $\phi(x, 0) = 0$, $\phi(0, t) = 1$
(b) $\phi_{tt} = [(1 + \varepsilon \sin x)\phi_x]_x$, $\phi(x, 0) = \phi_t(x, 0) = 0$, $\phi(0, t) = \sin \omega t$

in the region $0 < x < \infty$. Here ε and ω, satisfying $0 < \varepsilon \ll 1$ and $\omega > 0$, are constants. Is there any difficulty arising from the singularity at the origin of the (x, t) plane? [*Hint:* a preliminary Laplace transform may be useful.]

8.2.4 Discuss the feasibility of a perturbation approach (for $0 < \varepsilon \ll 1$) for each of the following problems:

(a) $\phi_{xx} + \varepsilon\phi_{yy} = 0$ in $0 < x < 1, 0 < y < 1$, ϕ specified on boundary,
(b) $\varepsilon\Delta\phi + \phi_x = 0$ in $0 < x < 1, 0 < y < 1$, ϕ specified on boundary,
(c) $\phi_{xx} + \varepsilon\phi_{yy} = \phi_t$ in $0 < x < 1, 0 < y < 1$, $\phi(x, y, 0) = 0$, ϕ specified as a function of t on the boundary.

What features of these three problems are crucial?

8.3 TWO EXAMPLES FROM FLUID MECHANICS

We will first use a perturbation approach to obtain "linearized" equations describing the small departure from equilibrium of a gas whose pressure p, density ρ, and velocity components (u_1, u_2, u_3) satisfy Eqs. (7.5)–(7.7). The equilibrium state is one in which p and ρ have the constant values p_0 and ρ_0, and in which $u_1 = U$ (a constant), $u_2 = u_3 = 0$. We now inquire what the response of this system to a small disturbance will be, and for that purpose write

$$p = p_0(1 + p'), \qquad \rho = \rho_0(1 + \rho')$$

$$u_1 = U(1 + u_1'), \qquad u_2 = Uu_2', \qquad u_3 = Uu_3'$$

where the dimensionless quantities $| p' |$, $| \rho' |$, and $| u_i' |$ are all much less than one. Substituting into Eqs. (7.5)–(7.7), we obtain (using $p_0 = A\rho_0{}^\gamma$)

$$p' = (1 + \rho')^\gamma - 1$$

$$\frac{\partial \rho'}{\partial t} + U \sum_{j=1}^{3} \frac{\partial}{\partial x_j} [(1 + \rho')(\delta_{1j} + u_j')] = 0 \qquad (8.6)$$

$$\frac{\rho_0 U}{p_0}(1 + \rho')\left[\frac{\partial u_i'}{\partial t} + U \sum_{j=1}^{3} \frac{\partial u_i'}{\partial x_j}(\delta_{1j} + u_j')\right] = -\frac{\partial p'}{\partial x_i}, \qquad i = 1, 2, 3$$

Here we have introduced Cartesian notation, via $x = x_1$, $y = x_2$, $z = x_3$, in order to obtain some economy in writing. Also, the Kronecker delta δ_{pq} is defined to equal zero unless the indices p and q are the same, in which case it equals unity; thus $\delta_{11} = 1$, $\delta_{12} = \delta_{13} = 0$, etc.

By hypothesis, each of p', ρ', u_i' is small compared to unity. (We assume that there are no "shock waves" or other steep gradients, so that the various partial derivatives of those three quantities are also small.) In consequence, it is reasonable to neglect terms in which products of these small quantities (or of their derivatives) occur, so that Eqs. (8.6) reduce to

$$p' = \gamma\rho'$$

$$\frac{\partial \rho'}{\partial t} + U \frac{\partial \rho'}{\partial x_1} + U \sum_{j=1}^{3} \frac{\partial u_j'}{\partial x_j} = 0 \qquad (8.7)$$

$$\frac{\rho_0 U}{p_0}\left[\frac{\partial u_i'}{\partial t} + U \frac{\partial u_i'}{\partial x_1}\right] = -\frac{\partial p'}{\partial x_i}, \qquad i = 1, 2, 3$$

These equations are linear, a fact that represents a great simplification over the original equation set. As one example of this simplification, we observe that after some straightforward algebra the pressure fluctuation p' satisfies the equation

$$\frac{\gamma p_0}{\rho_0}\left(\frac{\partial^2}{\partial x_1{}^2} + \frac{\partial^2}{\partial x_2{}^2} + \frac{\partial^2}{\partial x_3{}^2}\right)p' = \frac{\partial^2 p'}{\partial t^2} + 2U \frac{\partial p'}{\partial x_1 \partial t} + U^2 \frac{\partial^2 p'}{\partial x_1{}^2} \qquad (8.8)$$

In terms of a coordinate system that moves along with the undisturbed flow and is defined by $\xi_1 = x_1 - Ut$, $\xi_2 = x_2$, $\xi_3 = x_3$, $\tau = t$, Eq. (8.8) takes the form

$$\frac{\gamma p_0}{\rho_0} \left(\frac{\partial^2}{\partial \xi_1^2} + \frac{\partial^2}{\partial \xi_2^2} + \frac{\partial^2}{\partial \xi_3^2} \right) p' = \frac{\partial^2 p'}{\partial \tau^2} \tag{8.9}$$

which is the standard three-dimensional wave equation with propagation speed c defined by

$$c^2 = \frac{\gamma p_0}{\rho_0} = \gamma A \rho_0{}^{\gamma-1} = \frac{d}{d\rho_0} \left(A \rho_0{}^\gamma \right) = \frac{dp_0}{d\rho_0}$$

This equation governs the propagation of acoustic signals relative to the moving gas.

Our second example, associated historically with the names of Rayleigh and Janzen, concerns the two-dimensional flow of a compressible fluid around a cylinder (Fig. 8.1). Using Problem 7.2.9 and polar coordinates (r, θ), the mathematical problem can be formulated as follows: In the region $r > R$ of the (r, θ) plane, find $\Phi(r, \theta)$ such that

$$\Delta\Phi \left[1 + \frac{\gamma - 1}{2} M_\infty{}^2 \left(1 - \Phi_r{}^2 - \frac{1}{r^2} \Phi_\theta{}^2 \right) \right]$$

$$= M_\infty{}^2 \left[\Phi_r{}^2 \Phi_{rr} + \frac{2}{r^2} \Phi_r \Phi_\theta \Phi_{r\theta} - \frac{1}{r^3} \Phi_\theta{}^2 \Phi_r + \frac{1}{r^4} \Phi_\theta{}^2 \Phi_{\theta\theta} \right] \tag{8.10}$$

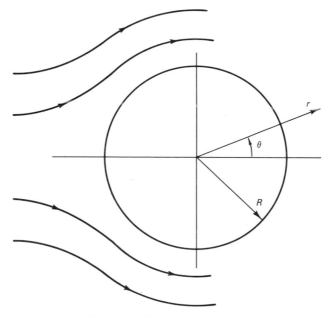

Fig. 8.1 Flow around cylinder.

where $\Delta\Phi = \Phi_{rr} + (1/r)\Phi_r + (1/r^2)\Phi_{\theta\theta}$, $\gamma = \text{const} > 1$ ($\gamma \cong 1.4$ for air), where $M_\infty = U_\infty/c_\infty$, and where U_∞ and c_∞ are, respectively, the prescribed fluid velocity and sound speed infinitely far from the cylinder.[†] The boundary conditions are

$$\Phi(r, \theta) \to r\cos\theta \qquad \text{as} \quad r \to \infty$$

$$\Phi_r(R, \theta) = 0$$

(8.11)

The constant M_∞ is termed the free-stream Mach number. If it is small, then Eq. (8.10) is approximated by $\Delta\Phi = 0$, which is easily solved; this suggests that a perturbation series

$$\Phi = \Phi^{(0)} + M_\infty^2\Phi^{(1)} + M_\infty^4\Phi^{(2)} + \cdots$$

(8.12)

might be useful. We try this, and find easily that

$$\Phi^{(0)} = \left(r + \frac{R^2}{r}\right)\cos\theta$$

$$\Phi^{(1)} = \frac{13}{12}\frac{R^2}{r}\cos\theta + \frac{R^4}{12r^3}\cos 3\theta - \frac{R^4}{2r^3}\cos\theta + \frac{R^6}{12r^5}\cos\theta - \frac{R^2}{4r}\cos 3\theta$$

.

.

.

The terms in this series are fairly complicated, and as far as convergence is concerned, the only proof available is that there exists some sufficiently small value of M_∞ for which it converges. Numerical computations indicate that the series converges up to about $M_\infty \cong 0.45$.

The maximum stream velocity occurs at $r = R$, $\theta = \pm\pi/2$, and has the value

$$U_{\max} = U_\infty\left[2 + \frac{7}{6}M_\infty^2 + \left\{\frac{281}{120} + \frac{71}{120}(\gamma - 1)\right\}M_\infty^4 + \cdots\right]$$

$$= U_\infty[2 + 0.187 + 0.066 + 0.031 + \cdots] \qquad \text{for} \quad M_\infty = 0.4, \ \gamma = 1.4$$

(8.13)

Along any streamline, the results of Problem 7.2.9 imply that

$$\frac{\gamma - 1}{2}q^2 + c^2 = \frac{\gamma - 1}{2}U_\infty^2 + c_\infty^2$$

[†] The physical conditions are that the velocity at ∞ is in the direction of the positive x axis and has uniform value U_∞ and that no fluid enters or leaves the cylinder. We replace K in Problem 7.2.9 by $\frac{1}{2}(\gamma - 1)U_\infty^2 + c_\infty^2$, and, for scaling purposes, set $\phi = U_\infty\Phi$.

where q is the local stream velocity and c the local speed of sound $\gamma p/\rho$, so that a value of c corresponds to any value of q as determined by use of Eq. (8.12). The dimensionless ratio q/c, termed the local Mach number M, is of fundamental importance. If $M > 1$, the flow is said to be supersonic; since a disturbance propagates in the fluid with velocity c while the fluid is moving past a stationary observer with velocity greater than c, any disturbances originating near the observer will be swept downstream and cannot be felt upstream of the observer. If $M < 1$, however, disturbances will propagate upstream of the observer, and we say the flow is locally subsonic.

For our cylinder problem, if terms up to order (M_∞^4) are included in the series (8.12), the flow at $r = R$, $\theta = \pm\pi/2$ becomes locally supersonic at $M_\infty \cong 0.404$. For $0.404 < M_\infty < 0.45$, there are regions just above and just below the cylinder in which the flow is supersonic and there is an apparently smooth transition from the subsonic flow elsewhere. For $M_\infty > 0.45$, the series appears to diverge. Experimentally, it is found that shock wave discontinuities (which we will discuss in a subsequent section) appear at about $M_\infty = 0.45$.

A further discussion of the Rayleigh–Janzen process, together with a comparison between it and alternative methods (such as that of Problem 8.5.7), will be found in Lighthill (1960).

8.4   BOUNDARY PERTURBATIONS

Let us consider Laplace's equation for the interior of a region bounded by an "almost" circle—in fact, by $r = R + \varepsilon \cos \theta$, where (r, θ) are polar coordinates. Here $R > 0$ and $0 < \varepsilon \ll R$ are constants. In the interior of this region we want to find $\phi(r, \theta)$ such that

$$\Delta\phi = \phi_{rr} + \frac{1}{r}\phi_r + \frac{1}{r^2}\phi_{\theta\theta} = 0 \tag{8.14}$$

with

$$\phi(R + \varepsilon \cos \theta, \theta) = f(\theta)$$

where $f(\theta)$ is prescribed.

We notice that the boundary condition can be rewritten by the use of a Taylor series as

$$\phi(R, \theta) + (\varepsilon \cos \theta)\cdot\phi_r(R, \theta) + \tfrac{1}{2}(\varepsilon \cos \theta)^2\phi_{rr}(R, \theta) + \cdots = f(\theta) \tag{8.15}$$

and this equation suggests the perturbation procedure

$$\phi(r, \theta) = \phi^{(0)}(r, \theta) + \varepsilon\phi^{(1)}(r, \theta) + \varepsilon^2\phi^{(2)}(r, \theta) + \cdots \tag{8.16}$$

Substitution into Eqs. (8.14) and (8.15) yields

$$\Delta\phi^{(j)} = 0, \qquad j = 0, 1, 2, \ldots$$

$$\phi^{(0)}(R, \theta) = f(\theta)$$

$$\phi^{(1)}(R, \theta) = -\cos\theta \cdot \phi_r^{(0)}(R, \theta)$$

$$\phi^{(2)}(R, \theta) = -\cos\theta \cdot \phi_r^{(1)}(R, \theta) - \tfrac{1}{2}\cos^2\theta \cdot \phi_{rr}^{(0)}(R, \theta)$$

(8.17)

$$\vdots$$

Thus we have a sequence of problems for the undeformed circular boundary, and these are easy to solve.

To obtain an explicit example, let us set $f(\theta) = \sin\theta$. Then $\phi^{(0)} = (r/R)\sin\theta$ by inspection, and to find $\phi^{(1)}$ we must solve

$$\Delta\phi^{(1)} = 0, \qquad \phi^{(1)}(R, \theta) = -(1/2R)\sin 2\theta$$

so that $\phi^{(1)} = -(r^2/2R^3)\sin 2\theta$. Continuing, we obtain

$$\phi = \frac{r}{R}\sin\theta - \varepsilon\frac{r^2}{2R^3}\sin 2\theta + \varepsilon^2\left[\frac{r^3}{2R^5}\sin 3\theta + \frac{r}{2R^3}\sin\theta\right] + \cdots$$

As a check on the calculation, we can let $r = R + \varepsilon\cos\theta$, and we should then obtain $\phi = \sin\theta$. This particular illustrative problem is simpler than most practical ones because here each term in the series can be obtained by inspection.

8.5 PROBLEMS

8.5.1 The perturbation series (8.12) was based on the prior reduction of the equations of Problem 7.2.9 to the form of Eq. (8.10). Once that it is known—or guessed—that M_∞^2 is an appropriate expansion parameter, an alternative approach would be to expand each of the quantities p, ρ, u_1, u_2, u_3 in such a series and to use the equations of Section 7.1 directly. Do this, and show that the final results are compatible with those obtained in Section 8.4. Compare the amount of labor involved in these two approaches.

8.5.2 Use the results of Problem 7.2.9 to obtain the equation replacing Eq. (8.10) for the case of axially symmetric compressible flow past a sphere. Use zonal harmonics (Section 7.7) to show that the perturbation expansion leads to

$$\Phi = \left(r + \frac{R^3}{2r^2}\right)\cos\theta + M_\infty^2 \left\{\left[\frac{1}{2}\frac{R^3}{r^2} - \frac{1}{5}\frac{R^6}{r^5} + \frac{1}{24}\frac{R^9}{r^8}\right]\cdot P_1\,(\cos\theta)\right.$$

$$\left. + \left[-\frac{3}{10}\frac{R^3}{r^2} + \frac{27}{55}\frac{R^5}{r^4} - \frac{3}{10}\frac{R^6}{r^5} + \frac{3}{176}\frac{R^9}{r^8}\right]\cdot P_3\,(\cos\theta)\right\} + \cdots$$

8.5.3   Let $\phi(x, t)$ satisfy the diffusion equation $\phi_t = a^2\phi_{xx}$ for $0 < x < \infty$, $t > 0$, where a is a constant. Let $\phi(x, 0) = A$, a positive constant, and let the boundary condition at $x = 0$ be given by the radiation condition

$$\phi_x(0, t) = C[\phi^4(0, t) - B^4]$$

where C and B are positive constants, with $0 < A - B \ll A$. Explain in detail how a perturbation approach might be applied to this problem so as to result in a sequence of linear problems.

8.5.4   Let $\Delta\phi = 0$ in the region of the (x, y) plane defined by $y > 0$, $\varepsilon \sin \omega y < x < \pi$, where ε and ω are positive constants with $\varepsilon \ll 1$. Find $\phi(x, y)$ satisfying the boundary conditions $\phi(\varepsilon \sin \omega y, y) = 0$, $\phi(\pi, y) = 0$, $\phi(x, 0) = \sin x$. [At ∞, ϕ is to be bounded.]

8.5.5   Let $\Delta\phi = 0$ in the (x, y) plane region defined by $0 < x < \pi$, $y > \varepsilon x$, $0 < \varepsilon \ll 1$. Let $\phi = 0$ on the vertical boundaries, and let $\phi(x, \varepsilon x) = \sin x$ for $0 < x < \pi$. Use two methods to determine ϕ: (a) a perturbation approach and (b) the change in independent variables $z = y - \varepsilon x$, followed by a series or transform solution. Compare results. Here the boundary condition on ϕ is continuous; what would happen to your results if the boundary condition on the line $y = \varepsilon x$ were replaced by $\phi(x, \varepsilon x) = x$ for $0 < x < \pi$?

8.5.6   In terms of spherical coordinates, let $\psi(r, \theta, \phi)$ satisfy the equation $\Delta\psi = 0$ in the region between two concentric spheres of radii R_1 and R_2, $R_1 < R_2$. Let $\psi(R_1, \theta, \phi) = 0$, $\psi(R_2, \theta, \phi) = 1$. Now displace the inner sphere along the z axis by an amount $\varepsilon \ll R_1$ while maintaining the prescribed ψ values on the two spheres. Determine ψ in the region between the spheres. Relate your results to the problem of determining the capacity of a nonperfect spherical condenser; is the capacity a linear or quadratic function of ε, for small ε?

8.5.7   Discuss the following outline of a "slender body" approximation to the two-dimensional flow of a compressible gas along the x axis (cf. Problem 7.2.9):

(a)   In Cartesian coordinates with $u_1 = U_\infty(1 + \Phi_x)$, $u_2 = U_\infty\Phi_y$, Eq.

(8.10) is replaced by

$$\Delta\Phi\left[1 + \frac{\gamma - 1}{2} M_\infty^2(-2\Phi_x - \Phi_x^2 - \Phi_y^2)\right]$$

$$= M_\infty^2\{(1 + \Phi_x)^2\Phi_{xx} + 2(1 + \Phi_x)\Phi_y\Phi_{xy} + \Phi_y^2\Phi_{yy}\}$$

so that $(1 - M_\infty^2)\Phi_{xx} + \Phi_{yy} \cong 0$.

(b) If the body occupies the portion $(0, b)$ of the x axis, with the upper and lower surfaces defined by $y = f_+(x)$ and $y = f_-(x)$, respectively (with the thickness $|f_\pm| \ll b$), then the boundary conditions have the form $u_2/u_1 = f'(x)$, which may be approximated by $\Phi_y(x, 0) = f'(x)$.

As a special case, can you analyze the flow along a slightly wavy wall? How do the results depend on M_∞? Is the nature of the governing equation compatible with these results?

8.5.8 Let $u_t = u_{xx}$ in the (x, t) plane region defined by $t > 0$, $\varepsilon \sin \omega t < x < \pi$, where ε and ω are positive constants with $\varepsilon \ll 1$. Find $u(x, t)$ satisfying the boundary conditions $u(\varepsilon \sin \omega t, t) = 0$, $u(\pi, t) = 0$, $u(x, 0) = \sin x$. Repeat this problem for the case in which u satisfies the equation $u_{tt} = u_{xx}$, with the added initial condition $u_t(x, 0) = 0$. In each case, compare the extent to which the perturbation is felt over the whole region with the corresponding result for Problem 8.5.4.

8.5.9 Let $\phi(x, t)$ satisfy the equation $\phi_{tt} = c^2\phi_{xx}$ in the region $x > 0$, $t > 0$, and let the motion be harmonic in the sense $\phi(x, t) = \text{Im}\{e^{i\omega t}\psi(x)\}$, where ω is a real constant. For $x = 0$, choose $\psi(0) = 1$. Then $\psi_{xx} + k^2\psi = 0$, where $k^2 = \omega^2/c^2$. If k were constant, we would obtain $\psi = e^{-ikx}$, where account has been taken of the requirement of outgoing waves at $x = \infty$. Consider, however, $k^2 = k_0^2[1 + \varepsilon \sin \lambda x]$ where k_0, ε, and ω are positive constants, with $\varepsilon \ll 1$. For what values of λ are the perturbation effects on ψ no longer small? Does the corresponding initial value problem (say $\phi(x, 0) = \phi_t(x, 0) = 0$, $\phi(0, t) = \sin \omega t$) exhibit a similar sensitivity with respect to the parameter λ?

8.5.10 Let $\phi(x, y)$ satisfy the equation

$$\phi_{xx} + \phi_{yy} - (k^2 + \varepsilon e^{-\alpha x})\phi = 0$$

in the region $0 < y < \pi$, $0 < x < \infty$, with k constant and with $|\varepsilon| \ll k^2$. As boundary conditions, choose $\phi(0, y) = \sin y$, $\phi_x(0, y) = -(1 + k^2)^{1/2} \times \sin y$, $\phi(x, 0) = \phi(x, \pi) = 0$. Use a perturbation method to find $\phi(x, y)$, and in particular, discuss the effect of a nonzero value of ε on the behavior of ϕ for large values of x. [Note that here we have not required any boundedness of ϕ at ∞]. How much of a change in the boundary conditions at $x = 0$ is required to avoid unboundedness at ∞ for nonzero ε? Does this mathematical problem seem to be "well posed" in the sense that it could arise from a physically reasonable problem?

8.5.11   Let $u(x, y)$ satisfy

$$xu_x + (y + \varepsilon u)u_y = 0$$

in the (x, y) plane with $u(x, 1) = \sin x$. Obtain an exact solution (valid in what region?), and attempt also to obtain a solution via perturbations. Make appropriate remarks.

8.5.12   There are a number of fluid mechanics problems, for example in geophysics, in which the stability of a basic flow pattern is examined by analyzing the growth or decay of a superimposed perturbation motion. A useful discussion of such problems, together with an extensive bibliography, is given by Lin (1955).

A prototype problem of this kind is the following. Restricting attention to two-dimensional motions, let u and v denote the x- and y-component velocities of an incompressible viscous fluid in the region $-\infty < x < \infty$, $0 < y < b$, where b is a constant. The incompressibility condition $u_x + v_y = 0$ implies the existence of a stream function $\psi(x, y)$ such that $u = \psi_y$, $v = -\psi_x$. The equations of motion may then be manipulated to yield

$$\Delta\psi_t + \psi_y \,\Delta\psi_x - \psi_x \,\Delta\psi_y = \nu \,\Delta \,\Delta\psi$$

where Δ is the Laplacian operator $(\partial^2/\partial x^2 + \partial^2/\partial y^2)$ and where ν is the coefficient of kinematic viscosity.

Let the basic flow be one between two plates in which $u = f'(y)$, $v = 0$. Write $\psi = f(y) + \psi'(x, y)$, where ψ' is a perturbation term, and obtain an approximate linear equation satisfied by ψ'. Explain why suitable boundary conditions would be $\psi_x' = \psi_y' = 0$ at $y = 0, b$. Consider next a perturbation of the form $\psi' = \phi(y)e^{i\alpha x}e^{\beta t}$, and obtain an ODE for ϕ. The boundary conditions for ϕ are homogeneous, so one solution is always $\phi(y) \equiv 0$. In general it turns out that, for given α, a nontrivial solution for ϕ can be found only if β has one of a certain set of allowed values, termed eigenvalues.[†] If any such β has a positive real part, the corresponding perturbation motion would grow in size and we would therefore expect the basic motion to be unstable.

---

† We discuss eigenvalue problems in Chapter 11.

9

GREEN'S FUNCTIONS

In Section 2.5 the solution of the diffusion equation problem $\phi_t = a^2\phi_{xx}$, $\phi(x, 0) = f(x)$, for $-\infty < x < \infty$, $t > 0$, was written in the form

$$\phi(x, t) = \int_{-\infty}^{\infty} g(x - \xi, t) f(\xi) \, d\xi$$

where the auxiliary function g was independent of the initial data described by f. We will see in this chapter that there are many linear PDE problems in which it is possible to write the solution in an analogous form, i.e., as an integral over some region of the product of an auxiliary function and the given data (which may be boundary or initial data or a nonhomogeneous term in the differential operator).

In honor of George Green,† such an auxiliary function—which will depend on the differential operator and on the geometry of the problem, but not on the particular data given—is termed a *Green's function*.

9.1 SOME CONSEQUENCES OF THE DIVERGENCE THEOREM

Let (A_1, A_2, A_3) be the three components of a vector field defined over some portion of three-dimensional space in terms of a Cartesian coordinate

---

† Green was a largely self-taught English mathematician who in 1828 published (and sold by private subscription) an essay entitled "On the Application of Mathematical Analysis to the Theories of Electricity and Magnetism." In this essay he derived the integral identities of Section 9.1 and used them to obtain integral representations for the solutions of problems involving the Laplacian operator.

system (x_1, x_2, x_3). The divergence theorem requires that, for continuously differentiable functions $A_i(x_1, x_2, x_3)$, the integral of the divergence of $\mathbf{A}$ over any volume V be equal to the outward flux of $\mathbf{A}$ over the surface S of that volume; i.e.,

$$\int_V \left(\frac{\partial A_1}{\partial x_1} + \frac{\partial A_2}{\partial x_2} + \frac{\partial A_3}{\partial x_3} \right) dV = \int_S (A_1 n_1 + A_2 n_2 + A_3 n_3) \, dS \quad (9.1)$$

where (n_1, n_2, n_3) are the three components of the outward unit normal vector for the surface element dS. (We assume S to possess a tangent plane at each point.)

As a special case, let ϕ be a scalar and define $\mathbf{A}$ to be the gradient of ϕ; i.e., $A_i = \partial \phi / \partial x_i$ for $i = 1, 2, 3$. Then Eq. (9.1) becomes

$$\int_V \Delta \phi \, dV = \int_S \frac{\partial \phi}{\partial n} \, dS \quad (9.2)$$

where $\partial \phi / \partial n$ is the directional derivative of ϕ in the outward normal direction We have already encountered Eq. (9.2) in Chapter 7 (and in two-dimensional form in Chapter 4); it was used there (cf. Problem 7.2.4) to derive the mean value theorem for harmonic functions and to show that the maximum or minimum of a harmonic function over a region must occur on the boundary of that region.

A second special case results from the choice $A_i = \phi \, \partial \psi / \partial x_i$, for $i = 1, 2, 3$, where ϕ and ψ are scalars. Equation (9.1) becomes

$$\int_V \phi \, \Delta \psi \, dV + \int_V \left(\frac{\partial \phi}{\partial x_1} \frac{\partial \psi}{\partial x_1} + \frac{\partial \phi}{\partial x_2} \frac{\partial \psi}{\partial x_2} + \frac{\partial \phi}{\partial x_3} \frac{\partial \psi}{\partial x_3} \right) dV = \int_S \phi \frac{\partial \psi}{\partial n} \, dS \quad (9.3)$$

As a variant of Eq. (9.3) we can rewrite the equation with ϕ and ψ interchanged and subtract the two results to obtain

$$\int_V (\phi \, \Delta \psi - \psi \, \Delta \phi) \, dV = \int_S \left(\phi \frac{\partial \psi}{\partial n} - \psi \frac{\partial \phi}{\partial n} \right) dS \quad (9.4)$$

Equations (9.3) and (9.4) are referred to as *Green's first and second identities*, respectively. The corresponding two-dimensional forms are also valid; the volume and surface integrals are merely replaced by area and contour integrals, respectively.

If we set $\phi = \psi$ and $\Delta \phi = 0$ in Eq. (9.3), we obtain

$$\int_V \left[\left(\frac{\partial \phi}{\partial x_1} \right)^2 + \left(\frac{\partial \phi}{\partial x_2} \right)^2 + \left(\frac{\partial \phi}{\partial x_3} \right)^2 \right] dV = \int_S \phi \frac{\partial \phi}{\partial n} \, dS$$

and the reader will easily verify that this result leads to another proof of the uniqueness thereom for the Dirichlet or Neumann problems for harmonic functions, as previously noted in Problem 7.2.4.

Next, let (ξ_1, ξ_2, ξ_3) be some fixed point inside our region V, and let r denote the distance between this point and any other point (x_1, x_2, x_3), i.e.,

$$r^2 = (x_1 - \xi_1)^2 + (x_2 - \xi_2)^2 + (x_3 - \xi_3)^2$$

Let ε be a small positive parameter $(0 < \varepsilon \ll 1)$, and define

$$\psi_\varepsilon(x_1, x_2, x_3; \xi_1, \xi_2, \xi_3) = 1/(r + \varepsilon) \qquad (9.5)$$

Replacing ψ in Eq. (9.4) by ψ_ε, and interpreting Δ as the Laplacian operator with respect to x_1, x_2, x_3 [remember that (ξ_1, ξ_2, ξ_3) is a fixed point], we obtain

$$\int_V \left[\frac{-2\varepsilon\phi}{r(r + \varepsilon)^3} - \frac{\Delta\phi}{r + \varepsilon} \right] dV = \int_S \left[\phi \frac{\partial}{\partial n} \left(\frac{1}{r + \varepsilon} \right) - \frac{1}{r + \varepsilon} \frac{\partial\phi}{\partial n} \right] dS \qquad (9.6)$$

where we have used the fact that

$$\Delta \left(\frac{1}{r + \varepsilon} \right) = \left(\frac{\partial^2}{\partial r^2} + \frac{2}{r} \frac{\partial}{\partial r} \right) \left(\frac{1}{r + \varepsilon} \right) = \frac{-2\varepsilon}{r(r + \varepsilon)^3} \qquad (9.7)$$

The function $(-2\varepsilon)/[r(r + \varepsilon)^3]$ on the right-hand side of Eq. (9.7) has a volume integral independent of ε, since

$$\int_0^\infty \frac{-2\varepsilon}{r(r + \varepsilon)^3} 4\pi r^2 \, dr = -4\pi$$

Moreover, for every set of fixed values of (x_1, x_2, x_3), except that at which $r = 0$, $(2\varepsilon)/[r(r + \varepsilon)^3] \to 0$ as $\varepsilon \to 0$; at $r = 0$, it is unbounded as $\varepsilon \to 0$. It is clear that this function is essentially a three-dimensional version of the functions of Eq. (2.21) used in the interpretation of the one-dimensional δ function. Thus we could write

$$\Delta \left(\frac{1}{r} \right) = \lim_{\varepsilon \to 0} \Delta \left(\frac{1}{r + \varepsilon} \right) = \lim_{\varepsilon \to 0} \left(\frac{-2\varepsilon}{r(r + \varepsilon)^3} \right)$$

$$\stackrel{?}{=} -4\pi\delta(x_1 - \xi_1) \cdot \delta(x_2 - \xi_2) \cdot \delta(x_3 - \xi_3) \qquad (9.8)$$

where the sign $\stackrel{?}{=}$ has the same interpretation as in Eq. (2.20).

Letting $\varepsilon \to 0$ and using Eq. (9.8), Eq. (9.6) leads to *Green's third identity*:

$$4\pi\phi(\xi_1, \xi_2, \xi_3) = \int_S \left[\frac{1}{r} \frac{\partial\phi}{\partial n} - \phi \frac{\partial}{\partial n} \left(\frac{1}{r} \right) \right] dS - \int_V \left(\frac{1}{r} \right) \Delta\phi \, dV \qquad (9.9)$$

The integrations range over appropriate values of the coordinates (x_1, x_2, x_3);

the point (ξ_1, ξ_2, ξ_3) is fixed within V. The volume integral does not "blow up" at the point $r = 0$ since dV near that point is proportional to r^2, so that the $1/r$ factor gets canceled out.

With the help of the function $\psi_\varepsilon = \ln(r + \varepsilon)$, where $r^2 = (x_1 - \xi_1)^2 + (x_2 - \xi_2)^2$, the two-dimensional analog of Eq. (9.9) is easily obtained:

$$2\pi\phi(\xi_1, \xi_2) = \int_C \left[\ln\left(\frac{1}{r}\right)\cdot\frac{\partial\phi}{\partial n} - \phi\frac{\partial}{\partial n}\ln\left(\frac{1}{r}\right) \right] dl - \int_A \ln\left(\frac{1}{r}\right)\cdot\Delta\phi\, dA \quad (9.10)$$

Here dl is the element of arc length of the curve C enclosing the area A. Again, the point (ξ_1, ξ_2) is held fixed in the integrations on the right-hand side.

We note the interesting fact that Eq. (9.9) relates the value of any function ϕ at a point inside V to the combination of an integral over V of its Laplacian (which vanishes if ϕ happens to be harmonic) and integrals over S of its value and of its normal derivative.

9.2 THE LAPLACIAN OPERATOR

Let (ξ_1, ξ_2, ξ_3) be a fixed point, and as in Section 9.1, let $r^2 = (x_1 - \xi_1)^2 + (x_2 - \xi_2)^2 + (x_3 - \xi_3)^2$. Let $p(x_1, x_2, x_3; \xi_1, \xi_2, \xi_3)$ be a function of (x_1, x_2, x_3), which will in general depend on the choice of the fixed point (ξ_1, ξ_2, ξ_3), but which in any event is harmonic in (x_1, x_2, x_3), so that $\Delta p = (\partial^2/\partial x_1^2 + \partial^2/\partial x_2^2 + \partial^2/\partial x_3^2)p = 0$. Define now

$$g(x, \xi) = g(x_1, x_2, x_3; \xi_1, \xi_2, \xi_3) = -(1/r) + p \quad (9.11)$$

We will write

$$\Delta g = \left(\frac{\partial^2}{\partial x_1^2} + \frac{\partial^2}{\partial x_2^2} + \frac{\partial^2}{\partial x_3^2} \right) g$$

$$= 4\pi\delta(x_1 - \xi_1)\cdot\delta(x_2 - \xi_2)\cdot\delta(x_3 - \xi_3)$$

with the usual understanding that this equation is symbolic only and is a shorthand representation for a process in which r is replaced by $r + \varepsilon$ (as in Section 9.1), with $\varepsilon \to 0$ only after all integrations have been completed.

If we now replace ψ in Eq. (9.4) by g, we obtain

$$4\pi\phi(\xi_1, \xi_2, \xi_3) = \int_S \left[\phi\frac{\partial g}{\partial n} - g\frac{\partial\phi}{\partial n} \right] dS + \int_V g\,\Delta\phi\, dV \quad (9.12)$$

So far the function p occurring in the definition of g has been an arbitrary harmonic function. We will now choose p in such a way as to simplify Eq. (9.12). Let us record, at each boundary point on S, the value of $(1/r)$— i.e., the reciprocal of the distance from the fixed interior point (ξ_1, ξ_2, ξ_3). In principle,† we can determine p so that it is (1) harmonic within V and (2) equal to the value of $(1/r)$ at each boundary point on S as recorded above. With this choice for p we have $g = 0$ whenever (x_1, x_2, x_3) lies on S. Equation (9.12) becomes

$$4\pi\phi(\xi_1, \xi_2, \xi_3) = \int_S \phi \frac{\partial g}{\partial n} \, dS + \int_V g \, \Delta\phi \, dV \qquad (9.13)$$

and so provides a formal solution to the Dirichlet problem: find ϕ if $\Delta\phi$ is a specified function of position within V, and if the boundary values of ϕ are specified on S. Of course we have to find g (essentially, p) as a function of (x_1, x_2, x_3) for each choice of (ξ_1, ξ_2, ξ_3) and this may not be easy. However, we will encounter cases in which it is not difficult to find g, and in any event we observe that once we have found g we can continue to use the same function g in Eq. (9.13), even if we change the specified boundary values of ϕ or the specified interior values for $\Delta\phi$. Observe that Eq. (9.13) gives $\phi(\xi_1, \xi_2, \xi_3)$ as a superposition of volume and surface contributions, where $\partial g/\partial n$ and g represent, respectively, the effects of concentrated "impulses" on the boundary or interior.

The function g defined as above is termed a *Green's function*. We note that it satisfies the conditions‡

$$\Delta g = 4\pi\delta(x_1 - \xi_1) \cdot \delta(x_2 - \xi_2) \cdot \delta(x_3 - \xi_3) \qquad \text{in} \quad V$$

$$g = 0 \qquad \text{for} \quad (x_1, x_2, x_3) \qquad \text{on} \quad S \qquad (9.14)$$

where Δ means $\partial^2/\partial x_1^2 + \partial^2/\partial x_2^2 + \partial^2/\partial x_3^3$.

Let us find g for the interior of a sphere of radius R. We will use spherical coordinates (r, θ, ψ) (with $x_1 = r \sin\theta \cos\psi$, etc.); denote the coordinates of the fixed point by (ρ, α, β). With reference to the inversion process discussed in Problem 7.2.5, the "image point" of (ρ, α, β) has coordinates $(R^2/\rho, \alpha, \beta)$ and so lies outside the sphere, since $\rho < R$. We now claim

---

† We will assume that a solution always exists for the problem of finding a harmonic function that takes on prescribed boundary values. If the region V is sufficiently regular (no infinitely sharp spikes, for example), rigorous mathematical proofs to this effect are available. One can also rely on physical intuition and consider, for example, the problem of steady-state heat conduction in the region, which leads to the same mathematical problem. Physically, a solution for this heat conduction problem must exist.

‡ For the special case $\Delta\phi = 0$ we would still obtain Eq. (9.13) even if we altered the second of conditions (9.14) to read $g = A$ for (x_1, x_2, x_3) on S, where A is an arbitrary constant. The reason for this is, of course, that $A \int (\partial\phi/\partial n) \, dS = 0$ if ϕ is harmonic.

that a suitable Green's function is given by

$$g(r, \theta, \psi; \rho, \alpha, \beta) = (R/\rho r_2) - (1/r_1) \qquad (9.15)$$

where r_1 and r_2 are the distances from (r, θ, ψ) to (ρ, α, β) and to $(R^2/\rho, \alpha, \beta)$, respectively; it is in fact a simple matter to verify that g as given by Eq. (9.15) satisfies Eq. (9.14), where the Laplacian operator is with respect to (r, θ, ψ). [Remember that $1/r_2$ is harmonic for any point (r, θ, ϕ) inside the sphere.] Equation (9.13), with g as defined in Eq. (9.15), now provides the solution for the Dirichlet problem for a sphere (cf. Problem 9.3.1).

9.3 PROBLEMS

9.3.1 If $\Delta\phi = 0$ inside a sphere of radius R, with $\phi(R, \theta, \psi)$ specified, show that

$$\phi(\rho, \alpha, \beta) = \frac{1}{4\pi} \int_0^{2\pi} d\psi \int_0^{\pi} d\theta$$

$$\times \frac{R(R^2 - \rho^2)\phi(R, \theta, \psi)\sin\theta}{[R^2 + \rho^2 - 2R\rho\{\sin\theta\sin\alpha\cos(\psi - \beta) + \cos\theta\cos\alpha\}]^{3/2}}$$

$$(9.16)$$

for $\rho < R$. Note that the quantity in curly brackets in the denominator is the cosine of the angle between the radius vectors to the observation point and to the surface element.

Second, use the Green's function approach to re-derive Eq. (4.12).

9.3.2 In Green's second identity, Eq. (9.4), replace ϕ by $g(x_1, x_2, x_3; \xi_1, \xi_2, \xi_3)$, as defined by Eq. (9.14), and ψ by $g(x_1, x_2, x_3; \eta_1, \eta_2, \eta_3)$, where the points (ξ_1, ξ_2, ξ_3) and (η_1, η_2, η_3) lie within V, to deduce that

$$g(\xi_1, \xi_2, \xi_3; \eta_1, \eta_2, \eta_3) = g(\eta_1, \eta_2, \eta_3; \xi_1, \xi_2, \xi_3)$$

so that the Green's function for the Laplacian is *symmetric* with respect to an interchange of its two sets of arguments.

Prove that g as defined by Eq. (9.14) is unique and nonpositive within V.

9.3.3 Consider a sphere of radius R lying inside a region within which ϕ is harmonic and positive. Choose a spherical polar coordinate system (ρ, α, β) whose origin coincides with the center of the sphere, and denote by ϕ_0 the value of ϕ at this origin. By replacing the quantity in curly brackets in the denominator of Eq. (9.16) by its upper and lower bounds

$+1$ and -1, respectively, and by using the mean value theorem (Problem 7.2.4), show that *Harnack's inequality* holds:

$$\frac{R(R-\rho)}{(R+\rho)^2}\,\phi_0 \leq \phi(\rho, \alpha, \beta) \leq \frac{R(R+\rho)}{(R-\rho)^2}\,\phi_0 \qquad (9.17)$$

This result provides a constraint on the rate at which the value of ϕ can alter as one moves away from the origin.

Obtain the corresponding result in two dimensions, and explain why (in both the two- and three-dimensional cases) Harnack's inequality implies that a function that is harmonic and bounded throughout all space is necessarily a constant.

9.3.4 In Eq. (9.16), let ρ/R be sufficiently small that the denominator can be expanded in a series of powers of (ρ/R). Do this, and obtain the first few terms of a power series expansion for a harmonic function valid in the neighborhood of the origin of a sphere within which the function is harmonic:

$$\phi(\rho, \alpha, \beta) = \phi(0, \alpha, \beta) + \frac{3}{4\pi}\frac{\rho}{R}\left[\sin\alpha\cos\beta\left(\iint \phi \sin^2\theta \cos\psi\, d\theta\, d\psi\right)\right.$$

$$+ \sin\alpha\sin\beta\left(\iint \phi \sin^2\theta \sin\psi\, d\theta\, d\psi\right)$$

$$\left. + \cos\alpha\left(\iint \phi \sin\theta \cos\theta\, d\theta\, d\psi\right)\right] + \left(\frac{\rho}{R}\right)^2 [\cdots] + \cdots$$

where the integrations are over the surface of the sphere of radius R.

Observe that if (x_1, x_2, x_3) are Cartesian coordinates with origin at the center of the sphere, this calculation leads to the result:

$$\phi(x_1, x_2, x_3) = \phi(0, 0, 0) + a_1 x_1 + a_2 x_2 + a_3 x_3 + a_{11} x_1^2 + \cdots$$

where $a_1, a_2, a_3, a_{11}, \ldots$, are constants.† We conclude that a function that is harmonic in a region is analytic at each point of that region—i.e., it can be expanded in a power series valid in some neighborhood of each interior point.

9.3.5 (a) Show that if $u(x_1, x_2, x_3)$ is continuous in some region V and satisfies the mean value theorem in V (i.e., for any point P chosen in V, the value of u at P is equal to the mean value of u over any sphere, center P, lying wholly in V; cf. Problem 7.2.4), then u is harmonic through-

---

† To obtain this series, it is perhaps more direct to replace ρ^2 by $x_1^2 + x_2^2 + x_3^2$, $\rho \sin\alpha \cos\beta$ by x_1, etc. in Eq. (9.16) prior to the expansion.

out V. [*Hint:* let v be a harmonic function whose boundary values on the sphere coincide with those of u. Then $u - v$ satisfies the mean value theorem, and so the maximum theorem, etc.)

(b)   Show that if u is harmonic in a region whose bounding surface includes a portion of a plane, and if $u = 0$ on that planar portion, then u may be continued harmonically across that planar portion by reflection.† Using Problem 7.2.5, state and prove a similar result for a region whose bounding surface includes a portion of a sphere.

(c)   Show that if a sequence of harmonic functions converges uniformly in a region the limit function is harmonic.

9.3.6   (a)   Let $\phi(x_1, x_2, x_3)$ be harmonic in the half space $x_1 > 0$, and let $\phi(0, x_2, x_3)$ be prescribed. Moreover, let $\phi \to C$ uniformly as $x_1^2 + x_2^2 + x_3^2 \to \infty$ in the half space, where C is some constant. A student derives the formula

$$\phi(\xi_1, \xi_2, \xi_3) = \frac{\xi_1}{2\pi} \iint\limits_{-\infty}^{\infty} \frac{\phi(0, x_2, x_3)\, dx_2\, dx_3}{[\xi_1^2 + (x_2 - \xi_2)^2 + (x_3 - \xi_3)^2]^{3/2}}$$

Is he correct? If not, modify the result appropriately. What happens as $\xi_1 \to 0$?

(b)   Discuss Green's function for Laplace's equation for the region exterior to a sphere.

9.3.7   Consider a region V, with surface S, and the problem: find ϕ harmonic in V, with $\partial\phi/\partial n$ prescribed on S. Assuming that a solution of this problem exists, provided that $\int \partial\phi/\partial n \, dS = 0$ (physically, think again of a heat conduction problem), show that an appropriate Green's function is defined by

$$h(x_1, x_2, x_3; \xi_1, \xi_2, \xi_3) = -(1/r) + p$$

where r is defined as before, and where p is a harmonic function satisfying

$$\frac{\partial p}{\partial n} = \frac{\partial}{\partial n}\left(\frac{1}{r}\right) + K$$

on S, where K is a suitable constant. Why can we not set $K = 0$? With h defined as above, Green's second identity leads to

$$4\pi\phi(\xi_1, \xi_2, \xi_3) = K \int_S \phi \, dS - \int_S h \frac{\partial\phi}{\partial n} \, dS$$

---

† To illustrate, let $u(x_1, x_2, x_3)$ be defined and harmonic for $x_3 > 0$, with $u(x_1, x_2, 0) = 0$. Then for $x_3 < 0$, define $u(x_1, x_2, x_3) = -u(x_1, x_2, -x_3)$.

Why is our inability to calculate $\int \phi \, dS$ of no importance? Is h symmetric in its two sets of arguments? If not so in general, can it be made so by a suitable choice of the arbitrary function of (ξ_1, ξ_2, ξ_3) necessarily forming part of p? Suppose, finally, that our problem was to find ϕ, harmonic in V and satisfying a mixed boundary condition on S (e.g., $\alpha\phi + \beta \, \partial\phi/\partial n$ prescribed, where α and β are functions of position). Is it feasible, at least in principle, to construct a Green's function for this problem?

9.3.8 Re-derive Eq. (9.9), without the use of the δ-function approach, by the following method. Delete from V the interior of a small sphere K of radius ε and centered on the point (ξ_1, ξ_2, ξ_3). Apply Green's second identity to the modified region (i.e., to the region interior to S and exterior to K), setting $\psi = 1/r$ as before, using the fact that on the inner spherical surface, $(\partial/\partial n)(1/r) = -(\partial/\partial r)(1/r) = 1/r^2 = 1/\varepsilon^2$, and observing that $(1/r)$ is well behaved and harmonic throughout the modified region. Let $\varepsilon \to 0$, and obtain again Eq. (9.9). Carry out a similar calculation for the case in which the point (ξ_1, ξ_2, ξ_3) is on S, to show that (if S has a tangent plane at this point) the factor 4π in Eq. (9.9) will be replaced by 2π. Finally, observe that if the point (ξ_i) is exterior to V the factor becomes 0. Carry out similar computations for the two-dimensional case. What happens if (ξ_i) is a corner point on the boundary?

9.3.9 A two-dimensional Green's function $g(x, y; \xi, \eta)$ for the Laplacian operator may be defined by

$$\left(\frac{\partial^2}{\partial x^2} + \frac{\partial^2}{\partial y^2}\right) g = \delta(x - \xi) \, \delta(y - \eta)$$

$$g = 0 \qquad \text{on boundary}$$

Show that for the case of the square region $0 < x < l, 0 < y < l$, g is given by

$$g = \sum_{m=1}^{\infty} \frac{\sin(m\pi\xi/l)}{m\pi \sinh m\pi} \sin \frac{m\pi x}{l}$$

$$\times \left\{\cosh \frac{m\pi}{l} [l - (y + \eta)] - \cosh \frac{m\pi}{l} [l - |y - \eta|]\right\}$$

9.4 POTENTIALS OF VOLUME AND SURFACE DISTRIBUTIONS

The discussion of Green's functions for the Laplacian may be instructively rephrased in terms of electrostatic fields; it will also turn out that this physical interpretation suggests some extensions.

We remind the reader that the electrostatic potential of a charge of Q esu, as measured at a point P a distance of r cm from that charge, is given by

$$\phi(r) = (Q/r) \text{ erg/esu}$$

It represents the work required to bring a unit positive test charge from ∞ to P. The superposition axiom of electrostatics permits us to obtain the potential of a number of such charges by simply adding together the individual potentials. In particular, let (x_1, x_2, x_3) be the Cartesian coordinates of a point P, and let there be a distribution of charge of intensity ρ esu/cm^2 throughout some region V. Then the potential resulting from this distribution is given by

$$\phi(x_1, x_2, x_3) = \int_V \frac{\rho(\xi_1, \xi_2, \xi_3)}{r} \, dV \tag{9.18}$$

where r measures the distance between the volume element dV at the point (ξ_1, ξ_2, ξ_3) and the point P—i.e., $r = [(x_1 - \xi_1)^2 + (x_2 - \xi_2)^2 + (x_3 - \xi_3)^2]^{1/2}$. The point P can be either outside or inside V; if the latter, we note that the volume integral exists, since the aparent singularity at $r = 0$ is canceled out by the r^2 dependence of dV in its neighborhood. Similarly, the potential of a charge distribution of σ esu/cm^2 over a surface Γ (which need not be a closed surface) is given by

$$\phi(x_1, x_2, x_3) = \int_\Gamma \frac{\sigma}{r} \, dS \tag{9.19}$$

where r denotes the distance from (x_1, x_2, x_3) to the surface element dS on Γ.

We can interpret the Green's function for a sphere, given by Eq. (9.15), as the potential arising from two concentrated charges—a negative unit charge at an interior point and a positive charge of magnitude (R/ρ) at the exterior image point. The construction of g by this device is one example of the "method of images."

Because $(1/r)$ is harmonic except at $r = 0$, we deduce that the potential function arising from any charge distribution is harmonic at any point not occupied by a charge.

We next define the field strength vector whose ith component is denoted by $E_i = -\partial\phi/\partial x_i$; physically, it represents the force vector acting on a unit positive test charge. At any point in free space, the fact that $\Delta\phi = 0$ implies that $\partial E_1/\partial x_1 + \partial E_2/\partial x_2 + \partial E_3/\partial x_3 = 0$. For any closed surface S enclosing no charges in its interior the divergence theorem therefore requires that the *flux integral* vanish, i.e., that

$$\int_S (E_1 n_1 + E_2 n_2 + E_3 n_3) \, dS = 0 \tag{9.20}$$

where n_i is, as usual, the ith component of the outward unit normal. For a single isolated charge contained within S, the right-hand side of Eq. (9.20) becomes $4\pi q$ [since the flux integral is the same for all surfaces enclosing q, simply calculate its value for a spherical surface centered on q; the effect of charges outside S need not be considered because their contribution to the flux integral is zero, from Eq. (9.20)]. For the case of a number of charges $q_1, q_2, \ldots$ contained within S, superposition yields

$$\int_S (E_1n_1 + E_2n_2 + E_3n_3)\, dS = 4\pi \sum q_i = 4\pi \int_V \rho\, dV \qquad (9.21)$$

where the second result holds for a distribution of charge of density ρ throughout the volume V contained within S. Now one way of computing the divergence of the vector field (E_i) is to take the limit of the ratio of the flux integral to the volume as $V \to 0$; for the charge distribution case, the preceding equation thereby yields

$$\frac{\partial E_1}{\partial x_1} + \frac{\partial E_2}{\partial x_2} + \frac{\partial E_3}{\partial x_3} = 4\pi\rho$$

or, since $E_i = -\partial\phi/\partial x_i$,

$$\Delta\phi = -4\pi\rho \qquad (9.22)$$

We observe that Eq. (9.22) is the same as the result we would have obtained by taking directly the Laplacian of Eq. (9.18), using Eq. (9.8). The reader may find these physical considerations a useful supplement to the δ-function approach of Section 9.1.

We turn now to the idea of a surface distribution of "dipoles." Consider a charge $(-q)$ at position (ξ_1, ξ_2, ξ_3), and a charge $(+q)$ at position $(\xi_1 + \delta\xi_1, \xi_2 + \delta\xi_2, \xi_3 + \delta\xi_3)$, where the $\delta\xi_i$ are small. If (n_i) is a unit vector in the direction of $(\delta\xi_i)$, we can write $\delta\xi_i = \varepsilon n_i$, where ε is the magnitude of the vector $(\delta\xi_i)$. The potential of these two charges at the point (x_1, x_2, x_3), a distance r from ξ_i, is given by

$$-\frac{q}{r} + q\left\{\frac{1}{r} + \sum_{i=1}^{3} \frac{\partial}{\partial\xi_i}\left(\frac{1}{r}\right) \cdot \varepsilon n_i + \text{higher-order terms}\right\}$$

If we now let $\varepsilon \to 0$ and simultaneously let q grow so that $q\varepsilon$ remains constant, the limiting value of this "dipole potential" becomes

$$(q\varepsilon) \sum_{i=1}^{3} \frac{\partial}{\partial\xi_i}\left(\frac{1}{r}\right) \cdot n_i = (q\varepsilon) \sum_{i=1}^{3} \frac{(x_i - \xi_i)n_i}{r^3}$$

$$= (q\varepsilon)\frac{\cos\omega}{r^2}$$

Here ω is the angle between the vector n_i (which is in the direction of the displacement of the positive charge away from the negative charge) and the vector from the dipole position to the observation point. The value of the constant $(q\varepsilon)$ is termed the strength of the dipole.

Green's third identity, Eq. (9.9), is now seen to be equivalent to the statement that any (adequately differentiable) function may be expressed as the potential of three kinds of charge distribution—sources over a volume, sources over a surface, and dipoles over a surface. This fact suggests that dipole distributions over a surface might play a useful role in potential theory, and the exercises of Section 9.5 will show this to be in fact the case. One result (cf. Problem 9.5.4) is that certain PDE problems in potential theory can be rephrased as integral equations, a fact that provides further tools for dealing with such problems.

9.5   PROBLEMS

9.5.1   Let dipoles, each perpendicular to the surface, be distributed over a finite (in general nonclosed) surface Γ. Let the dipole distribution intensity be constant and have value k per unit area. Let A be a point not on Γ; show that the potential at A is equal to $\pm k\Omega$ where Ω is the size of the solid angle subtended by the boundary of Γ at A, with the $+$ or $-$ sign corresponding to whether A is on the "positive" or "negative" side of the dipole charge distribution. Can Γ have wrinkles? Can it be closed about some volume V? Deduce that there is a potential discontinuity of size $4\pi k$ across the surface.

9.5.2   Section 9.4 dealt with three-dimensional situations. Carry out an analogous discussion for the two-dimensional case.

9.5.3   (a)   Let a sphere of radius R be filled with a uniform density of charge with intensity σ. Compute the field strength vector $\mathbf{E}$ and potential ϕ at any point located at distance r from the center (for $r < R$ and for $r > R$), and show that each is continuous at $r = R$. [*Hint:* use symmetry, together with the flux integral equation (9.21).]

(b)   Let the spherical shell $r = R$ be given a surface charge distribution of intensity σ. Compute $\mathbf{E}$ and ϕ for all values of r, and state what discontinuities, if any, exist across $r = R$.

(c)   Let the spherical shell $r = R$ carry a uniform dipole distribution, normal to the surface, of intensity k (positive outwards). Compute $\mathbf{E}$ and ϕ for $r < R$ and for $r > R$; describe any discontinuities at $r = R$. [*Note:* Let the dipole distribution be approached as a limiting case of two spherical shells of charge, at $r = R \pm \varepsilon$, as the distance 2ε between them approaches 0. The answer to this problem must be compatible with the result of Problem 9.5.1.] Explain in what sense one could state that $\phi(R) = -2\pi k$.

9.5.4 The above results can be expected to carry over in a natural way to non-spherically-symmetric, nonconstant charge or dipole distributions. Consider, for example, a portion of a surface, Γ, and let there be a position-dependent normal dipole distribution of intensity k on Γ. Then the potential at any point P is given by

$$\phi(P) = \int_\Gamma k \frac{\partial}{\partial n}\left(\frac{1}{r}\right) dS$$

where the normal derivative direction is the same direction as that in which the positive sense for the dipole is chosen. Let P_0 be a point on the surface and P_+, P_- two points infinitesimally close to P_0 on the "positive" and "negative" sides of Γ, respectively. Show that†

$$\phi(P_+) - \phi(P_0) = \phi(P_0) - \phi(P_-) = 2\pi k(P_0)$$

9.5.5 Let S be a closed surface enclosing a volume V. Let the intensity of a normal dipole distribution over S be denoted by k, positive if directed outwards. The potential $\phi(P)$ at any point in space is given by

$$\phi(P) = \int_S k \frac{\partial}{\partial n}\left(\frac{1}{r}\right) dS$$

Except for P on S, $\phi(P)$ is harmonic. Let P_0 be any point on S at which the value of k is $k(P_0)$, and denote by $u(P_0)$ the limit of $\phi(P_-)$ as $P_- \to P_0$, where P_- is in V. From Problem 9.5.3 we know that $u(P_0) = \phi(P_0) - 2\pi k(P_0)$. Deduce now that, if a function u, harmonic in V with $u(P_0)$ specified for P_0 on S, is to be representable as the potential of some dipole distribution on S of intensity k, then the function k must satisfy the *integral equation*

$$u(P_0) + 2\pi k(P_0) = \int_S k \frac{\partial}{\partial n}\left(\frac{1}{r}\right) dS \tag{9.23}$$

† Γ is supposed to have a tangent plane at each point. One approach to this problem is to approximate Γ in the neighborhood of P_0 by a portion of a spherical surface with constant dipole intensity $k(P_0)$, and to use the results of Problem 9.5.3, realizing that any discontinuity in ϕ can arise only from that part of Γ immediately neighboring P_0. Another approach, mathematically more elegant, is to "extend" the function k continuously into the region bordering one side of Γ in such a way that $\partial k/\partial n = 0$ on that side, and to apply the modified Green's identity of Problem 9.3.8 to this function k.

where r is the distance from dS to P_0. Explain exactly what the integral means. Obtain an analogous result for the two-dimensional case. If S is a sphere (or circle, in two dimensions), write the integral equation in as compact a form as you can. [*Note:* From integral equation theory, it can be proved—at least for suitable surfaces S—that Eq. (9.23) does have a solution, k; consequently, any harmonic function can be represented as the potential of a dipole distribution over such an S.]

Finally, obtain a similar integral equation equivalent to the Neumann problem ($\partial u/\partial n$ specified on S) in two or three dimensions, using now a single layer on S instead of a dipole distribution. If this equation is solvable, would its solution be unique? What would you expect to be a necessary condition for solvability?

Integral equation theory may be used very efficiently to prove formal existence theorems in potential theory. An informative treatment is given in Garabedian (1964, Chap. 9) and also in Petrovsky (1964, Sec. 34).

9.6   MODIFIED LAPLACIAN

Let (x_1, x_2) be Cartesian coordinates in the plane, and consider a simple closed contour C in the plane, enclosing a region R. We want to find a function $u(x_1, x_2)$ satisfying an equation involving a modified Laplacian operator:

$$\Delta u - \alpha^2 u = f \tag{9.24}$$

in R, where α is a constant, $\Delta = \partial^2/\partial x_1{}^2 + \partial_2/\partial x_2{}^2$, and where $f(x_1, x_2)$ is prescribed in R. On the boundary C of R, values of u are prescribed.

This kind of problem would arise in the analysis of the small displacement of a tightly stretched membrane, as discussed in Section 3.7 and Problem 4.2.2, if one adds the condition that the membrane be elastically restrained. The Green's function to be defined below will correspond to a concentrated unit force applied to a point (ξ, ξ_2) on the membrane, and the final expression for u in terms of g and f, as given in Eq. (9.27), can be interpreted physically in terms of a superposition of loadings.

In analogy with Eq. (9.14), we now define a Green's function $g(x_1, x_2; \xi_1, \xi_2)$ by

$$\Delta g - \alpha^2 g = \delta(x_1 - \xi_1) \cdot \delta(x_2 - \xi_2) \tag{9.25}$$

$$g = 0 \quad \text{for} \quad (x_1, x_2) \quad \text{on} \quad C \tag{9.26}$$

Here (ξ_1, ξ_2) is a fixed point in R, and the Δ operator applies to the (x_1, x_2) variables. The above-mentioned physical interpretation of g suggests that such a function exists, and we will shortly turn to its actual construction. First, however, let us assume that we have found a function g satisfying Eqs. (9.25) and (9.26); we now want to use it to obtain a formal solution of Eq. (9.24).

The technique to be used is generally called the "multiply-and-subtract" method; it occurs repeatedly in analysis. We multiply Eq. (9.24) by g, Eq. (9.25) by u, and subtract to obtain

$$g \, \Delta u - u \, \Delta g = fg - u \, \delta(x_1 - \xi_1) \, \delta(x_2 - \xi_2)$$

Integration over the region R enclosed by C and the use of Green's second identity [Eq. (9.4)] gives

$$\int_C \left(g \frac{\partial u}{\partial n} - u \frac{\partial g}{\partial n} \right) dl = \int_R fg \, dA - u(\xi_1, \xi_2)$$

so that (using $g = 0$ on C)

$$u(\xi_1, \xi_2) = \int_R fg \, dA + \int_C u \frac{\partial g}{\partial n} \, dl \qquad (9.27)$$

In this equation ξ_1 and ξ_2 are fixed parameters, and integrations are with respect to the (x_1, x_2) variables. Thus if g can be determined, Eq. (9.27) will provide the desired formal solution for u for any given function f in R and for any given boundary values of u on C.

To construct the function g, we first look for a relatively simple function g_1 having the correct kind of singularity at the source point, and then satisfy the boundary condition by adjoining an appropriate nonsingular function p to g_1. Let $g_1(x_1, x_2; \xi_1, \xi_2)$ depend only on the radial distance r between (x_1, x_2) and (ξ, ξ_2) and satisfy

$$\Delta g_1 - \alpha^2 g_1 = \left(\frac{d^2}{dr^2} + \frac{1}{r} \frac{d}{dr} - \alpha^2 \right) g_1 = 0, \qquad r > 0$$
$$\qquad (9.28)$$
$$g_1 \to 0 \qquad \text{as} \quad r \to \infty$$

The solution of this differential equation is given in terms of the modified Bessel functions I_0 and K_0 by

$$g_1 = AK_0(\alpha r) + BI_0(\alpha r)$$

where A and B are constants. Since $K_0 \to 0$ and $I_0 \to \infty$ as the argument $(\alpha r) \to \infty$, we must set $B = 0$, so

$$g_1 = AK_0(\alpha r) \qquad (9.29)$$

The function $K_0(\alpha r)$ has a logarithmic singularity at the origin described by $K_0(\alpha r) \sim \ln[1/(\alpha r)]$ as $r \to 0$. We next want to choose A so that g

satisfies Eq. (9.25); if we use our previous technique and replace r by $(r + \varepsilon)$, $\varepsilon > 0$, and then let $\varepsilon \to 0$, it follows easily that $A = -1/(2\pi)$. (For an alternative technique, see Problem 9.7.1.) Returning now to g, we are thus led to

$$g = -(1/2\pi)K_0(\alpha r) + p \qquad (9.30)$$

where $r^2 = (x_1 - \xi_1)^2 + (x_2 - \xi_2)^2$, and where p is a function satisfying $\Delta p - \alpha^2 p = 0$ in R, $p = 1/2\pi K_0(\alpha r)$ for (x_1, x_2) on C. Since the point (ξ_1, ξ_2) is not on C, r does not vanish for (x_1, x_2) on C, and thus there are no singularities in the conditions, in R or on C, to be satisfied by p. On physical grounds (e.g., the membrane problem mentioned earlier), we can anticipate that such a function p can be found—by conventional series expansions, transforms, or numerical methods—at least if the contour C is a reasonable one.

If the sign of the second term in Eq. (9.24) is reversed, we encounter the Bessel function $H_0^{(2)}$ instead of K_0; cf. Problem 9.7.2.

9.7 PROBLEMS

9.7.1 Replace conditions (9.28) by

$$\left(\frac{d^2}{dr^2} + \frac{1}{r}\frac{d}{dr} - \alpha^2\right) g_1 = \frac{1}{2\pi\varepsilon}\,\delta(r - \varepsilon)$$

$$\qquad (9.31)$$

$$g_1 \to 0 \qquad \text{as} \quad r \to \infty, \qquad g_1(0) = \text{finite}$$

where $\varepsilon > 0$, and find $g_1(r)$.† Let $\varepsilon \to 0$, and recover $g_1 = -(1/2\pi)K_0(\alpha r)$. Why is Eq. (9.31) an appropriate replacement for Eq. (9.25)?

9.7.2 Consider the two-dimensional wave equation for a function $\phi(x_1, x_2, t)$:

$$\phi_{tt} - c^2\,\Delta\phi = m$$

where c is constant and m a function of (x_1, x_2, t). Solutions depending sinusoidally on time can exist for appropriate m and boundary conditions; write $m = \text{Re}\{-fe^{i\omega t}\}$, $\phi = \text{Re}\{\psi e^{i\omega t}\}$, where ω is a positive constant and f, ψ functions of (x_1, x_2). Then

$$\Delta\psi + \frac{\omega^2}{c^2}\psi = \frac{1}{c^2}f \qquad (9.32)$$

---

† The ODE may be written $(rg_1')' - \alpha^2 rg_1 = (r/2\pi\varepsilon)\,\delta(r - \varepsilon)$. Requiring g_1 to be continuous at $r = \varepsilon$, this ODE requires that there be a jump in g_1' across $r = \varepsilon$ given by $g_1'(\varepsilon+) - g_1'(\varepsilon-) = 1/2\pi\varepsilon$.

for (x_1, x_2) within some contour C enclosing an area A. Show that one possible Green's function has the form

$$g = (i/4)H_0^{(2)}(r\omega/c) + p \qquad (9.33)$$

where $H_0^{(2)}$ is the Hankel function of the second kind, defined by $H_0^{(2)}(z) = J_0(z) - iY_0(z)$, and where p is well behaved. Here the singularity term has been chosen so as to (in conjunction with the time dependence $e^{i\omega t}$) represent outgoing waves at ∞, since for large r,

$$H_0^{(2)}\left(r\frac{\omega}{c}\right) \sim \left(\frac{2}{\pi(r\omega/c)}\right)^{1/2} e^{-i(r\omega/c - \pi/4)}$$

Comment on this choice. Assuming that Eq. (9.32) has a solution (we will see in Chapter 11 that, for special choices of ω, f, and boundary conditions, it may not), obtain a formal solution of this equation by making an appropriate choice for p and by using the Green's function method.

9.7.3 Show that the three-dimensional analogs of g_1 as it appears in Eqs. (9.30) and (9.33) are

$$g_1 = -(1/4\pi r)e^{-\alpha r}, \qquad g_1 = -(1/4\pi r)e^{-i(\omega/c)r}$$

respectively.

9.7.4 Discuss the possibility of a Green's function approach to the elliptic equation

$$L\phi = A\phi_{xx} + 2B\phi_{xy} + C\phi_{yy} + D\phi_x + E\phi_y + F\phi = G$$

where $A, B, \ldots$ are functions of x and y, by use of the Green's function $g(x, y; \xi, \eta)$ satisfying the "adjoint" equation

$$Mg = (Ag)_{xx} + (2Bg)_{xy} + (Cg)_{yy} - (Dg)_x - (Eg)_y + Fg$$

$$= \delta(x - \xi)\,\delta(y - \eta)$$

9.7.5 With $g_1 = -(1/4\pi r)e^{-i(\omega/c)r}$, show that the solution ψ of the three-dimensional equivalent of Eq. (9.32) satisfies

$$\psi(\xi_1, \xi_2, \xi_3) = \int_V \frac{1}{c^2}g_1 f\,dV + \int_S \left(\psi\frac{\partial g_1}{\partial n} - g_1\frac{\partial \psi}{\partial n}\right)dS \qquad (9.34)$$

an equation somewhat reminiscent of Huygens' principle in optics. Here S encloses a finite volume V. Show also that if V is the volume *outside* a closed surface S, then the same result holds, provided that the *Sommerfeld radiation condition*

$$R[(\partial \psi/\partial R) + i(\omega/c)\psi] \to 0 \qquad \text{as} \quad R \to \infty \qquad (9.35)$$

is satisfied by ψ as evaluated on the surface of a large sphere of radius R.

9.7.6   In Eq. (9.34), multiply through by $e^{i\omega t}$ and interpret quantities like $[\phi]_R = \psi e^{i\omega[t-(r/c)]}$ as the value of ϕ at a time earlier than t by that time interval necessary for signal propagation over a distance r (we call $[\phi]_R$ the *retarded value* of ϕ), to obtain

$$\phi(\xi_1, \xi_2, \xi_3) = \frac{1}{4\pi c^2} \int_V \frac{[m]_R}{r} dV$$

$$+ \frac{1}{4\pi} \int_S \left\{ \frac{1}{r} \left[\frac{\partial \phi}{\partial n} \right]_R - [\phi]_R \frac{\partial}{\partial n} \left(\frac{1}{r} \right) + \frac{1}{rc} \left[\frac{\partial \phi}{\partial t} \right]_R \frac{\partial r}{\partial n} \right\} dS \quad (9.36)$$

where $m = -fe^{i\omega t}$. This formula does not involve ω, and since it applies to each frequency component of an arbitrary solution of the time-dependent equation $\phi_{tt} - c^2 \Delta\phi = m$, it follows that this *Kirchhoff formula* must hold for any solution of this wave equation.

Use the function

$$g_2 = -\frac{1}{4\pi r} e^{ir\omega/c}$$

(which differs from g_1 only in the sign of the exponent) to derive a similar result involving *advanced* rather than retarded values of ϕ.

Simplify Eq. (9.36) as much as you can for the case in which S is a sphere.

9.7.7   (a)   Solve Eqs. (9.25) and (9.26) for the case in which the region of interest is that inside a circle.

(b)   Find the Green's function for the two-dimensional operator of Eq. (9.32) for the region $x_1 > 0$, and with $\partial g/\partial n = 0$ on $x_1 = 0$. [*Hint:* use an image source in the left half plane.]

(c)   Repeat part (a) for a region contained between concentric circles.

(d)   Repeat part (a) for the region inside a square (cf. Problem 9.3.9).

9.8   WAVE EQUATION

A modified strategy associated with the fact that time is now one of the independent variables is useful when dealing with wave equation problems. Consider the problem of determining a function $g(x_1, x_2, x_3, t; \xi_1, \xi_2, \xi_3, \tau)$ [denoted for short by $g(x, t; \xi, \tau)$] satisfying

$$g_{tt} - c^2 \Delta g = \delta(x_1 - \xi_1) \cdot \delta(x_2 - \xi_2) \cdot \delta(x_3 - \xi_3) \cdot \delta(t - \tau)$$

$$g = g_t = 0 \quad \text{for} \quad t < \tau \quad (9.37)$$

Here the Laplacian operator Δ is with respect to the (x_i) variables, the equation is to hold throughout all space, and c is a fixed positive constant. Taking a Laplace transform with respect to time, solving the resultant equation by the use of spherical symmetry, and inverting the transform, we obtain

$$g(x, t; \xi, \tau) = \frac{1}{4\pi c^2 r} \delta\left[(t - \tau) - \frac{r}{c}\right] \tag{9.38}$$

where $r^2 = (x_1 - \xi_1)^2 + (x_2 - \xi_2)^2 + (x_3 - \xi_3)^2$, as before. Thus we have a kind of radial delta function spreading out from the source point; note the $(1/r)$ amplitude decay.

An interesting use of Eq. (9.38) is in the derivation of the form of the disturbance β resulting from a unit source moving with fixed velocity V along the x_1 axis:

$$\beta_{tt} - c^2 \Delta\beta = \delta(x_1 - Vt) \cdot \delta(x_2) \cdot \delta(x_3) \tag{9.39}$$

We can rewrite the right-hand side as

$$\delta(x_1 - Vt) \cdot \delta(x_2) \cdot \delta(x_3)$$

$$= \delta(x_2) \cdot \delta(x_3) \int\limits_{-\infty}^{\infty}\!\!\! \delta(\xi_1 - V\tau) \cdot \delta(t - \tau)\, \delta(x_1 - \xi_1)\, d\xi_1\, d\tau$$

and consequently obtain the function β by a superposition of solutions of the form (9.38):

$$\beta = \frac{1}{4\pi c^2} \int\limits_{-\infty}^{\infty}\!\!\! d\xi_1\, d\tau\, \delta(\xi_1 - V\tau)\, \frac{\delta[(t - \tau) - (r/c)]}{r}$$

$$= \frac{1}{4\pi c^2} \int_{-\infty}^{\infty} d\tau\, \frac{\delta\{(t - \tau) - (1/c)[(x_1 - V\tau)^2 + x_2^2 + x_3^2]^{1/2}\}}{[(x_1 - V\tau)^2 + x_2^2 + x_3^2]^{1/2}}$$

Following the change in variable $\lambda = \tau + (1/c)[(x_1 - V\tau)^2 + x_2^2 + x_3^2]^{1/2}$, integration yields, for the case $V < c$,

$$\beta = \frac{1}{4\pi c^2}\left[(x_1 - Vt)^2 + \left(1 - \frac{V^2}{c^2}\right)(x_2^2 + x_3^2)\right]^{-1/2} \tag{9.40}$$

which is the desired solution.

Return now to Eq. (9.38) and note that g satisfies the symmetry relation

$$g(\xi, -\tau; x, -t) = g(x, t; \xi, \tau)$$

so that, denoting $(\partial^2/\partial\xi_1^2 + \partial^2/\partial\xi_2^2 + \partial^2/\partial\xi_3^2)$ by Δ_0, we have

$$g_{\tau\tau} - c^2 \Delta_0 g = \delta(x_1 - \xi_1)\, \delta(x_2 - \xi_2)\, \delta(x_3 - \xi_3)\, \delta(t - \tau) \tag{9.41}$$

Next, let a function $\phi(\xi_1, \xi_2, \xi_3, \tau)$ satisfy the equation

$$\phi_{\tau\tau} - c^2 \, \Delta_0 \phi = f(\xi_1, \xi_2, \xi_3, \tau) \tag{9.42}$$

within some region V_0, surface S_0, of ξ_i space. Multiply Eq. (9.41) by ϕ, Eq. (9.42) by g, subtract, and integrate over V_0 and also with respect to τ from 0 to $(t + 0)$ to give

$$\phi(x, t) = \int_0^{t+} d\tau \left[\int_{V_0} f(\xi, \tau) g(x, t; \xi, \tau) \, dV_0 \right.$$

$$\left. - c^2 \int_{S_0} \left\{ \phi \frac{\partial g}{\partial n_0} - g \frac{\partial \phi}{\partial n_0} \right\} dS_0 \right] - \int_{V_0} \left[\phi \frac{\partial g}{\partial \tau} - g \frac{\partial \phi}{\partial \tau} \right]\bigg|_{\tau=0} dV_0$$

$$\tag{9.43}$$

provided the point (x_1, x_2, x_3) lies within V_0. The reader may use Eq. (9.43) to re-derive the result of Problem 9.7.6.

In the next section we give some problems relating to a variety of Green's function situations. We should emphasize, however, that there are many problems in which the method of Fourier transforms is invaluable; in this book we content ourselves with the special case of Laplace transforms in order to avoid the requirement of a substantial background in complex analysis on the part of the reader. Many examples of the use of general Fourier transforms, together with a discussion of approximation and asymptotic methods, will be found in Carrier *et al.* (1966, Chap. 7).

9.9   PROBLEMS

9.9.1   Derive the results analogous to Eqs. (9.38), (9.40), and (9.43) for the cases of one and two space dimensions, comparing the one-dimensional result to previous formulas. For the traveling-source problem, consider both $V < c$ and $V > c$.

How much progress is possible if c is a function of position and/or time?

9.9.2   (a)   Let $\Delta \phi = 0$ in the (x, y)-plane region defined by $0 < y < l$, $-\infty < x < \infty$, with $\phi(x, 0) = \delta(x - \xi)$, $\phi(x, l) = 0$, and $\phi \to 0$ as $|x| \to \infty$. Find ϕ in the form of a series, and sum the series to obtain

$$\phi = \frac{1}{2l} \frac{\sin(\pi/l) y}{\cosh(\pi/l)(x - \xi) - \cos(\pi/l) y}$$

Note the special case $l \to \infty$, for which

$$\phi \to \frac{1}{\pi} \frac{y}{(x - \xi)^2 + y^2}$$

which clearly approaches $\delta(x - \xi)$ as $y \to 0$.

(b)   Use the result of part (a) to write at once the solution to the quarter-plane problem: $\Delta\phi = 0$ in the region defined by $x > 0$, $y > 0$, with $\phi(x, 0) = f(x)$, $\phi(0, y) = g(y)$, where $f(x)$ and $g(y)$ are prescribed. [*Hint*: Begin with the problem $\phi(x, 0) = f(x)$ for $x > 0$, $\phi(x, 0) = -f(-x)$ for $x < 0$, for the half-plane region $y > 0$.]

9.9.3   (a)   At time $t = 0$, a unit amount of heat is released at the origin of an infinite solid possessing constant thermal diffusivity. Find the temperature, at subsequent times, for each point in the region. How could this solution be used to solve a more general problem?

(b)   Find $\phi(x, y, t)$ satisfying

$$\phi_t - a^2 \Delta\phi = 0 \qquad \text{in} \quad -\infty < x < \infty, \quad y > 0$$

with $\phi(x, y, 0) = 0$ and with $\phi(x, 0, t) = \delta(x - \xi) \cdot \delta(t - [0+])$, where $[0+]$ means a time just after time zero. Here a and ξ are constants. Is this a reasonable problem? How much heat is released? Consider also the corresponding problem for a half space.

(c)   A moving heat source in a three-dimensional half space starts at time zero and proceeds along a straight line parallel to the free surface, which is maintained at zero temperature. Find the resulting temperature if the thermal diffusivity is constant.

9.9.4   Let $\phi(x, y)$ satisfy the equation

$$\Delta\phi = \delta(x - \xi) \cdot \delta(y - \eta)$$

in the region $-\infty < x < \infty$, $0 < y < l$, with $\phi(x, 0) = \phi(x, l) = 0$, $\phi \to 0$ as $|x| \to \infty$. Use a series expansion technique to determine ϕ, and explain how the result could alternatively have been obtained by appropriately locating positive and negative sources outside the region (the "method of images," which we have previously used in Section 9.2 and in a number of exercises). Sum the series. Consider also the case $l \to \infty$; what happens to the "images"?

Repeat this problem for the case in which the governing equation is

$$\Delta\phi - \alpha^2\phi = \delta(x - \xi) \cdot \delta(y - \eta)$$

where α is a constant.

For what kinds of problems—i.e., nature of differential operators (e.g., constant or variable coefficients, linear or nonlinear, order, etc.) and boundary conditions—would you expect the method of images to be useful?

9.9.5 Find the Green's function $g(x, y; \xi, \eta)$ satisfying the equation $\Delta g = \delta(x - \xi) \, \delta(y - \eta)$, $g = 0$ on C, where C is the "almost" circle defined by $r = R + \varepsilon f(\theta)$. Here (r, θ) are polar coordinates, $R > 0$ a given constant, $f(\theta)$ a given periodic function, and ε a small parameter. Use a perturbation method of the kind discussed in Chapter 8.

9.9.6 Let Eq. (9.32) be valid in the entire plane and let ψ satisfy the two-dimensional radiation condition

$$\sqrt{R}[(\partial \psi / \partial R) + i(\omega/c)\psi] \to 0 \qquad \text{as} \quad R \to \infty$$

when ψ is evaluated on the boundary of a circle of radius R. Then show that we can write

$$\psi(x, y) = \frac{1}{c^2} \iint\limits_{-\infty}^{\infty} f(\xi, \eta) \cdot \frac{i}{4} H_0^{(2)} \left(\frac{\omega}{c} r\right) d\xi \, d\eta,$$

where $r^2 = (x - \xi)^2 + (y - \eta)^2$, provided that $f \to 0$ with adequate rapidity as $x^2 + y^2 \to \infty$. Next, replace Eq. (9.32) by

$$\Delta \psi_1 + (\omega^2/c^2)[1 + \varepsilon J_0(\alpha(x^2 + y^2)^{1/2})]\psi_1 = (1/c^2)f$$

where α and ε are parameters. Treating ε as small, use a perturbation technique to determine ψ. Are there values of the parameter α that will make ψ_1 differ appreciably from the preceding ψ function?

9.9.7 A model equation that has been proposed† for the study of the diffusion of reproductive cells in a certain environment is

$$c_t = D(c_{xx} + c_{yy} + c_{zz}) - vc_z + Ac$$

for $c(x, y, z, t)$ in the half space $-\infty < x < \infty$, $-\infty < y < \infty$, $z > 0$, and for time $t > 0$. Here D, A, v are constants. The initial condition is $c(x, y, z, 0) = \delta(x) \, \delta(y) \, \delta(z)$, and the boundary conditions are $c_z(x, y, 0, t) = 0$ (which corresponds to zero flux at the boundary of the half space) and $c \to 0$ as $z \to \infty$. Obtain one solution for this equation by means of the substitution $c = f(x, y, t)g(z, t)$; does this solution provide a suitable Green's function? Discuss.

† Blumenson (1970, p. 273).

10

VARIATIONAL METHODS

10.1 A MINIMIZATION PROBLEM

Let a wire be bent into the form of a closed loop, and let the loop be twisted a bit so that it does not lie in a plane. If this loop is now dipped into a soap solution and removed, a soap film can be produced; this film will have the form of a surface in space whose boundary is the wire loop. Because of surface tension, we can expect that the soap film will contract as much as it can, and this suggests the following mathematical problem: given a closed curve in space, find that surface, bounded by the curve, whose area is a minimum. This problem differs from conventional maxima or minima problems in calculus in that we look not for a *point* at which a given function attains its extremum, but rather for a *function* (here that function describing the shape of the surface) for which a given property (in this example, the area) is extremized. Such a problem is said to be a problem in the *calculus of variations*.

The reason for our interest in such problems is that there is an equivalence between problems in the calculus of variations and problems involving partial differential equations. In particular, any PDE problem can be phrased in variational form (cf. Problem 10.2.7), and it often happens that the variational form permits a useful alternative approach to the solution of the PDE problem.

To illustrate the general technique used in demonstrating the equivalence, we consider now a variational problem that will lead to a PDE of potential type (we return to the minimum-area problem in Problem 10.2.1c). Let Γ be a closed smooth curve in the (x, y) plane, enclosing an area A; let a function $p(x, y)$ be defined within A, and let a function g be defined on Γ. Among all functions $u(x, y)$ that are C'' in A (i.e., whose second derivatives are continuous) and that satisfy the condition $u = g$ on Γ, we seek the one that minimizes the integral

$$I = \int_A \left[\tfrac{1}{2}(u_x{}^2 + u_y{}^2) - pu \right] dA \qquad (10.1)$$

Assume for the moment that such a minimizing function does exist, and denote it by $w(x, y)$. What can we deduce about w? To answer this question, we apply directly the condition that I attains its minimum when u is replaced by w. This means in particular that if we write

$$u = w + \varepsilon\eta$$

where $\eta(x, y)$ is any chosen (but fixed, once chosen) C'' function that vanishes on Γ, and where ε is a parameter, then

$$I(\varepsilon) = \int_A \left[\tfrac{1}{2}(w_x + \varepsilon\eta_x)^2 + \tfrac{1}{2}(w_y + \varepsilon\eta_y)^2 - p(w + \varepsilon\eta) \right] dA$$

must be a minimum for $\varepsilon = 0$. Thus $dI/d\varepsilon = 0$ at $\varepsilon = 0$, and conventional differentiation of the integrand with respect to ε, followed by the replacement of ε by 0, gives

$$\int_A \left[w_x\eta_x + w_y\eta_y - p\eta \right] dA = 0$$

By use of the divergence theorem, this equation becomes

$$\int_\Gamma \frac{\partial w}{\partial n}\eta \, ds - \int_A \left[w_{xx} + w_{yy} + p \right]\eta \, dA = 0$$

where ds is the element of arc length along Γ, and $\partial w/\partial n$ is the outward normal derivative of w. Since $\eta = 0$ on Γ, it follows that

$$\int_A \left[\Delta w + p \right]\eta \, dA = 0 \qquad (10.2)$$

But apart from the C'' condition and the condition $\eta = 0$ on Γ, η was entirely arbitrary; the only way in which Eq. (10.2) can be true for every such η function is if the integrand factor multiplying η is identically zero.†

† For if this factor is nonzero but, say, positive at some point, then by continuity it is also positive in some neighborhood of this point; a contradiction to Eq. (10.2) is then obtained by the choice of a C'' function η that is positive in an appropriate subneighborhood of the point and zero elsewhere.

Thus, if there is a solution $w(x, y)$ to the stated minimization problem, it must satisfy the PDE

$$\Delta w + p = 0 \qquad (10.3)$$

We know from potential theory that if Γ, p, and g are sufficiently well behaved, then Eq. (10.3), together with the boundary condition $w - g$ on Γ, has a solution and indeed a unique solution. However, we have so far only shown that *if* the variational problem has a solution, it must be given by the function w satisfying Eq. (10.3) and, of course, the condition $w = g$ on Γ. We now show that the function w defined by this PDE problem does indeed make I a minimum. For let $\phi(x, y)$ be any C'' function satisfying $\phi = g$ on Γ. Define $\psi = \phi - w$; then

$$\int_A \left[\tfrac{1}{2}(\phi_x{}^2 + \phi_y{}^2) - p\phi \right] dA = \int_A \left[\tfrac{1}{2}(w_x{}^2 + w_y{}^2) - pw \right] dA$$

$$+ \int_A \left[\psi_x w_x + \psi_y w_y - p\psi \right] dA + \int_A \left[\tfrac{1}{2}\psi_x{}^2 + \tfrac{1}{2}\psi_y{}^2 \right] dA$$

Use of the divergence theorem, Eq. (10.3), and the condition that $\psi = 0$ on Γ shows that the second integral on the right-hand side vanishes, and since the third integral is nonnegative, we have the desired result.

Thus in this example, I is indeed a minimum for the particular solution function w. It should be noticed, however, that the only condition we actually used in deriving Eq. (10.3) was that I be *stationary* for $u = w$ — i.e., that the first-order change in I (corresponding to a change $\varepsilon\eta$ in u) be zero. This is the usual situation in variational problems. A function that makes a desired integral stationary is sought, and whether the integral is a minimum, a maximum, or has an inflection point behavior for that solution function is frequently of secondary importance.

It is customary to abbreviate somewhat the calculation that led to Eq. (10.3). Let w be the choice of u that makes I of Eq. (10.1) stationary, and let $\delta w(x, y)$ represent a change in the function w. Since $w + \delta w$ is to continue to satisfy the boundary condition, we require $\delta w = 0$ on Γ. The corresponding first-order change in I, denoted by δI (and termed the "variation" in I) is given by

$$\delta I = \delta \int_A \left[\tfrac{1}{2}w_x{}^2 + \tfrac{1}{2}w_y{}^2 - pw \right] dA$$

$$= \int_A \left[w_x \, \delta w_x + w_y \, \delta w_y - p \, \delta w \right] dA$$

in exact analogy with the usual method whereby the differential of a function $h^2(x, y)$ is given by $2h\,dh$. We have not distinguished between $(\delta w)_x$ and $\delta(w_x)$, for example, since these two quantities are identical as a result of the linearity property of differentiation. Continuing, we use the divergence theorem to give

$$\delta I = \int_\Gamma \frac{\partial w}{\partial n} \delta w\,ds - \int_A [w_{xx} + w_{yy} + p]\,\delta w\,dA$$

Using $\delta w = 0$ on Γ, the first integral vanishes. We now use the condition that I be stationary (i.e., $\delta I = 0$) for any such δw, and deduce Eq. (10.3) just as before.

10.2 PROBLEMS

10.2.1 In each of the following integrals, $u(x, y)$ is to be adequately differentiable in a region A of the (x, y) plane and must satisfy the conditions prescribed for the boundary Γ of A. Show that if $w(x, y)$ is the choice of u that renders J stationary, then w must satisfy the stated PDE.

(a) $J = \int_A [(1 + x^2)u_x^2 + 3u_y^2 - e^y(u^2 + u)]\,dA$ $u = 0$ on Γ

Then $[(1 + x^2)w_x]_x + 3w_{yy} + e^y(w + \tfrac{1}{2}) = 0$

(b) $J = \int_A [uu_x + u_y^2 + xu_xu_y]\,dA$ u = prescribed function on Γ

Then $2w_{yy} + w_y + 2xw_{xy} = 0$

(c) $J = \int_A [1 + u_x^2 + u_y^2]^{1/2}\,dA$ u = prescribed function on Γ

Then $(1 + w_y^2)w_{xx} + (1 + w_x^2)w_{yy} - 2w_xw_yw_{xy} = 0$

Show that this is the minimal area problem ("Plateau's problem") discussed in the first paragraph of Section 10.1.

10.2.2 If second derivatives occur in a variational integral, then two applications of the divergence theorem are generally required. With the notation of Problem 10.2.1, solve $\delta J = 0$ for each of

(a) $J = \int_A [(u_{xx} + u_{yy})^2 + (x + y)u]\,dA$ $u = \partial u/\partial n = 0$ on Γ

Then $2\,\Delta\Delta w + (x+y) = 0$

(b) $$J = \int_A \left[u_{xx}u_{yy} - u_{xy}^2 \right] dA$$

with u and $\partial u/\partial n$ prescribed on Γ (so that all of δu, δu_x, δu_y must vanish on Γ). Then no PDE requirement on w ensues.†

10.2.3 Let (r, θ) be polar coordinates. Among all C'' functions $u(r, \theta)$ attaining prescribed values on the boundary Γ of a region A, find the PDE satisfied by the one that makes stationary the integral

$$I = \int_A \left[u_r^2 + \frac{1}{r^2} u_\theta^2 \right] dA$$

Does this process suggest a way in which one can find the expression for the Laplacian operator in curvilinear coordinates?

10.2.4 Let V be a region in three-dimensional space for which (x, y, z) are Cartesian coordinates. The boundary of V is denoted by S. Find the PDE satisfied by the function $\phi(x, y, z)$ that makes stationary the integral

$$I = \int_V \left[\phi_x^2 + \phi_y^2 + \phi_z^2 \right] dV$$

and for which $\phi = f$ on S, where f is some prescribed function. Show that I is a minimum for the solution function.

Formulate and solve the analogous problem in spherical coordinates (r, θ, ψ) where $x = r \sin\theta \cos\psi$, $y = r \sin\theta \sin\psi$, $z = r \cos\theta$.

10.2.5 A differential equation arising from a variational principle—such as those of Problem 10.2.1—is said to be the Euler equation of that variational principle. Find the Euler equation for each of $\delta I = 0$, $\delta J = 0$, $\delta K = 0$, where

(a) $$I = \int_A F(x, y, u, u_x, u_y)\, dA$$

(b) $$J = \int_V G(x, y, z, u, u_x, u_y, u_z)\, dV$$

(c) $$K = \int_A H(x, y, u, u_x, u_y, u_{xx}, u_{xy}, u_{yy})\, dA$$

† The reason for this is that the integrand can be written as $(u_x u_{yy})_x - (u_x u_{xy})_y$ so that an application of the divergence theorem yields $J = \int \Gamma u_x (d/ds) u_y\, ds$, where ds is the element of arc length along Γ. Thus the value of J depends only on the boundary values of u_x and u_y, i.e., only on u and $\partial u/\partial n$ as specified on Γ.

Here A and V are regions of two- and three-dimensional space, respectively, the quantities F, G, and H are functions of the listed arguments, and appropriate boundary conditions on u are assumed specified. [Answer for Part (a): $F_u - (\partial F/\partial u_x)_x - (\partial F/\partial u_y)_y = 0$.]

10.2.6 Solve the problem $\Delta u = 1$ in the square $0 < x < 1$, $0 < y < 1$, with $u = 0$ on the boundary, by writing $u = \sum\sum a_{mn} \sin m\pi x \sin n\pi y$ in Eq. (10.1) and by choosing the coefficients a_{mn} so as to minimize I. This is an example of the *direct method* of the calculus of variations.

As a second example of the direct method, let (r, θ) be polar coordinates and solve $\Delta \phi = 0$ for $r < 1$, for $\phi(1, \theta) = f(\theta)$, where $f(\theta)$ is a prescribed function, by use of the integral of Problem 10.2.3 and the series expansion

$$\phi = \tfrac{1}{2}a_0(r) + \sum_1^\infty [a_n(r) \cos n\theta + b_n(r) \sin n\theta]$$

where $a_j(r)$, $b_j(r)$ are to be chosen so as to minimize I.

10.2.7 It was remarked in Section 10.1 that any PDE problem can be exhibited in variational form. One way of doing this is as follows: Let $\phi(x, y)$ satisfy the equation $F(x, y, \phi, \phi_x, \phi_y, \phi_{xx}, \phi_{xy}, \phi_{yy}) = 0$, say, together with some kind of boundary conditions. Then simply define

$$I = \int_A [F(x, y, u, u_x, u_y, u_{xx}, u_{xy}, u_{yy}]^2 \, dA$$

where A is any region in (x, y) space over which the equation $F = 0$ is to be solved, and where u is any function satisfying the boundary conditions imposed on ϕ. It is clear that the integrand is always nonnegative and vanishes over A if $u \equiv \phi$; thus the minimum value (zero) of I will be obtained for $u \equiv \phi$.

As an example of this process, set $I = \int_A (u_{xx} + u_{yy} + p)^2 \, dA$, form $\delta I = 0$ for suitable boundary conditions on u, and compare the results with those of Section 10.1.

10.2.8 Consider the space interval $0 < x < l$, and the time interval $0 < t < T$. Find the PDE satisfied by that C'' function $u(x, t)$ for which $u(0, t) = u(l, t) = 0$, $u(x, 0) = f(x)$, $u(x, T) = g(x)$, where f and g are prescribed functions (vanishing at $x = 0$, $x = l$), and that renders stationary the integral

$$I = \int_0^T dt \int_0^l dx\{[u_t^2 - u_x^2] + h(x, t) \cdot u\}$$

Here $h(x, t)$ is a prescribed function of x and t.

10.3 NATURAL BOUNDARY CONDITIONS

Let us return to the integral (10.1):

$$I = \int_A [\tfrac{1}{2}(u_x^2 + u_y^2) - pu]\, dA \qquad (10.4)$$

We now enlarge the class of admissible functions by considering *all C''* functions *u*—i.e., we remove the restriction that $u = g$ on Γ. For which, if any, of these functions is *I* stationary?

The condition $\delta I = 0$ leads as before to

$$\int_\Gamma \frac{\partial u}{\partial n}\, \delta u\, ds - \int_A [\Delta u + p]\, \delta u\, dA = 0 \qquad (10.5)$$

for all admissible δu. Among the admissible δu functions are certainly those that vanish on Γ; for any such function, the area integral alone is left, and its vanishing implies (as we have already seen) that $\Delta u + p = 0$. Thus if there is a function *u* for which *I* is stationary, it must satisfy $\Delta u + p = 0$, and we can make use of this fact in Eq. (10.5) to reduce this equation to

$$\int_\Gamma \frac{\partial u}{\partial n}\, \delta u\, ds = 0 \qquad (10.6)$$

But the only way Eq. (10.6) can hold for all possible δu functions is if $\partial u/\partial n = 0$ on Γ. Thus if *u* is the function that makes *I* stationary, it must satisfy the conditions

$$\Delta u + p = 0 \quad \text{in } A, \qquad \frac{\partial u}{\partial n} = 0 \quad \text{on } \Gamma \qquad (10.7)$$

Of course, such a function *u* will not exist unless $\int p\, dA = 0$, so that the variational problem we have now posed in connection with Eq. (10.4) will not have a solution if this consistency condition is violated. However, the result of interest to us at the moment is that if we do not specify *u* on the boundary, then the variational principle leads to a condition that *u* must satisfy on the boundary; such a condition, which arises from the variational principle itself, is termed a *natural boundary condition*.

Next, let us find the function, among all *C''* functions *u*, that makes stationary the quantity

$$J = \int_A [\tfrac{1}{2}(u_x^2 + u_y^2) - pu]\, dA - \int_\Gamma f(s)u(s)\, ds \qquad (10.8)$$

where $f(s)$ is prescribed on Γ. A straightforward calculation yields

$$\int_\Gamma \left[\frac{\partial u}{\partial n} - f(s) \right] \delta u \, ds - \int_A [\Delta u + p] \delta u \, dA = 0$$

so that u must satisfy

$$\Delta u + p = 0 \quad \text{in } A, \qquad \partial u / \partial n = f(s) \quad \text{on } \Gamma$$

if it is to make J stationary. In comparison with the previous problem, we observe that we have altered the natural boundary conditions but not the Euler equation. As an example of a higher-order situation, the reader may verify that if we remove the restriction $u = \partial u / \partial n = 0$ on Γ in Problem 10.2.2a, the natural boundary conditions become $\Delta u = \partial / \partial n \, \Delta u = 0$ on Γ. If we remove only the restriction $\partial u / \partial n = 0$ on Γ but keep the restriction $u = 0$ on Γ, the natural boundary condition becomes $\Delta u = 0$ on Γ.

An example that is illustrative of a problem involving more than one dependent variable is the following: among all C'' functions $u(x, y)$ and $v(x, y)$, find those that make stationary the integral $\int_A [(u_x - v_y)^2 + (u_y + v_x)^2] \, dA$. We obtain $\Delta u = \Delta v = 0$ as Euler equations, and $\partial u / \partial n = \partial v / \partial s$, $\partial v / \partial n = -\partial u / \partial s$ on Γ as natural boundary conditions.

10.4 SUBSIDIARY CONDITIONS

We begin by recalling the idea of Lagrange multipliers in problems of constrained minimization. Suppose first that two independent variables x and y are related to one another by the condition $g(x, y) = 0$; subject to this constraint, we want to find that point at which $f(x, y)$ attains a minimum. Let (x_0, y_0) be such a point (supposed interior to the range of definition of f); if $g_y(x_0, y_0) \neq 0$, we can think of the condition $g(x, y) = 0$ as defining y as a function of x in the neighborhood of (x_0, y_0), so that $f(x, y)$ becomes $f(x, y(x))$. The condition that it be a minimum reduces to

$$f_x + f_y y' = 0 \tag{10.9}$$

where we can solve for y' from the condition $g_x + g_y y' = 0$. Let the numerical value of the ratio f_y / g_y, as evaluated at the point (x_0, y_0), be denoted by $(-\lambda)$; then $f_y = -\lambda g_y$, and Eq. (10.9) now implies that $f_x = \lambda g_y y' = -\lambda g_x$, so that

$$f_x + \lambda g_x = 0, \qquad f_y + \lambda g_y = 0$$

Thus there is some constant λ such that the two partial derivatives of the composite function $(f + \lambda g)$ vanish at (x_0, y_0). We assumed here that

$g_y(x_0, y_0) \neq 0$; if this is not the case, but instead $g_x(x_0, y_0) \neq 0$, we can obtain the same result by interchanging the roles of x and y. If both $g_x(x_0, y_0)$ and $g_y(x_0, y_0)$ vanish, then it is no longer necessarily true that such a constant λ exist.† We can, however, include this special case by saying that there will always be two constants, λ_0 and λ, such that at the minimum point

$$(\lambda_0 f + \lambda g)_x = (\lambda_0 f + \lambda g)_y - 0$$

In the exceptional case $\lambda_0 = 0$; otherwise, we can set $\lambda_0 = 1$.

Another approach may be instructive. Suppose that $f(x_1, x_2, \ldots, x_n)$ is stationary at the point $(x_1^0, x_2^0, \ldots, x_n^0)$, where the (x_i) are related by m conditions of the form $g_j(x_1, x_2, \ldots, x_n) = 0$, $j = 1, 2, \ldots, m$, with $m < n$. Then we require

$$\frac{\partial f}{\partial x_1} dx_1 + \frac{\partial f}{\partial x_2} dx_2 + \cdots + \frac{\partial f}{\partial x_n} dx_n = 0 \tag{10.10}$$

for each set of quantities $(dx_1, dx_2, \ldots, dx_n)$ satisfying the m conditions

$$\frac{\partial g_j}{dx_1} dx_1 + \frac{\partial g_j}{\partial x_2} dx_2 + \cdots + \frac{\partial g_j}{\partial x_n} dx_n = 0, \qquad j = 1, 2, \ldots, m \tag{10.11}$$

All partial derivatives in Eqs. (10.10) and (10.11) are to be evaluated at the point $(x_1^0, x_2^0, \ldots, x_n^0)$. Geometrically, Eq. (10.11) requires the n-dimensional vector whose components are $(dx_1, \ldots, dx_n)$ to be orthogonal to the gradient vector of each of the g_j functions; Eq. (10.10) states that the gradient vector of the f function is orthogonal to each such $(dx_1, \ldots, dx_n)$ vector. This will be the case if and only if the f-gradient vector is a linear combination of the g_j-gradient vectors, so that

$$\mathbf{grad}\, f = -\lambda_1 \,\mathbf{grad}\, g_1 - \lambda_2 \,\mathbf{grad}\, g_2 - \cdots - \lambda_m \,\mathbf{grad}\, g_m$$

where the λ_i are constants—i.e., each partial derivative of

$$f + \lambda_1 g_1 + \lambda_2 g_2 + \cdots + \lambda_m g_m$$

must vanish. Again, an exceptional case can arise (specifically, if the rank of the matrix of partial derivatives of the g_j is less than m), and to include this we state the result in the form that there must exist constants $\lambda_0, \lambda_1, \ldots, \lambda_m$ such that the partial derivatives of

$$\lambda_0 f + \lambda_1 g_1 + \lambda_2 g_2 + \cdots + \lambda_m g_m$$

all vanish.

---

† For setting the two partial derivatives of $f + \lambda g$ equal to zero would then yield $f_x = f_y = 0$, and this need not be the case. A very simple counterexample is $f = x$, $g = y^2 - x^5$; here f has a minimum at $(x, y) = (0, 0)$, subject to the condition $g = 0$, yet $f_x(0, 0) \neq 0$.

In subsequent applications of this Lagrange multiplier method, we will assume without discussion that $\lambda_0 \neq 0$, and will take it for granted that if trouble arises in a particular problem one will have to go back and include the λ_0 factor. Thus our procedure for making f stationary, subject to the constraints $g_j = 0, j = 1, 2, \ldots, m$, will be to form the composite function

$$F = f + \lambda_1 g_1 + \cdots + \lambda_m g_m \tag{10.12}$$

where the λ_i are unknown constants, to set the partial derivative of F with respect to each of the x_i equal to zero (i.e., treat F as a function to be made stationary without constraints), and then adjoin the conditions $g_1 = g_2 = \cdots g_m = 0$. This gives a total of $n + m$ equations to be solved for the $n + m$ unknowns x_i and λ_j. Of course the values of the λ_i are usually only of subsidiary interest; what we really want are the x_i values. To the reader unacquainted with Lagrange multipliers, the impression may be that the technique represents a difficult approach to a straightforward problem. However, it has an advantage that can be of great utility; namely, the technique does not require that one solve any algebraic equations during the process of obtaining all of the equations that must be satisfied at an extremal point. Incidentally, we note in passing that the conditions $g_j = 0$ are exactly those that would be obtained by formally requiring F to be stationary with respect to the λ_j also; thus the conditions that F in Eq. (10.12) be stationary with respect to all of the x_i and λ_i, without any subsidiary constraints being imposed, provides a compact and rather symmetrical formulation of the original problem.

We now extend these ideas to variational problems. As a first example, consider the problem of finding the function $\psi(x, y)$ defined in a region A of the (x, y) plane, with boundary Γ, and subject to the conditions $\psi = 0$ on Γ and $\int_A \psi^2 \, dA = 1$, that makes $\int_A [\psi_x{}^2 + \psi_y{}^2] \, dA$ stationary. A convenient approach to this problem is as follows. Let $\xi(x, y)$ and $\eta(x, y)$ be a pair of chosen functions, C'' in A and vanishing on Γ, and let $\varepsilon_1, \varepsilon_2$ be a pair of parameters. Define

$$I(\varepsilon_1, \varepsilon_2) = \int_A \left[(\psi_x + \varepsilon_1 \xi_x + \varepsilon_2 \eta_x)^2 + (\psi_y + \varepsilon_1 \xi_y + \varepsilon_2 \eta_y)^2 \right] dA$$

$$J(\varepsilon_1, \varepsilon_2) = \int_A \left[\psi + \varepsilon_1 \xi + \varepsilon_2 \eta \right]^2 dA$$

Then if ψ is to be the solution function, we can say that in terms of the variables ε_1 and ε_2, constrained by the condition $J(\varepsilon_1, \varepsilon_2) = 1$, we must have $I(\varepsilon_1, \varepsilon_2)$ stationary at the point $\varepsilon_1 = \varepsilon_2 = 0$. But in this form the problem falls within the scope of the usual Lagrange multiplier method, so we can say that some constant λ exists such that $I(\varepsilon_1, \varepsilon_2) + \lambda[J(\varepsilon_1, \varepsilon_2) - 1]$ is stationary at $\varepsilon_1 = \varepsilon_2 = 0$. Thus the partial derivative with respect to each of ε_1 and ε_2 of the quantity $I + \lambda(J - 1)$ vanishes at $\varepsilon_1 = \varepsilon_2 = 0$, and this gives

$$\int_A 2[\psi_x \xi_x + \psi_y \xi_y]\, dA + \lambda \int_A 2[\psi \xi]\, dA = 0$$

as well as a similar equation in which ξ is replaced by η. But ξ (or η) is any function that vanishes on Γ; denoting any such function by $\delta\psi$, we can therefore write

$$\int_A 2[\psi_x\, \delta\psi_x + \psi_y\, \delta\psi_y]\, dA + \lambda \int_A 2\psi\, \delta\psi\, dA = 0 \qquad (10.13)$$

Equation (10.13) is equivalent to the condition $\delta K = 0$ for any $\delta\psi$ vanishing on Γ, where

$$K = \int_A [\psi_x^2 + \psi_y^2]\, dA + \lambda\left[\int_A \psi^2\, dA - 1\right] \qquad (10.14)$$

and this now represents the form of the Lagrange multiplier method for the problem posed. Let us complete the solution. By use of the divergence theorem, as well as the condition $\delta\psi = 0$ on Γ, Eq. (10.13) is equivalent to

$$\int_A [-\Delta\psi + \lambda\psi]\, \delta\psi\, dA = 0$$

which requires

$$\Delta\psi = \lambda\psi \qquad (10.15)$$

Equation (10.15), together with the condition $\psi = 0$ on Γ, formulates an eigenvalue problem of the kind to be discussed in Chapter 11. We will find there that it has a non-identically-zero solution for certain discrete values of λ. It follows from the homogeneity of Eq. (10.15) that, if λ is given one of these values for which a solution exists, then any constant times this solution is also a solution, so that we can scale ψ up or down to satisfy the subsidiary condition $\int_A \psi^2\, dA = 1$. In any event, there are functions ψ (in fact, an infinite number) for which the variational problem has a solution.

In Problem 10.5.5 a less formal but very suggestive procedure leading to the same Lagrange multiplier formulation is described for this problem. This alternative approach helps make clear why, in a problem in which the constraint condition is applied at each point of a region, the appropriate Lagrange multiplier is not a constant but rather a function of position. An example of such a situation is given in Problem 10.5.7.

10.5 PROBLEMS

10.5.1   Let $\phi(x, y, z)$, defined (and C'') throughout a certain volume V of three-dimensional space, with boundary surface S, be such that $\delta I = 0$ for all functions $\delta\phi$, where

$$I = \int_V [\phi_x^2 + k\phi_y^2 + \phi_z^2]\, dV + \int_S f\phi\, dS$$

where f is a prescribed function of position on S. Determine the Euler equation and the natural boundary condition for the cases (a) $k = 1$, (b) $k = 2$, (c) $k = 0$, (d) $k = -1$. Are all of these cases reasonable?

10.5.2   Given an (x, y) plane region A with boundary Γ, let $\phi(x, y)$ be that C'' function for which $\delta J = 0$, where

$$J = \int_A [\phi_x^2 + \phi_y^2 + 2f\phi]\, dA + \int_\Gamma [g\phi - h\phi^2]\, ds$$

Here f, g, h are prescribed functions of position in A, on Γ, and on Γ, respectively. Find the Euler equation and the natural boundary condition. Comment on the limiting situation that ensues if h and g are made large compared to unity.

10.5.3   Let A denote a region in the (x, y) plane, with boundary Γ. Find the Euler equation and natural boundary condition for the problem of making stationary the expression

$$\int_A F[x, y, \phi, \phi_x, \phi_y]\, dA + \int_\Gamma G\left[x, \phi, \frac{d\phi}{ds}\right] ds$$

where F and G are prescribed functions of the arguments indicated.

10.5.4   Let A be a region in the (x, y) plane, with boundary Γ. Discuss the effect of the constant B on the Euler equation and natural boundary conditions of $\delta I = 0$, where

$$I = \int_A [(\phi_{xx} + \phi_{yy})^2 + B(\phi_{xx}\phi_{yy} - \phi_{xy}^2)]\, dA + \int_\Gamma f\phi\, ds$$

Here f is a prescribed function of position on Γ.

10.5.5   In the example that led to Eq. (10.14), the problem was to make $\int_A [\psi_x^2 + \psi_y^2]\, dA$ stationary, subject to $\psi = 0$ on Γ and $\int_A \psi^2\, dA = 1$. We require $\int_A [\psi_x\, \delta\psi_x + \psi_y\, \delta\psi_y]\, dA = 0$ for every permissible $\delta\psi$, and the divergence theorem leads to $\int_A \Delta\psi \cdot \delta\psi\, dA = 0$ for all such $\delta\psi$. Let A be divided into a large number n of area elements ΔA, and let the index i denote the value of the indicated quantity at a chosen point within the ith area element. Then this last area integral can be written (approximately) as

$$[(\Delta\psi)_1\, \delta\psi_1 + (\Delta\psi)_2\, \delta\psi_2 + \cdots + (\Delta\psi)_n\, \delta\psi_n]\, \Delta A = 0 \qquad (10.16)$$

Similarly, we can write the constraint condition $\delta[\int_A \psi^2\, dA - 1] = 0$ as (approximately)

$$[\psi_1\, \delta\psi_1 + \psi_2\, \delta\psi_2 + \cdots + \psi_n\, \delta\psi_n]\, \Delta A = 0 \qquad (10.17)$$

We can now think of a set of n mesh point values, such as $\{\psi_i\}$, as constituting a vector in n-dimensional space. Moreover, Eqs. (10.16) and (10.17) have exactly the form of Eqs. (10.10) and (10.11), provided that $\delta\psi_i$ is identified with dx_i.

Follow out this idea to show how Eq. (10.14) is the natural consequence of Eq. (10.12).

10.5.6 Let a function $\psi(x, y)$ be prescribed in an (x, y)-plane region A whose boundary is Γ. Use the Lagrange multiplier method to characterize the function ϕ that, subject to the three conditions $\phi = 0$ on Γ, $\int_A \phi^2\, dA = 1$, and $\int_A \phi\psi\, dA = 0$, makes stationary the integral $\int_A (\phi_x{}^2 + \phi_y{}^2)\, dA$.

10.5.7 Find two functions $u(x, y)$ and $v(x, y)$ related by the conditions $u_y = v_x$ in a region A, on whose boundary Γ both u and v are specified, that make stationary the integral $\int_A [u_x + v_y]^2\, dA$. In particular, show that

(a) The Lagrange multiplier technique (cf. Problem 10.5.6) leads to $\delta I = 0$ for unconstrained (except on Γ) δu and δv, where

$$I = \int_A \{(u_x + v_y)^2 + \lambda(x, y) \cdot (u_y - v_x)\}\, dA$$

and where the Lagrange multiplier $\lambda(x, y)$ is now a function of position within A.

(b) The results of the $\delta I = 0$ process (note that λ_x and λ_y will occur) are equivalent to those obtained by a direct approach to the original problem, in which the condition $u_y = v_x$ is used to imply the existence of a function ψ such that $u = \psi_x$, $v = \psi_y$, and in which the variational problem is reformulated in terms of ψ.

10.5.8 Let a membrane lying in the (x, y) plane (occupying a region A with boundary Γ) undergo a small displacement $w(x, y)$ in the z direction as a result of transverse loading [of intensity $F(x, y)$ per unit area] or of boundary motion. The uniform tension in the membrane is denoted by T (cf. Section 3.7). As a result of this displacement, an area element dA is increased in the ratio $[1 + w_x{}^2 + w_y{}^2]^{1/2} \cong 1 + \frac{1}{2}(w_x{}^2 + w_y{}^2)$. If an added displacement δw is contemplated, the increase in strain energy must equal the work done by external forces (principle of virtual work) so that

$$\delta \int_A \tfrac{1}{2} T(w_x{}^2 + w_y{}^2) \, dA = \int_A F \, \delta w \, dA + \int_\Gamma f \, \delta w \, ds$$

where f denotes the vertical component of the boundary traction. Show that the result of carrying out this variational process leads to the equation $T \, \Delta w = -F$ in A, with $T \, \partial w/\partial n = f$ as the natural boundary condition. Second, adopt a different point of view in which w is specified on the boundary, but in which this boundary constraint is treated by means of Lagrange multipliers; relate the resulting Lagrangian multiplier to the physical constraint force.

10.6 APPROXIMATE METHODS

One of the simplest and most natural approximate methods based on the calculus of variations is that of *Rayleigh and Ritz*. It can be used in any problem for which a variational principle exists; we will choose here an example involving Laplace's equation. Let it be required that $\phi(x, y)$ be found, harmonic in a region A of the (x, y) plane, so that ϕ takes on specified values, say $\phi = f$, on the boundary Γ of A. We have already seen that an equivalent formulation of this problem is to search for the function ϕ that makes the integral $I = \int_A [\phi_x{}^2 + \phi_y{}^2] \, dA$ stationary with respect to all functions satisfying $\phi = f$ on Γ. In this special case we can make a stronger statement to the effect that the solution ϕ of our problem makes I not only stationary, but actually a *minimum*—the proof of this statement proceeds just as in Section 10.1.

The idea of the Rayleigh–Ritz method is as follows. We choose certain functions $\psi^{(0)}(x, y)$, $\psi^{(1)}(x, y), \ldots, \psi^{(n)}(x, y)$ satisfying $\psi^{(0)} = f$ on Γ, all other $\psi^{(j)} = 0$ on Γ, but otherwise arbitrary. If $c_1, c_2, \ldots, c_n$ are any constants, the boundary conditions satisfied by the $\psi^{(j)}$ functions ensure that the linear combination

$$\psi = \psi^{(0)} + c_1 \psi^{(1)} + c_2 \psi^{(2)} + \cdots + c_n \psi^{(n)} \tag{10.18}$$

satisfies $\psi = f$ on Γ; we now hope that we can determine the constants c_i so that ψ is a good approximation to ϕ. Now the true solution ϕ makes $\int_A [\phi_x{}^2 + \phi_y{}^2] \, dA$ a minimum with respect to all functions satisfying the prescribed boundary conditions, and it is therefore very reasonable to use as a criterion for the choice of the c_i the condition that $\int_A [\psi_x{}^2 + \psi_y{}^2] \, dA$ be as small as possible. The integral

$$\int_A \{[\psi_x^{(0)} + c_1\psi_x^{(1)} + \cdots + c_n\psi_x^{(n)}]^2$$

$$+ [\psi_y^{(0)} + c_1\psi_y^{(1)} + \cdots + c_n\psi_y^{(n)}]^2\} \, dA \qquad (10.19)$$

must indeed have a minimum, and this can be obtained by differentiating this expression with respect to each of the c_i, and setting each result equal to zero. We obtain

$$\int_A \{[\psi_x^{(0)} + c_1\psi_x^{(1)} + \cdots + c_n\psi_x^{(n)}]\psi_x^{(j)}$$

$$+ [\psi_y^{(0)} + c_1\psi_y^{(1)} + \cdots + c_n\psi_y^{(n)}]\psi_y^{(j)}\} \, dA = 0 \qquad (10.20)$$

$$\text{for} \quad j = 1, 2, \ldots, n$$

This set of n linear algebraic equations determines the n constants $c_1, \ldots, c_n$.

In a moment we will consider an example. Note first, however, that the method can only be as good as the choice of the $\psi^{(j)}$ functions permits—unless the $\psi^{(j)}$ are such that Eq. (10.18) has the capability of representing ϕ within a good degree of approximation, we cannot hope to obtain good results. Thus some effort, involving a combination of intuition and experience (exploitation of anticipated symmetries, for example), must be applied to the appropriate choice of the $\psi^{(j)}$ functions. We generally get better accuracy by increasing the number n (in the limit, we could always get an exact solution by letting n be infinite, and using for the $\psi^{(j)}$ a complete set of functions), but the amount of computational labor increases. We may be able to reduce this labor by using $\psi^{(j)}$ functions that are orthogonal to one another in the sense that

$$\int_A [\psi_x^{(p)}\psi_x^{(q)} + \psi_y^{(p)}\psi_y^{(q)}] \, dA = 0 \qquad \text{for} \quad p \neq q \qquad (10.21)$$

but the added complications introduced by this restriction may outweigh the advantages.

An important feature of the Rayleigh–Ritz method in this example is that only first derivatives of the approximating function ψ occur in the calculations, despite the fact that the Laplace equation satisfied by the true solution ϕ involves second derivatives. This means that the adequacy of our approximate solution is generally less sensitive to the detailed nature of the $\psi^{(j)}$ functions; on the other hand, it could mean that ψ approximates ϕ very well, but that, for example, ψ_{xx} approximates ϕ_{xx} very badly.

Although the Rayleigh–Ritz method was here applied to an example in which the variational principle represented a minimum—a fact which made its reasonableness more evident—it can also be used in situations in which the variational integral is merely stationary. The argument here is that this is a feature of the exact solution that we make the approximate solution share, so that one can hope that this is as good a way as any in which to choose the c_i—and in any event, by using a variational principle, we are using a criterion that involves all of the region A. Finally, the idea of the Rayleigh–Ritz method can easily be extended to the construction of functions ψ that depend nonlinearly on the parameters c_i, or that may even involve undetermined functions rather than constants; moreover, we can also avoid the restriction $\psi = f$ on Γ by incorporating the boundary conditions into the variational principle. We will consider some such possibilities in the exercises; let us turn now, however, to an example involving a choice of ψ in the form (10.18).

Let ϕ be harmonic in the triangular region A defined by $x > 0$, $y > 0$, $y + \frac{1}{2}x < 1$ (Fig. 10.1), and let the boundary values be $\phi(x, 0) = \sin(\pi x/2)$, $\phi(0, y) = 0$, $\phi(x, 1 - \frac{1}{2}x) = 0$. We take $n = 1$ and choose

$$\psi^{(0)} = (2y + x - 2) \frac{\sin(\pi x/2)}{x - 2}$$

$$\psi^{(1)} = (2y + x - 2) xy \tag{10.22}$$

Note that $\psi^{(0)}$, and all of its derivatives, are well behaved at $x = 2$. The integrals arising in Eq. (10.20) become tedious, however, so that if we want to obtain an analytical rather than a numerical result, it is useful to express

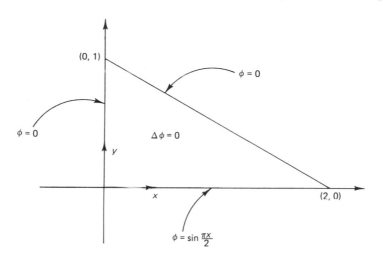

Fig. 10.1   A Dirichlet problem.

$\psi^{(0)}$ via a Taylor series:

$$\psi^{(0)} = (2y + x - 2)x\{-1 + (\tfrac{1}{8}\pi^2 - 1)(x - 1)^2 + \cdots\}$$

With $n = 1$ it is probably not worthwhile to use more than the first term for $\psi^{(0)}$. A straightforward solution of Eq. (10.20) then gives $c_1 \cong \tfrac{3}{5}$, so that

$$\psi \cong -x(2y + x - 2)[1 - \tfrac{3}{5}y] \qquad (10.23)$$

In incorporating more $\psi^{(j)}$ functions, a convenient way in which to ensure the vanishing of each such term on the boundary is to include a factor $xy(2y + x - 2)$ in each. It is clear that a computer soon becomes worthwhile.

If the reader has carried through the calculation leading to Eq. (10.23), he has probably already noticed that the integrals arising in Eq. (10.20) can be simplified by use of the divergence theorem. In fact, Eq. (10.20) can be written

$$\int_A [\Delta(\psi^0 + c_1\psi^{(1)} + \cdots + c_n\psi^{(n)})]\psi^{(j)}\, dA = 0$$

$$\text{for} \quad j = 1, 2, \ldots, n \qquad (10.24)$$

The quantity in square brackets is simply the result of applying the differential operator Δ for our basic problem to the approximation function ψ; Eq. (10.24) requires that the c_j be chosen so as to make $\Delta\psi$ orthogonal to each $\psi^{(j)}$. This observation suggests the *Galerkin method*, which may be applied to a problem in which no simple variational principle exists. Let the problem be described by $L\phi = 0$ in A, $\phi = f$ on Γ, where L is a differential operator and f a given function. Then construct an approximation function ψ as in Eq. (10.18) and determine the c_i so that $\int_A (L\psi)\psi^{(j)}\, dA = 0$ for each choice of (j). The operator L need not be linear; moreover, as in the generalization of the Rayleigh–Ritz method mentioned above, the c_i need not appear linearly in the expression for ψ. We can also increase the flexibility by using undetermined functions of one or more variables. We do not even have to make $L\psi$ orthogonal to each of the approximation functions $\psi^{(j)}$; we can choose any set of convenient functions $\beta^{(j)}(x, y)$ and find the c_i by the condition

$$\int_A \beta^{(j)} \cdot L\{\psi^{(0)} + \sum_1^n c_k\psi^{(k)}\}\, dA = 0, \qquad \text{for} \quad j = 1, 2, \ldots, n \quad (10.25)$$

A special case of this process arises if we choose δ functions for the $\beta^{(j)}$, each having the form

$$\beta^{(j)} = \delta(x - x_j) \cdot \delta(y - y_j)$$

where (x_j, y_j) is a chosen point inside A. The effect of Eq. (10.25) is then to require that $L\psi$ vanish exactly at each of n so-called *collocation points* (x_j, y_j) inside A. Another possibility is to divide A into n subregions and to choose $\beta^{(j)}$ to vanish everywhere except in the jth subregion, where $\beta^{(j)} = 1$.

It is clear that one can invent many variants of the Rayleigh–Ritz and Galerkin procedures. Useful examples of such ideas will be found in Kantorovich and Krylov (1958), in Mikhlin (1964), and in Finlayson (1972).

10.7 PROBLEMS

10.7.1   Carry out the example of Eq. (10.22) for $n = 2$ with an appropriate choice for $\psi^{(2)}$. Compute ψ, ψ_x, ψ_{xy}, $\Delta\psi$ at the midpoint of the triangle for each of $n = 0, 1, 2$.

For the case $n = 1$, use some of the other methods described in the preceding section to determine c_1 (e.g., collocation, or use of subregions) and compare results.

10.7.2   An obvious variational principle leading to $L\phi = 0$ in a region A is $\delta I = 0$, where $I = \int_A (L\phi)^2 \, dA$. Construct an approximation method based on this least-mean-square idea and discuss it in the framework of Section 10.6.

10.7.3   Devise an example that will enable you to illustrate the effect of a discontinuity in the boundary data on the effectiveness of the Rayleigh–Ritz and Galerkin methods. If a discontinuity is present in a particular problem, would it be useful to subtract an appropriate singular solution before using the approximation method? Consider also the more subtle corner singularity arising in the problem $\Delta\phi = 1$ inside the square $0 < x < 1$, $0 < y < 1$, with $\phi = 0$ on the boundary.

10.7.4   Let $\Delta\phi = 0$ in the region $-1 < x < 1$, $-1 < y < 1$, $z > 0$, with $\phi = 0$ on the lateral sides $x = \pm 1$, $y = \pm 1$, and with $\phi(x, y, 0) = (1 - x^2)(1 - y^2)$. Use the Rayleigh–Ritz method to approximate ϕ by a function ψ having the form

$$\psi = (1 - x^2)(1 - y^2)\{ f_0(z) + x f_1(z) \}$$

and determine appropriate choices for the differential equations and boundary conditions satisfied by the $f_i(z)$. Solve for the $f_i(z)$ and compare your results with the exact solution to this problem.

10.7.5   Invent a simple example involving either the diffusion or wave equation (in one space dimension and time-dependent) for which you can compare an exact solution with that obtained by either the Rayleigh–Ritz or Galerkin procedure.

10.7.6 Modify the problem associated with Eqs. (10.22) so as to include the boundary conditions in the variational principle—i.e., so as to be able to drop the explicit requirement that $\psi = f$ on Γ. Choose a new $\psi^{(0)}$ and $\psi^{(1)}$, write $\psi = c_0\psi^{(0)} + c_1\psi^{(1)}$, and determine c_0 and c_1 by use of your modified principle. Compare results.

10.7.7 Let $\psi_1, \psi_2, \ldots, \psi_n$ be n chosen functions, each harmonic in a region A of the (x, y) plane. Define $\psi = a_1\psi_1 + a_2\psi_2 + \cdots + a_n\psi_n$, where the a_n are constants determined from the conditions

$$\int_\Gamma \left[f - \sum_1^n a_j\psi_j \right] \frac{\partial\psi_k}{\partial n}\, ds = 0$$

where f is a prescribed function on the boundary Γ of A. As before, $\partial/\partial n$ denotes the outward normal derivative. Also, let ϕ be the solution of $\Delta\phi = 0$ in A, $\phi = f$ on Γ.

Show (Trefftz) that this method of choosing the a_j coefficients minimizes the integral

$$\int_A \left[(\phi_x - \psi_x)^2 + (\phi_y - \psi_y)^2 \right] dA$$

and also that

$$\int \left[\psi_x^2 + \psi_y^2 \right] dA \le \int \left[\phi_x^2 + \phi_y^2 \right] dA \qquad (10.26)$$

The approximating function obtained in the Rayleigh–Ritz method of Section 10.6 gave a value for the Dirichlet integral that was always greater than or equal to the true value; from Eq. (10.26) we now obtain a lower bound† for the Dirichlet integral. The closeness with which a guessed solution approximates the true value of the Dirichlet integral is of course one measure of the adequacy of that guessed solution.

10.7.8 A wire loop is described by the parametric equations $x = R \cos\theta$, $y = R \sin\theta$, $z = \frac{1}{4}R \sin\theta$ where $R > 0$ is a constant, and $0 \le \theta \le 2\pi$. Use a variational method to approximate the equation $z = f(x, y)$ describing the surface of minimal area whose boundary is this loop. Does this problem involve any new features? How adequate is your solution?

---

† A general discussion of techniques for obtaining both upper and lower bounds by variational principles is given by Courant and Hilbert (1953, Chap. 4).

10.8 FINITE-ELEMENT METHOD

The widely used finite-element method is a special case of the Rayleigh–Ritz and Galerkin methods. It is based on the subdivision of the region of interest into a number of subregions and on the use of a different analytic expression for the approximate solution function within each such subregion; some form of continuity across subregion boundaries is usually imposed.

As an example to illustrate the idea, consider again the problem of finding the function $\phi(x, y)$ that satisfies $\Delta\phi = 0$ in a region A of the (x, y) plane, with $\phi = f$ on the boundary Γ, where f is some prescribed function. We divide the region A into a number of triangular subregions, as in Fig. 10.2, and number the nodes from 1 to n in any manner, as shown. We now try to determine a function $\psi(x, y)$ that is an approximation to the solution ϕ; denote the value of ψ at the jth nodal point by ψ_j. We require ψ to be a linear function of x and y in each triangle, and to be continuous across each triangle boundary. An easy way in which to impose this continuity condition is to think of the nodal ψ_j values as the unknowns to be determined, with the coefficients of the linear expression for ψ in each triangle being expressed in terms of the ψ_j values at the three nodes of that triangle. Thus inside the triangle whose nodes are at 1, 4, and 5, we will have

$$\psi = a + bx + cy$$

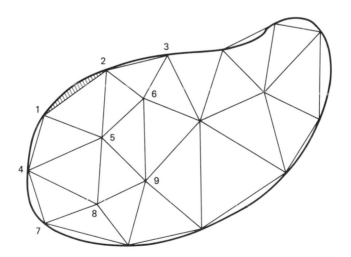

Fig. 10.2 Finite elements.

where a, b, c are related to ψ_1, ψ_4, ψ_5 by the conditions

$$a + bx_4 + cy_4 = \psi_4$$

$$a + bx_5 + cy_5 = \psi_5$$

$$a + bx_1 + cy_1 = \psi_1$$

Here (x_j, y_j) are the Cartesian coordinates of node j. Note that the coefficients a, b, c are linear expressions in ψ_4, ψ_5, ψ_1. When ψ is constructed in this manner for each triangle, then no matter what values the ψ_j quantities are assigned, ψ will clearly be continuous across triangle boundaries.

At those nodes lying on the boundary, such as node 4 in the figure, we assign $\psi_4 = f_4$, where f_4 is the value at that point of the specified boundary function f. The values of ψ_j at all other nodes are unknowns to be determined by use of a variational principle. We choose these ψ_j values by requiring

$$I = \int_A (\psi_x{}^2 + \psi_y{}^2) \, dA \qquad (10.27)$$

to be stationary. In particular, ψ_x and ψ_y are computed within each triangle (where they have constant values, each such constant value being a linear function of the nodal ψ_j values for that triangle), and the I integral constructed as a sum of integrals over triangles. The fact that the boundary Γ does not quite correspond with the boundary triangle edges is neglected; in this calculation no contribution to I is exacted from a region such as the shaded area in Fig. 10.2. The quantity I is thus a quadratic function of those ψ_j values corresponding to internal nodes, and we now set the derivative of I with respect to each such ψ_j value equal to zero. This gives a set of linear equations, equal in number to the number of unknown ψ_j values, which may be solved for the ψ_j values. The structure of this set of linear equations is interesting. In Fig. 10.2, the value of ψ_5 can affect ψ only inside those triangles that have node 5 as one of their vertices, and so that contribution to I involving ψ_5 will also involve ψ_8, ψ_9, ψ_6, but not the other unknowns. This situation holds generally, so that any one of the linear equations to be solved will involve only a small number of the ψ_j; thus the coefficient matrix of the equation set is sparse. (For an example, see Problem 10.9.4). For this problem, it also turns out that—unless some of the triangles contain small angles—the coefficient matrix is well conditioned and so is relatively easily solved by such standard methods as Gaussian elimination.

If a particular triangular decomposition as in Fig. 10.2 has been chosen and the resulting set of linear equations solved, an immediate question is how close the ψ_j values are to the exact solution values ϕ_j at corresponding nodes. In practice, one might re-solve the problem with a finer mesh (i.e., smaller triangles) and see if the solution values change appreciably, where one's interpretation of the term "appreciably" involves a combination of experience and judgment. We expect that as the mesh becomes finer and finer, ψ will approach ϕ, and this brings up the technical matter of how rapid the asymptotic rate of convergence is. In the sample problem just discussed, let h denote the length of the largest side of any triangle; then (if Γ and f are adequately smooth) it turns out that $|\phi - \psi| = O(h^2)$ as $h \to 0$. This result is not unreasonable, since this is the kind of accuracy with which a function of the ψ class (i.e., linear over triangles) can be made to interpolate the solution function ϕ (for small h this follows from Taylor's theorem), and the ψ_j values determined as above correspond to a ψ function that fits ϕ at least as well as any other ψ function [in the sense of a minimization of I defined in Eq. (10.27)], and in particular, at least as well as the interpolating ψ function that agrees with ϕ at each nodal point. A more formal discussion of the convergence rate problem is given by Strang and Fix.† One is frequently more interested in derivative quantities, say ϕ_x or ϕ_{xy}, than in ϕ itself; we might then expect (again from interpolation theory) that the convergence to their exact values of derivatives as computed from the ψ_j values is at a slower rate than of $O(h^2)$; however, this matter is somewhat fuzzy since it depends on the way in which derivatives are calculated from ψ_j values.

The illustrative example we have just discussed is a relatively simple one, and in practice a number of generalizations can be made. We list some of these.

(1) The approximating function ψ need not be linear over each triangle; it can for example be quadratic, or of more complex form. Note that if the variational principle involves second derivatives of ϕ (as for the case in which ϕ is biharmonic) then ψ_x and ψ_y should be continuous across subdomain boundaries. See Problem 10.9.5.

(2) The subdomains need not be triangular in shape; quadrilaterals or more complicated shapes are feasible. Near the boundary, curvilinear shapes can even be used. See Problem 10.9.7.

(3) The differential equation and boundary conditions satisfied by ϕ may be more complicated (and the dimensionality of the problem may be higher); the problem need not be elliptic or even linear.

(4) A variational principle need not be available; the Galerkin idea can be used. In this case, it is useful to manipulate the integrals so as to reduce the derivative order in the integrand. See Problem 10.9.8.

† Strang and Fix (1973, pp. 39ff and 136ff).

We remark finally that computer programming considerations play an important role in practical applications of the finite-element method, since the number of equations to be solved tends to become rather large. A good discussion of these considerations is given by Strang and Fix.† Much of the development of finite element technology has been in connection with structural problems; the text by Zienkiewicz (1971) may be referred to.

10.9 SUPPLEMENTARY PROBLEMS

10.9.1 Let $f(x)$ and $K(x, t)$ be prescribed functions, with $K(x, t) = K(t, x)$, in the range $a \leq x \leq b$, $a \leq t \leq b$. Let a function $\phi(x)$ satisfy the Fredholm integral equation of the second kind:

$$\phi(x) = f(x) + \int_a^b K(x, t)\phi(t)\, dt$$

for x in $[a, b]$. Show that a variational principle equivalent to this integral equation can be obtained by use of the integrals

$$\int_a^b \int_a^b K(x, t)\phi(x)\phi(t)\, dx\, dt, \qquad \int_a^b [\phi(t)]^2\, dt, \qquad \int_a^b f(x)\phi(x)\, dx$$

10.9.2 Relate the problem $\delta I = 0$, for

$$I = \int_V \left[\tfrac{1}{2}r(x, y, z)\{\phi_x^2 + \phi_y^2 + \phi_z^2\} + \int_0^\phi g(x, y, z, \xi)\, d\xi \right] dV$$

$$+ \int_{S_2} h\phi\, dS + \int_{S_3} m(\phi - \phi_0)^2\, dS$$

to a problem in steady-state heat conduction. Here V is a region in (x, y, z) space whose surface S consists of three portions S_1, S_2, S_3; r, g, h are functions of position; m, ϕ_0 are constants; and $\delta\phi = 0$ on S_1.

10.9.3 Let $\phi(x, t)$ satisfy $\phi_t = a^2\phi_{xx}$ for $0 < x < l$, $0 < t$; let $\phi(x, 0)$ be specified and $\phi(0, t) = \phi(l, t) = 0$, for all t. Obtain a variational principle equivalent to this problem by first taking a Laplace transform to yield

$$-\phi(x, 0) + s\Phi(x, s) = a^2\Phi_{xx}(x, s)$$

---

† Strang and Fix (1973). This reference contains an extensive bibliography.

by noting that a variational integral leading to this equation has the form

$$\int_0^l \left[\tfrac{1}{2} a^2 \Phi_x{}^2 + \tfrac{1}{2} s \Phi^2 - \phi(x, 0) \Phi(x, s) \right] dx$$

and that each term can be inverted by use of the convolution theorem. Verify directly that the result does provide the desired principle.

10.9.4   In the sample problem of Section 10.8, suppose that a portion of the region A is covered by uniform triangles produced by bisecting rectangles of sides h and k, as in Fig. 10.3. Observing that in the triangle whose nodes are numbered 2, 5, 3 the value of ψ_x is constant and equal to $(\psi_3 - \psi_2)/h$, for example, it is easy to write the contribution to the integral I of Eq. (10.27) that arises from the triangles influenced by node 2. Complete this process, and show that the equation obtained from $\partial I / \partial \psi_2 = 0$ is

$$2\psi_2[1 + (k^2/h^2)] = \psi_5 + \psi_8 + (k^2/h^2)(\psi_1 + \psi_3)$$

10.9.5   In the triangular decomposition of Fig. 10.2, insert one new node at the midpoint of each triangle side (Fig. 10.4). Inside each triangle, let the function ψ of Section 10.8 have the form

$$\psi = a + bx + cy + dx^2 + exy + fy^2$$

where $a, b, \ldots, f$ are constants that vary from triangle to triangle. Let these six constants be determined so that ψ has chosen values, say $\psi_1, \psi_2, \ldots, \psi_6,$

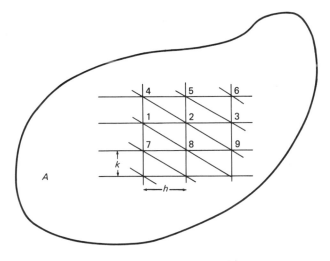

Fig. 10.3   Regular decomposition.

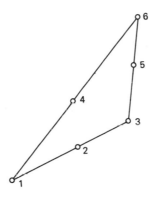

Fig. 10.4 Added nodes.

at the nodes of Fig. 10.4. In the finite-element method we now treat all (internal) corner and midpoint nodes as unknowns. Why is it immediately clear that, for any choice of ψ_j values, ψ is continuous across triangle boundaries? How many ψ_j values arise in a typical member of the set of linear equations that must eventually be solved? Obtain an actual typical equation for the particular decomposition of Fig. 10.3, with appropriately added nodes. Give a plausible reason why the convergence rate could now be expected to be $O(h^3)$ in favorable circumstances.

Show that an extension to cubics is feasible if we insert two nodes along each edge rather than only one, as well as a node at the centroid of the triangle. Can the equation corresponding to each such centroidal node be easily removed from the final equation set? Show that an alternative specification of the coefficients of the cubic can be made in terms of ψ, ψ_x, ψ_y at each corner node, together with ψ at the centroidal node. Is interface continuity maintained? This latter technique has the advantage of bringing derivative values into the calculation, and of ensuring derivative continuity at nodes (but not across edges).

10.9.6   As an extension of the results of Problem 10.9.5, let ψ be a quintic (a 5th-degree polynomial in x and y; a typical term is $cx^\alpha y^\beta$, where $\alpha + \beta \leq 5$) with ψ, ψ_x, ψ_y, ψ_{xx}, ψ_{xy}, ψ_{yy} specified at each node and with the normal derivative of ψ specified at the midpoint of each side. Show that ψ and its first derivatives are continuous across triangle sides. Explain how such a ψ could be used in connection with the biharmonic operator, and construct an example using the mesh of Fig. 10.3.

10.9.7   Instead of triangles, consider the subdivision of a region into equal rectangles with sides parallel to the axes, perhaps merged with triangles near the boundary. Discuss the advantages or disadvantages of using, for

ψ in each rectangle,

    (a)   a bilinear function $\psi = a + bx + cy + dxy$

    (b)   a biquadratic function $\psi = a + bx + cy + dxy + ex^2 + fx^2y + gy^2 + hxy^2 + ix^2y^2$ with one extra node at the midpoint of each side and also at the center.

Show that in three dimensions the equivalent of a bilinear function would be a trilinear function: $\psi = a + bx + cy + dz + exy + fxz + gyz + hxyz$.

10.9.8   Let $\phi(x, y, t)$ satisfy the heat conduction equation

$$\phi_t = \Delta\phi + g(x, y, t)$$

for (x, y) in a region A of the (x, y) plane, and for $t > 0$; here g is a specified function of (x, y, t). The initial condition is that $\phi(x, y, 0)$ is some specified function of x and y, and the boundary condition is that $\phi = 0$ on the boundary Γ of A, for all t.

We divide A into triangles as in Fig. 10.2 and define a set of "basis functions" $\eta(x, y)$ for each internal node j by the requirement that $\eta_j = 1$ when $x = x_j$ and $y = y_j$, that $\eta_j = 0$ when (x, y) corresponds to any other nodal point, and that η_j vary linearly with x and y between the nodal values in any triangle. (Thus η_j vanishes outside of those triangles that share node j.) Let $\psi(x, y, t)$ be an approximating function to ϕ, defined by

$$\psi = \sum_j c_j(t)\eta_j(x, y)$$

where the summation is over all internal nodes.

Show first that, if $\omega(x, y)$ is any function vanishing on Γ, the solution ϕ satisfies the Galerkin equation

$$\int_A \left[\phi_t\omega + \phi_x\omega_x + \phi_y\omega_y - g\omega\right] dA = 0$$

which suggests the set of equations

$$\int_A \left[\psi_t\eta_j + \psi_x(\eta_j)_x + \psi_y(\eta_j)_y - g\eta_j\right] dA = 0$$

to be satisfied by ψ. This is a set of linear differential equations for the $c_j(t)$; what is their specific form for the case of Fig. 10.3? How could initial values be determined? Drawing on your knowledge of numerical methods for ordinary differential equations, how would you recommend solving them? If the boundary condition on Γ were that ϕ was to be a specified function of position (and perhaps of time), what modifications in procedure would be appropriate?

10.9.9   Show that the Euler equation that emerges from the variational principle

$$\delta \int_{V_n} [(\mathbf{grad}\ \phi)^2 - \lambda\phi^2]\, dx_1\, dx_2 \cdots dx_n = 0 \qquad (10.28)$$

is

$$\phi_{x_1 x_1} + \phi_{x_2 x_2} + \cdots + \phi_{x_n x_n} + \lambda\phi = 0 \qquad (10.29)$$

The integration in (10.28) is over an n-dimensional region with Cartesian coordinates $(x_1, x_2, \ldots, x_n)$; λ is a constant. Use this result, together with appropriate descriptions of $\mathbf{grad}\ \phi$ in cylindrical and spherical coordinates in three dimensions, to derive the appropriate forms for the Helmholtz equation (10.29). Compare the difficulty of this derivation with that of the direct conversion of Eq. (10.29) to these coordinates systems.

10.9.10   Let V be a region of three-dimensional space, with boundary S; let f be a function prescribed in V, and g a function prescribed on S. Let $u(x, y, z)$ be any function satisfying $\Delta u = f$ in V, and let $v(x, y, z)$ be any function satisfying $v = g$ on S. Finally, let $w(x, y, z)$ be the solution of the problem $\Delta w = f$ in V, $w = g$ on S.

Defining the pseudo scalar product (α, β) of any two functions α and β defined in V by

$$(\alpha, \beta) = \int_V [\alpha_x \beta_x + \alpha_y \beta_y + \alpha_z \beta_z]\, dV$$

with $\| \alpha \|^2 = (\alpha, \alpha)$, show that $(w - u, w - v) = 0$, and consequently that

$$\| w - \tfrac{1}{2}(u + v) \|^2 = \tfrac{1}{4} \| u - v \|^2$$

which is the equation of a "hypercircle" in this kind of function space. This equation provides a measure of the error resulting from an approximation of w by $\tfrac{1}{2}(u + v)$. If one now modifies u by adding a linear combination of harmonic functions to it, how should one choose the coefficients of this linear combination so as to improve the approximation to w? How could one alter v so as to proceed similarly? [This hypercircle property has been exploited by Synge in a number of applications. For references, and for a discussion of the relationship to reciprocal variational principles, see Courant and Hilbert (1953, p. 252).]

10.9.11   (a)   The equation governing incompressible viscous flow in a two-dimensional region R of the (x, y) plane is

$$\nu\Delta\Delta\psi + \psi_x\Delta\psi_y - \psi_y\Delta\psi_x = 0$$

where ψ is the stream function, Δ the Laplacian operator, and ν the coefficient of viscosity. Multiply by $\delta\psi$ and integrate over R to show that an equivalent variational formulation is given by

$$\int_R \left[\delta\{\tfrac{1}{2}(\Delta\psi)^2\} + \psi_y\Delta\psi\delta\psi_x - \psi_x\Delta\psi\delta\psi_y \right] dA = 0$$

for all $\delta\psi$ such that $\delta\psi = (\partial/\partial n)(\delta\psi) = 0$ on the boundary of R. If one or both of these boundary constraints is omitted, what would the natural boundary conditions become?

(b) Defining

$$I = \int_R \left[\frac{\gamma-1}{2} M^2(\phi_x^2 + \phi_y^2) - \left(1 + \frac{\gamma-1}{2} M^2\right) \right]^{\gamma/\gamma-1} dA$$

where $\gamma > 1$ and M are constants, show that the condition $\delta I = 0$ for all $\delta\phi$ leads to the Euler condition

$$\Delta\phi[1 + \tfrac{1}{2}(\gamma - 1)M^2(1 - \phi_x^2 - \phi_y^2)]$$

$$- M^2[\phi_x^2\phi_{xx} + \phi_y^2\phi_{yy} + 2\phi_x\phi_y\phi_{xy}] = 0$$

(which is the equation governing irrotational incompressible flow in two dimensions; ϕ is the potential function, γ the ratio of specific heats, and M the Mach number).

11

EIGENVALUE PROBLEMS

11.1 A PROTOTYPE PROBLEM

Let C be a closed contour enclosing a region R in the (x, y) plane. Let $f(x, y)$ be a given function defined in R, satisfying $f > 0$ in R, and let λ be a given constant. Then one solution $u(x, y)$ of the problem

$$\Delta u + \lambda f u = 0 \quad \text{in } R, \qquad u = 0 \quad \text{on } C \qquad (11.1)$$

(where $\Delta = \partial^2/\partial x^2 + \partial^2/\partial y^2$) is given by the so-called *trivial solution* $u \equiv 0$; we ask now whether we can also find a nontrivial solution to this problem. We note that if there does exist such a nontrivial solution, then because of the homogeneity of the equation and of the boundary condition, any multiple of that solution is also a solution.

For simple choices of R and f, the question just raised can be answered very easily. For example, let R be the rectangle $0 < x < a, 0 < y < b$, and let $f \equiv 1$. A conventional series expansion technique then shows that a nontrivial solution will exist only if λ has one of the values $\lambda_{mn} = \pi^2[m^2/a^2 + n^2/b^2]$, where m and n are positive integers; the solution corresponding to the choice $\lambda = \lambda_{mn}$ is then given by any multiple of

$$u_{mn} = \sin \frac{m\pi x}{a} \sin \frac{n\pi y}{b}$$

Similarly, if R is the circular region defined in terms of polar coordinates (r, θ) by $r < a$, and if $f \equiv 1$, the reader may show that a nontrivial solution will exist only if λ has one of the values $\lambda_{nj} = \xi_{nj}^2/a^2$, where ξ_{nj} is the jth zero of the Bessel function of first kind and of integral order n [i.e., $J_n(\xi_{nj}) = 0$].

From consideration of special cases such as these, and also from experience with the corresponding situation for ODEs,[†] we might expect that a nontrivial solution of the problem of Eq. (11.1) will exist only for certain discrete values of λ. That this is indeed the case for a rather wide class of bounded regions R and positive functions f may be proved formally by use of integral equation theory (cf. Problem 11.3.2), and we now accept this result. A value of λ for which a nontrivial solution exists is termed an *eigenvalue*, and the corresponding solution is termed an *eigenfunction*.

Eigenvalue problems—of which Eq. (11.1) is a relatively simple representative—frequently arise in oscillation problems and in stability analysis. An example of the latter, for a biological system, will be found in Problem 11.3.1; here we choose an example involving the oscillation of a membrane. Let the membrane occupy a region R, with boundary C, in the (x, y) plane. As in Section 3.7, let the (constant) tension per unit length be T, and let the small displacement be $w(x, y, t)$, where t is time. We fix the edge of the membrane, so $w = 0$ on C. Denoting the mass per unit area by $\rho(x, y)$, and the transverse force per unit area by $h(x, y, t)$, the equation of motion is easily found to be

$$w_{xx} + w_{yy} - (1/c^2)w_{tt} = p \qquad (11.2)$$

where $p(x, y, t) = -h(x, y, t)/T$, and $c^2(x, y) = T/\rho\,(x, y)$.

In the absence of external force, it is known[‡] that the time-dependent small displacement from equilibrium of a conservative mechanical system may be expressed as a superposition of so-called *normal modes*. Each normal mode is a motion in which all particles move harmonically at a fixed frequency and in phase with one another. In the present example we try $w(x, y, t) = u(x, y) \sin \omega t$, where ω is a constant, and where $u = 0$ on C. With $p = 0$, substitution into Eq. (11.2) shows that u must satisfy Eq. (11.1), where $\lambda = \omega^2$ and $f = 1/c^2$. Thus the fact that a nontrivial solution can be found for the problem of Eq. (11.1) is in accordance with vibration theory and, incidentally, also with musical experience.

11.2 SOME EIGENVALUE PROPERTIES

We consider again Eq. (11.1), with $u = 0$ on C. Let λ_m and λ_n be two different eigenvalues, and let u_m, u_n denote corresponding eigenfunctions.

[†] An analogous ODE problem would be to find $y(x)$ in $a < x < b$ such that $y'' + \lambda f y = 0$, with $y(a) = y(b) = 0$.

[‡] Goldstein (1950, p. 332).

Then each of u_m and u_n vanishes on C, and in R we have

$$\Delta u_m + \lambda_m f u_m = 0 \tag{11.3}$$

$$\Delta u_n + \lambda_n f u_n = 0 \tag{11.4}$$

Multiply each term of Eq. (11.3) by u_n, each term of Eq. (11.4) by u_m, subtract the results, and integrate over R to obtain

$$\int_R [u_n \, \Delta u_m - u_m \, \Delta u_n] \, dA + (\lambda_m - \lambda_n) \int_R f u_m u_n \, dA = 0$$

The first term vanishes [by Eq. (9.4), using $u_m = u_n = 0$ on C], and since $\lambda_m \neq \lambda_n$, we deduce

$$\int_R f u_m u_n \, dA = 0 \tag{11.5}$$

We interpret the integral in Eq. (11.5) as a generalized scalar product of the two functions u_m and u_n, with respect to the weight function f, and say that eigenfunctions belonging to different eigenvalues are *orthogonal* to one another.

A consequence of Eq. (11.5) is that all eigenvalues of the problem of Eq. (11.1) are real. For if λ_m, say, were complex, then we could take the complex conjugate of Eq. (11.3) to show that $\lambda_m{}^*$, the complex conjugate of λ_m, is also an eigenvalue, with the corresponding eigenfunction $u_m{}^*(x, y)$. We then set $\lambda_n = \lambda_m{}^*$, $u_n = u_m{}^*$ in the above discussion, and Eq. (11.5) leads to

$$\int_R f u_m u_m{}^* \, dA = 0$$

But $u_m u_m{}^*$ is the sum of the squares of the real and imaginary parts of u_m, and since $f > 0$ in R, the only way in which this integral can vanish is if $u_m \equiv 0$ in R. Thus λ_m cannot be complex. Moreover, we can always find a real eigenfunction corresponding to any λ_m, for since λ_m is now known to be real in Eq. (11.3), it follows that if u_m were complex each of its real and imaginary parts would separately satisfy Eq. (11.3), so that each would be a real eigenfunction. Consequently, we can restrict our attention to real eigenfunctions.

We can next show that each eigenvalue λ_m must be positive. In fact, if we multiply Eq. (11.3) by u_m and integrate over R, we obtain

$$\int_R u_m \, \Delta u_m \, dA + \lambda_m \int_R f u_m{}^2 \, dA = 0$$

Using Eq. (9.3) (with $\phi = \psi = u_m$) and the fact that $u_m = 0$ on C, this equation yields

$$\lambda_m = \frac{\int_R \left[(\partial u_m/\partial x)^2 + (\partial u_m/\partial y)^2 \right] dA}{\int_R f u_m^2 \, dA} \tag{11.6}$$

and since this is the quotient of two positive quantities, it follows that $\lambda_m > 0$.

We have remarked that if an eigenfunction is multiplied by any constant, the result is still an eigenfunction. It is often convenient to remove this indeterminancy by *normalizing* with respect to the scalar product occurring in Eq. (11.5); in future we will assume that each eigenfunction u_m has been scaled so that $\int_R f u_m^2 \, dA = 1$.

It may be that two or more quite different eigenfunctions correspond to the same eigenvalue (one then says that the eigenvalue is *degenerate*). For example, in the rectangular region problem of Section 11.1 let $a = b$; then corresponding to the eigenvalue $\lambda = 5\pi^2/a^2$ we have the two eigenfunctions $\sin(\pi x/a) \sin(2\pi y/a)$ and $\sin(2\pi x/a) \sin(\pi y/a)$. Other examples may lead to a larger number of different eigenfunctions associated with a particular eigenvalue. However, the number of linearly independent eigenfunctions associated with each distinct eigenvalue must be finite (cf. Problem 11.3.2). Also, by the Gram–Schmidt process of Problem 11.3.3, it is possible to choose linear combinations of these eigenfunctions that are orthogonal to one another. The final result is that we can list the eigenvalues, in order of increasing size (where each one occurs in the list a number of times equal to the number of linearly independent eigenfunctions associated with it), as $\lambda_1, \lambda_2, \lambda_3, \ldots$; the corresponding eigenfunctions $u_1, u_2, u_3, \ldots$ may then be considered as orthogonal to one another and normalized (we say they are *orthonormal*).

If the region R is not bounded, or if Eq. (11.1) is modified so as to have a more complicated structure, then it may be that the eigenvalues no longer form a discrete set. For example, let Eq. (11.1), with $f \equiv 1$, apply to the region R defined by $0 < y < b$, $-\infty < x < \infty$. If u is to be bounded, and to satisfy $u(x, 0) = u(x, b) = 0$, then the function $u = \sin(n\pi y/b) \cdot \sin\{[\lambda - (n^2\pi^2/b^2)]^{1/2}x\}$ provides an eigenfunction for any choice of $\lambda > n^2\pi^2/b^2$, where n is a chosen integer. As another example let (r, θ, ϕ) be spherical polar coordinates, k a positive constant, and Δ the three-dimensional Laplacian operator. Then Schrödinger's equation

$$\Delta\psi + \left(\lambda + \frac{k}{r}\right)\psi = 0 \tag{11.7}$$

for $\psi(r, \theta, \phi)$, where ψ is to be finite at $r = \infty$, has the property that every positive value of λ is an eigenvalue, but that the negative eigenvalues form a discrete set. This result is obtainable by use of a product series expansion involving trigonometric functions of ϕ, associated Legendre functions of θ, and Laguerre functions of r.

11.3 PROBLEMS

11.3.1 A chemical substance known as acrasin, secreted by certain amoebae, has the property of inducing amoebae to move towards an increasing concentration of this substance. One result of this effect is that a uniform distribution of amoebae may be unstable, in that a small perturbation of the distribution may lead to aggregation of the amoebae into clusters.

A simple model that can be used to investigate this kind of instability is the following:† Let $a(x, y, t)$ and $\rho(x, y, t)$ denote the densities of amoebae and acrasin, respectively, as functions of time t and position (x, y) within some plane region R with boundary C. The rate of change with respect to time of each of these quantities is influenced by diffusive effects, and also by the concentration of the other quantity, according to

$$a_t = -(c_1\rho_x)_x - (c_1\rho_y)_y + (c_2a_x)_x + (c_2a_y)_y$$

$$\rho_t = -c_3\rho/(1 + c_4\rho) + af(\rho) + c_5(\rho_{xx} + \rho_{yy})$$

where c_1 and c_2 are given functions of (x, y), f a prescribed function of ρ, and c_4, c_5, c_6 are given constants. The boundary conditions (corresponding to zero flux) are that the normal derivatives $\partial a/\partial n$ and $\partial \rho/\partial n$ are to vanish on C.

The solution corresponding to a uniform state is given by $a = a_0 = $ const, $\rho = \rho_0 = $ const, with

$$a_0 f(\rho_0) = c_3\rho_0/(1 + c_4\rho_0)$$

Consider a perturbation from this uniform state, defined by $a = a_0 + a'$, $\rho = \rho_0 + \rho'$, and use the methods of Section 8.3 to obtain the linearized equations (and boundary conditions) satisfied by a' and ρ'. Next, consider a perturbation pattern of the special form (involving perhaps complex

---

† We modify slightly a problem studied by Keller and Segel (1970, p. 399). A useful discussion of similar stability problems in fluid dynamics has been given by Lin (1955).

quantities)

$$a' = \alpha(x, y)e^{\sigma t}, \qquad \rho' = \beta(x, y)e^{\sigma t}$$

and formulate the equations to be satisfied by α and β, together with the boundary conditions. Explain why this is an eigenvalue problem for σ. This kind of perturbation would be unstable if some eigenvalue σ had a positive real part.

11.3.2 Let $g(x, y; \xi, \eta)$ be the Green's function for the Laplacian operator, satisfying $\Delta g = \delta(x - \xi)\,\delta(y - \eta)$ in R and $g = 0$ on C. Show that Eq. (11.1) may be transformed into the integral equation

$$u(\xi, \eta) = -\lambda \int_R f(x, y)u(x, y)g(x, y; \xi, \eta)\,dx\,dy \qquad (11.8)$$

Show that, because of the symmetry of g (cf. Chapter 9) and the fact that $f > 0$ in R, the function $v(x, y) = [f(x, y)]^{1/2} \cdot u(x, y)$ satisfies an integral equation of the form

$$v(\xi, \eta) = \lambda \int_R K(x, y; \xi, \eta)v(x, y)\,dx\,dy \qquad (11.9)$$

where K is symmetric in the sense $K(x, y; \xi, \eta) = K(\xi, \eta; x, y)$, and where K is square integrable, in that $\int_R K^2\,dA$ exists.

It is true of homogeneous integral equations of the form (11.9) that nontrivial solutions (eigenfunctions) will exist only for special values (eigenvalues) of the parameter λ, that there is at least one such eigenvalue, and that the totality of eigenvalues forms a denumerable set without limit points.†

Accepting these results, let a particular value of λ, say λ_m, be degenerate, in that there are q orthonormal eigenfunctions corresponding to it, viz., $v_1, v_2, \ldots, v_q$ (cf. Problem 11.3.3). By evaluating

$$\int_R dx\,dy \int_R d\xi\,d\eta \left[K - \frac{v_1(x, y)v_1(\xi, \eta)}{\lambda_m} - \cdots - \frac{v_q(x, y)v_q(\xi, \eta)}{\lambda_m} \right]^2$$

which is necessarily greater than or equal to zero, show that the number q of such eigenfunctions corresponding to λ_m must be finite, and obtain a bound for q.

11.3.3 Let $\phi_1(x, y)$, $\phi_2(x, y)$, ..., $\phi_n(x, y)$ be n functions defined over a region R of the (x, y) plane. Let these functions be linearly independent, in that no linear combination $a_1\phi_1 + a_2\phi_2 + \cdots + a_n\phi_n$ (where the a_j are

---

† See Tricomi (1957) or Garabedian (1964, Chap. 10).

constants) vanishes identically in R. Let $f(x, y)$ be a positive function defined over R, and define the scalar product of any two functions $u(x, y)$ and $v(x, y)$ to be

$$(u, v) = \int_R fuv \, dA$$

Show that the following *Gram–Schmidt* process yields a set of linearly independent and orthonormal functions $\psi_1, \psi_2, \ldots, \psi_n$ having the property that each function ϕ_j is a linear combination of the ψ_j functions and vice versa:

$$\psi_1 = \phi_1/[(\phi_1, \phi_1)]^{1/2}$$

$$\psi_2' = \phi_2 - (\phi_2, \psi_1)\psi_1$$

$$\psi_2 = \psi_2'/[(\psi_2', \psi_2')]^{1/2}$$

$$\psi_3' = \phi_3 - (\phi_3, \psi_1)\psi_1 - (\phi_3, \psi_2)\psi_2$$

$$\psi_3 = \psi_3'/[(\psi_3', \psi_3')]^{1/2}$$

$$\cdot$$
$$\cdot$$
$$\cdot$$

11.3.4 Another result of integral equation theory (cf. Problem 11.3.2) is that the set of orthonormal eigenfunctions $u_1, u_2, \ldots$ of Eq. (11.1) is complete in the sense that any function $s(x, y)$ defined in R and satisfying mild smoothness conditions can be expanded in terms of them. Consider now Eq. (11.2), and write $w = \Sigma c_n(t)u_n$; let w and w_t be prescribed for $t = 0$, and let $w = 0$ on C. Show that the equation satisfied by $c_j(t)$ is

$$c_j'' + \lambda_j c_j = p_j(t) \qquad (11.10)$$

where $p_j(t)$ is defined by

$$\frac{p}{f} = - \sum p_j(t)u_j$$

and explain how the initial conditions permit Eq. (11.10) to be solved.

11.3.5 Denote the eigenvalues of the problem of Eq. (11.1) by $\lambda_1, \lambda_2, \ldots$ and the associated orthonormal eigenfunctions by $u_1, u_2, \ldots$. Consider now a function $\psi(x, y)$ satisfying

$$\Delta\psi + \lambda f\psi = s \quad \text{in } R, \qquad \psi = 0 \quad \text{on } C$$

where $s(x, y)$ is a function defined in R, and where λ is a constant. Defining

$s_n = \int_R s u_n \, dA$, $\psi_n = \int_R f \psi u_n \, dA$, show that

$$\psi_n = s_n / (\lambda - \lambda_n)$$

and so obtain a formal series solution for ψ. What happens if $s = \delta(x - \xi) \, \delta(y - \eta)$, where (ξ, η) is a point in R? What condition must s_3 satisfy if a solution ψ for the case $\lambda = \lambda_3$ is to exist, and what is the solution in that case? Interpret the case $\lambda = \lambda_3$ in terms of resonance in the membrane problem of Section 11.1. What alterations in the above discussion would a prescription of nonzero values for ψ on C entail?

11.3.6 Find all eigenfunctions satisfying Eq. (11.1), with $f \equiv 1$, in the following regions, and with the assigned boundary conditions:

(a) $0 < x < 1, 0 < y < \infty$, $u = 0$ on C and bounded at ∞ ;
(b) $0 \leq x^2 + y^2 < 1$, $\partial u / \partial n = 0$ on C;
(c) $\frac{1}{2} < x^2 + y^2 < 1$, $u = 0$ on C;
(d) $\frac{1}{2} < x^2 + y^2 < 1, 0 < \tan^{-1}(y/x) < \pi/2$, $u = 0$ on C;
(e) $0 \leq x^2 + y^2 < 1$, $u + k \, \partial u / \partial n = 0$ on C, (k a constant);
(f) $1 < x^2 + y^2 < \infty$, $u = 0$ on $x^2 + y^2 = 1$ and finite at ∞.

11.3.7 Let (r, θ, ϕ) be spherical polar coordinates, and let $u(r, \theta, \phi)$ satisfy the equation

$$\Delta u + k u = 0$$

in the sphere $r < 1$, with $u(1, \theta, \phi) = 0$. Find all eigenvalues and eigenfunctions. Solve also the eigenvalue problem defined by the same equation, valid now inside the cylinder $r < 1, 0 < z < 1$, where (r, θ, z) are cylindrical coordinates; we require $u = 0$ on the surface of the cylinder.

11.3.8 Let $p(x, y)$, $r(x, y)$, $g(x, y)$ be given functions in a bounded region R of the (x, y) plane, with boundary C. Let p and r be positive in R. Discuss the eigenvalue problem described by

$$(p u_x)_x + (p u_y)_y + [\lambda r - g]u = 0 \quad \text{in} \quad R$$
$$\alpha u + \beta \, \partial u / \partial n = 0 \quad \text{on} \quad C \tag{11.11}$$

where α and β are prescribed functions on C. Devise a diffusion equation problem in which separation of variables leads to this kind of problem.

11.3.9 Let L be a linear operator defined by

$$L\phi = a\phi_{xx} + 2b\phi_{xy} + c\phi_{yy} + d\phi_x + e\phi_y + f\phi$$

where $a, b, \ldots$ are functions of (x, y) defined in a region R with boundary C. The operator M defined by

$$M\phi = (a\phi)_{xx} + 2(b\phi)_{xy} + (c\phi)_{yy} - (d\phi)_x - (e\phi)_y + f\phi$$

is said to be *adjoint* to L. Show that if $u(x, y)$ and $v(x, y)$ are defined in R, then the integral

$$I = \int_R (uLv - vMu) \, dA$$

is expressible in terms of values on C of u, v and their first partial derivatives. Show in particular that if $u = 0$ on C, $v = 0$ on C, then $I = 0$. Show next that if L is elliptic in R, and if

$$Lu + \lambda fu = 0 \quad \text{in } R, \qquad u = 0 \quad \text{on } C$$

$$Mv + kfv = 0 \quad \text{in } R, \qquad v = 0 \quad \text{on } C$$

where $f(x, y) > 0$ in R, and where λ and k are constants, then each eigenfunction u_j is orthogonal to each eigenfunction v_i, if $\lambda_j \neq k_i$.

Consider next the problem $L\psi + \lambda f\psi = p$ in R, $\psi = 0$ on C, where $p(x, y)$ is a given function. Show that if λ coincides with one of the eigenvalues k_j, then this problem cannot have a solution unless $\int_R pv_j \, dA = 0$.

Since $L = M$ for the operators of Eqs. (11.1) and (11.11), we say these operators are *self-adjoint*.

11.3.10   Discuss the eigenvalue problem for $u(x, y)$ defined by $\Delta\Delta u + \lambda u = 0$ in a region R of the (x, y) plane, where Δ is the Laplacian operator, and where $u = \partial u/\partial n = 0$ on the boundary C of R.

11.4   PERTURBATIONS

Let (r, θ) be polar coordinates, and let $\phi(r, \theta)$ satisfy the equation

$$\phi_{rr} + (1/r)\phi_r + (1/r^2)\phi_{\theta\theta} + (\lambda - \varepsilon \sin^2 \theta)\phi = 0 \qquad (11.12)$$

in the unit circle region $r < 1$, with $\phi(1, \theta) = 0$ and with ϕ finite at the origin. Here ε is a small parameter, with $|\varepsilon| \ll 1$. We investigate the dependence of the lowest eigenvalue on ε, and write

$$\phi = \phi_0 + \varepsilon\phi_1 + \varepsilon^2\phi_2 + \cdots$$

$$\lambda = \lambda_0 + \varepsilon\lambda_1 + \varepsilon^2\lambda_2 + \cdots$$

Substitution into Eq. (11.12) yields the sequence of problems

$$\Delta\phi_0 + \lambda_0\phi_0 = 0, \qquad\qquad\qquad \phi_0(1, \theta) = 0$$

$$\Delta\phi_1 + \lambda_0\phi_1 = (\sin^2 \theta - \lambda_1)\phi_0, \qquad \phi_1(1, \theta) = 0 \qquad (11.13)$$

$$\Delta\phi_2 + \lambda_0\phi_2 = (\sin^2 \theta - \lambda_1)\phi_1 - \lambda_2\phi_0, \qquad \phi_2(1, \theta) = 0$$

.

.

.

The first of these problems gives (since we are interested in the lowest eigenvalue)

$$\phi_0 = J_0(\xi r), \qquad \lambda_0 = \xi^2$$

where ξ is the smallest zero of J_0. The second problem can have a solution only if the nonhomogeneous term is orthogonal to ϕ_0, as is easily seen by multiplying the first equation by ϕ_1, the second by ϕ_0, subtracting, and integrating over the area of the unit disk. This condition requires

$$\int_0^{2\pi} \int_0^1 (\sin^2 \theta - \lambda_1) J_0^2(\xi r) r \, dr \, d\theta = 0$$

whence $\lambda_1 = \frac{1}{2}$. With this choice for λ_1, we can now solve the second equation for ϕ_1 (by the method of Problem 11.3.5, for example). The solution of the second equation will contain an arbitrary multiple of ϕ_0, but—just as in the case of perturbation methods for ordinary differential equations†— there is no loss in disregarding this term since it merely affects the arbitrary constant by which ϕ may always be multiplied in any event. A continuation of this process yields the remaining λ_i and ϕ_i.

A perturbation process is also effective for some problems involving a modification of the boundary of a region. Consider for example the problem

$$\Delta\phi + \lambda\phi = 0 \quad \text{in } R, \qquad \phi = 0 \quad \text{on } C \qquad (11.14)$$

where R is a region in the (x, y) plane whose boundary C is a smooth curve (possessing a tangent everywhere). We now deform C slightly and ask how a particular eigenvalue is affected. We content ourselves with a first-order calculation.

Let each boundary point be displaced along the outward normal by an amount $\varepsilon f(s)$, where f is a prescribed function of contour arc length s. Writing $\phi = \phi_0 + \varepsilon\phi_1 + \cdots$, the condition that $\phi = 0$ on the deformed boundary C^* becomes

$$[\phi_0 + (\partial\phi_0/\partial n)\varepsilon f(s) + \cdots] + \varepsilon[\phi_1 + \cdots] + \cdots = 0$$

where each of ϕ_0, $\partial\phi_0/\partial n$, and ϕ_1 is evaluated on the "old" boundary C. Thus $\phi_0 = 0$ on C, and $\phi_1 + (\partial\phi_0/\partial n)f(s) = 0$ on C. Moreover, $\Delta\phi_0 + \lambda_0\phi_0 = 0$ in R, and $\Delta\phi_1 + \lambda_0\phi_1 = -\lambda_1\phi_0$ in R, where $\lambda = \lambda_0 + \varepsilon\lambda_1 + \cdots$. Thus λ_0 and ϕ_0 are the same as the eigenvalue and eigenfunction of interest for the problem with the undeformed boundary; to determine λ_1, we must solve

$$\Delta\phi_1 + \lambda_0\phi_1 = -\lambda_1\phi_0 \quad \text{in } R, \qquad \phi_1 = -(\partial\phi_0/\partial n)f(s) \quad \text{on } C$$

† Carrier and Pearson, (1968, Sec. (9.2)).

The usual "multiply-and-subtract" technique now shows that this problem can have a solution only if λ_1 is such that

$$\int_C \left[\phi_0 \frac{\partial \phi_1}{\partial n} - \phi_1 \frac{\partial \phi_0}{\partial n} \right] ds = -\lambda_1 \int_R \phi_0^2 \, dA$$

i.e.,

$$\lambda_1 = - \frac{\left(\int_C (\partial \phi_0/\partial n)^2 f(s) \, ds \right)}{\int_R \phi_0^2 \, dA} \tag{11.15}$$

Since the change in λ is $\varepsilon \lambda_1$, we observe that a positive value for $\varepsilon f(s)$ tends to decrease λ, a result which is physically reasonable.

11.5 APPROXIMATIONS

Let p, r, g be positive functions defined in a bounded region R of the (x, y) plane, with boundary C. All eigenvalues of the problem

$$(pu_x)_x + (pu_y)_y + (\lambda r - g)u = 0 \qquad \text{in } R$$
$$u = 0 \qquad \text{on } C \tag{11.16}$$

are real and positive; denote them in order of increasing size by $\lambda_1, \lambda_2, \ldots$, and let the corresponding eigenfunctions be $u_1, u_2, \ldots$. By the method of Section 11.2, we obtain

$$\lambda_n = \frac{\int [p\{(u_n)_x^2 + (u_n)_y^2\} + gu_n^2] \, dA}{\int ru_n^2 \, dA} \tag{11.17}$$

where the integrals are over R. Consider now *any* function $\phi(x, y)$, vanishing on C, and form a similar expression for ϕ:

$$I(\phi) = \frac{\int [p\{\phi_x^2 + \phi_y^2\} + g\phi^2] \, dA}{\int r\phi^2 \, dA} \tag{11.18}$$

This expression is always positive. We ask if there is some function ϕ (vanishing on C) that makes it stationary with respect to variations $\delta\phi$ satisfying $\delta\phi = 0$ on C. Taking the variation of this quotient expression with respect to ϕ (i.e., using the methods of Chapter 10), we find that such a function ϕ must satisfy

$$(p\phi_x)_x + (p\phi_y)_y + (Ir - g)\phi = 0$$

where I is the value of the quotient corresponding to the extremizing function ϕ. But this is again Eq. (11.16), with λ replaced by I. We therefore conclude that the quotient (11.18) is stationary for $\phi = u_n$, $n = 1, 2, 3, \ldots$, where u_n is the nth eigenfunction of Eq. (11.16) and the corresponding value of the quotient is just the nth eigenvalue λ_n of Eq. (11.16). This means that we can use the Rayleigh–Ritz method of Chapter 10 to approximate eigenvalues and eigenfunctions.

In the present example, we can make a stronger statement. Let the $\{u_n\}$ be normalized, so that $\int r u_m u_n \, dA = \delta_{mn}$. Then for any (twice differentiable) function ϕ satisfying $\phi = 0$ on C, write (cf. Problem 11.3.4)

$$\phi = \sum_1^\infty c_j u_j$$

where the c_j are constants, and substitute into Eq. (11.18). After some algebra, we obtain

$$I(\phi) = \frac{\lambda_1 c_1^2 + \lambda_2 c_2^2 + \cdots}{c_1^2 + c_2^2 + \cdots} \tag{11.19}$$

and since $\lambda_1 \leq \lambda_2 \leq \lambda_3 \leq \cdots$, it is clear that

(1) the minimum of $I(\phi)$ over all admissible ϕ is λ_1,

(2) the minimum of $I(\phi)$ over all admissible ϕ that are orthogonal to $u_1, u_2, \ldots, u_k$ is λ_{k+1},

(3) if $k - 1$ continuous functions $\psi_1, \psi_2, \ldots, \psi_{k-1}$ are chosen arbitrarily (*not* necessarily vanishing on C), and if $I(\phi)$ is minimized over all differentiable functions ϕ satisfying $\phi = 0$ on C and $\int r\phi\psi_j \, dA = 0$ for $j = 1, 2, \ldots, k - 1$, then the maximum of such minima, over all sets $(\psi_1, \ldots, \psi_{k-1})$, is λ_k. [For there is some linear combination of $u_1, \ldots, u_{k-1}, u_k$, that is orthogonal to each of $\psi_1, \ldots, \psi_{k-1}$, and a choice of ϕ equal to this linear combination will yield $I(\phi) \leq \lambda_k$.] This result is known as *Courant's maximum–minimum principle*.

We observe that an alternative (Lagrange multiplier) formulation of the variational problem (11.18) is to ask for that function ϕ, vanishing on C and satisfying $\int r\phi^2 \, dA = 1$, that makes the numerator of the quotient expression stationary. Of more practical interest is the fact that the first-order change in $I(\phi)$ vanishes at the stationary point, so that $I(\phi)$ is not very sensitive to changes in ϕ when near such a point; thus the extremization process is apt to yield the eigenvalues more accurately than the eigenfunctions.

If a simple variational formulation of a particular eigenvalue problem is not available, a Galerkin procedure may be feasible. If the problem has the form $L(u) = \lambda u$, say, where L is some linear differential operator, then we can choose a number n of independent functions ϕ_j satisfying whatever homogeneous boundary conditions are prescribed, and choose constants c_j in

$$\psi = c_1\phi_1 + \cdots + c_j\phi_j$$

so as to make $(L - \lambda)\psi$ orthogonal to each of the ϕ_i (or to any other n functions, for that matter). This yields a set of linear homogeneous equations in the c_j, which have solutions only if their coefficient determinant (which involves λ) vanishes; with some luck, the roots of this determinantal equation should approximate the first few λ_j. As in our discussions of the Galerkin method in Chapter 10, it is not necessary that either L or the ψ expansion be linear, and undetermined functions rather than undetermined constants may be used.

11.6 PROBLEMS

11.6.1 Use a perturbation method to discuss the eigenvalues of the problem

$$[(1 + \varepsilon x)u_x]_x + (1 + \varepsilon x)u_{yy} + \lambda u = 0$$

inside a square region, with $u = 0$ on the boundary and with $|\varepsilon| \ll 1$. Does anything interesting happen if the factor $(1 + \varepsilon x)$ in front of the term u_{yy} is replaced by unity? Can you verify your results by means of an exact solution?

11.6.2 Obtain an approximation for the lowest eigenvalue for the case in which the region R of Eq. (11.1), with $f \equiv 1$, is elliptical, the major axis being twice the minor. Determine, also approximately, the change in this lowest eigenvalue resulting from a contour deformation that makes the ellipse slightly fatter.

11.6.3 Discuss—with examples where feasible—perturbation and variational methods for the problem $\Delta\Delta u + \lambda u = 0$ in R, $u = \partial u/\partial n = 0$ on C. Here R is a region in the (x, y) plane, having boundary C.

11.6.4 Let p, r, g be positive functions of (x, y) in a region R, and let α be a positive function of position on the boundary C of R. Show that the problem of minimizing

$$J(\phi) = \int_R [p(\phi_x{}^2 + \phi_y{}^2) + g\phi^2] \, dA + \int_C p\alpha\phi^2 \, ds$$

subject to the condition $\int_R r\phi^2 \, dA = 1$ leads to an eigenvalue problem with the natural homogeneous boundary condition $\partial u/\partial n + \alpha u = 0$ on C.

11.6.5 Show that each eigenvalue λ_j of the problem of Eq. (11.16) is not increased if the region R is enlarged. State and prove similar results concerning the dependence of each λ_j on the functions p, q, r; comment on the physical reasonableness of these results for the vibrating membrane problem of Section 11.1.

11.6.6 Let the region R of Eq. (11.1) have the "starlike" property—i.e., there is some interior point [which we choose as the origin of a polar coordinate system (r, θ)] from which each boundary point is visible. Take $f = 1$, and write the equation of the boundary curve as $r = g(\theta)$. In Rayleigh's quotient [Eq. (11.18)], use the function $\phi(r, \theta) = w(\xi)$, where $\xi = r/g(\theta)$, with $w(1) = 0$, and choose w so as to minimize the quotient. Show that this technique yields the upper bound

$$\lambda_1 \leq \frac{k^2}{2A} \int_C \frac{ds}{h} \tag{11.20}$$

where k is the smallest root of $J_0(z) = 0$, and where h is the perpendicular distance from the origin to the tangent drawn to the boundary curve at ds. Consider also the special case of a circle.

A number of additional results are available for the membrane problem. In particular, $\lambda_1 \geq \pi k^2/A$ (where k is the same as in Eq. (11.20), and $\lim_{n\to\infty}(\lambda_n/n) = 4\pi/A$. These and other results, as well as a further discussion of approximation techniques, will be found in Garabedian (1964, Chap. 11), Courant and Hilbert (1953, Chap. 6), Collatz (1948), and Polya and Szego (1951).

12

MORE ON FIRST-ORDER EQUATIONS

12.1 ENVELOPES

We will begin our discussion of general first-order equations with a review of the generation of envelopes by families of lines or surfaces.

A sketch shows that the family of straight lines given by $y = \alpha x + (1/\alpha)$, where $\alpha > 0$ is a parameter, possesses an *envelope*, i.e., a curve that at each point is tangent to some member of the family. To find the equation of this envelope, we note that two adjoining members of the family (for which the parameters are α and $\alpha + d\alpha$) will intersect at a common point (x, y) satisfying the two equations

$$y = \alpha x + (1/\alpha), \qquad y = (\alpha + d\alpha)x + 1/(\alpha + d\alpha)$$

Subtracting, dividing by $d\alpha$, and letting $d\alpha \to 0$, we can replace these two equations by

$$y = \alpha x + (1/\alpha), \qquad 0 = x - (1/\alpha^2)$$

from which α may be eliminated to yield $y = 2\sqrt{x}$ as the equation of the envelope.

More generally, a similar derivation shows that if a family of curves $F(x, y, \alpha) = 0$ (where α is again a parameter) has an envelope, that envelope may be found by eliminating α from the pair of equations

$$F(x, y, \alpha) = 0, \qquad F_\alpha(x, y, \alpha) = 0 \qquad (12.1)$$

It may happen that this elimination cannot be carried out explicitly, in which case we think of the coordinates x and y of the envelope curve as being defined parametrically by means of Eqs. (12.1).

For a given family of curves, say $F(x, y, \alpha) = 0$, it is generally possible—at least in principle—to find a differential equation satisfied by each member of the family, where this differential equation does not involve the parameter α. For along any one curve (on which α is fixed), we have $F_x + F_y y' = 0$, and α may be eliminated between this equation and the equation $F = 0$ to obtain an equation involving x, y, and y' only. We will write this equation as $f(x, y, p) = 0$, where $p = y'$. It is clear that if the family $F(x, y, \alpha) = 0$ has an envelope, then this envelope curve also satisfies the differential equation $f(x, y, p) = 0$, since the envelope curve is tangent at each point to some member of the family, so that the values of x, y, and y' at each point on the envelope are just those of some member of the family at that point. We observe moreover that if each member of a family of curves satisfies $f(x, y, p) = 0$, and if we know that this family of curves has an envelope, then adjoining members of the family will intersect at a common point (x, y) satisfying $f(x, y, p) = 0$ and $f(x, y, p + dp) = 0$, so that—going to the limit—the equation of the envelope can be obtained directly from the differential equation by eliminating p from the pair

$$f(x, y, p) = 0, \qquad f_p(x, y, p) = 0 \qquad (12.2)$$

Such an envelope-type solution of an ordinary differential equation is often termed a *singular solution*.

It may happen that our family of curves is given in parametric form. Let t be a parameter that is constant for any one member of the family, but that varies as we go from one curve to another; let s be a parameter varying along each curve. Then we can write the equations for the family in the form

$$x = x(s, t), \qquad y = y(s, t) \qquad (12.3)$$

If this family has an envelope, then adjoining members of the family (corresponding to t, $t + dt$, say) will intersect at some common point (x, y). Writing the values of s corresponding to this intersection point as s and $s + ds$, we have

$$x(s, t) = x(s + ds, t + dt), \qquad y(s, t) = y(s + ds, t + dt)$$

In the limit,

$$x_s \, ds + x_t \, dt = 0, \qquad y_s \, ds + y_t \, dt = 0$$

But our envelope hypothesis implies that there is a nontrivial solution for ds and dt; thus

$$x_s y_t - x_t y_s = 0 \qquad (12.4)$$

is the condition that—together of course with Eqs. (12.3)—defines the envelope. As a simple example, the reader may verify that the family $x = st$, $y = s + t$ has the same envelope as the previous example.

An extension of this discussion to the case of an envelope of a family of surfaces—i.e., a surface that at each point is tangent to some member of the family—is straightforward and will be left to the reader. The basic result for three dimensions is that if a family of surfaces $F(x, y, z, \alpha) = 0$ has an envelope, where α is a parameter, then the equation of the envelope is obtained from the pair of simultaneous equations $F = F_\alpha = 0$. If the equation is in parametric form, say $x = x(s, t, \alpha)$, $y = y(s, t, \alpha)$, $z = z(s, t, \alpha)$, then the envelope equation satisfies these three equations and also

$$
J\left(\frac{x, y, z}{s, t, \alpha}\right) = \begin{vmatrix} x_s & x_t & x_\alpha \\ y_s & y_t & y_\alpha \\ z_s & z_t & z_\alpha \end{vmatrix} = 0 \tag{12.5}
$$

In Section 12.3 we will consider the relationship between the partial differential equation satisfied by each member of a family of surfaces and that satisfied by the envelope of such a family.

12.2 CHARACTERISTIC STRIPS

In Chapter 6 we considered first-order equations of quasilinear form:

$$
f(x, y, u)p + g(x, y, u)q = h(x, y, u) \tag{12.6}
$$

where $p = u_x$, $q = u_y$, and where f, g, h are prescribed functions of x, y, u. We found that the set of characteristic curves passing through some arbitrarily prescribed initial curve in (x, y, u) space would (except in pathological situations) generate a solution surface $u = u(x, y)$ for Eq. (12.6) satisfying the initial conditions.

We now turn our attention to the general first-order equation

$$
F(x, y, u, p, q) = 0 \tag{12.7}
$$

where F is some given function—perhaps highly nonlinear—of x, y, u, p, and q, and where, as before, $p = u_x$ and $q = u_y$. From our experience with Eq. (12.6) we might anticipate that the general solution of Eq. (12.7) will involve an arbitrary function—e.g., the value of u prescribed along some chosen curve in the (x, y) plane—rather than an arbitrary constant, as in the case of an ordinary differential equation. Moreover, we might hope that the idea of characteristics will continue to play a useful role in the solution of a problem in which such an initial condition is prescribed.

The reader will recall that a characteristic curve has been defined in the past as a curve along which Cauchy data is not sufficient to permit extrapolation of u values into adjoining regions, or alternatively, as a curve across which there could be a discontinuity in some high-order derivative. In finding such characteristic curves, it turned out to be useful to introduce a new local (ξ, η) coordinate system.

If the initial condition for Eq. (12.7) consists in the specification of u along some curve Γ in the (x, y) plane, then the tangential derivative of u along Γ is easily computed. By use of Eq. (12.7) we would then also expect to be able to determine the derivative of u in a direction normal to Γ. However, a complication arises in that Eq. (12.7) is nonlinear, and so there may be several possible values for the normal derivative at any point on Γ. It is therefore conventional to require that both u and $\partial u / \partial n$ be specified on Γ as initial data; these values of u and $\partial u / \partial n$ imply certain values of p and q at each point of Γ, and we require of course that these values be compatible with Eq. (12.7).

We now ask: is there some curve Γ, and associated data for u and $\partial u / \partial n$ on Γ, such that we cannot extrapolate u values into adjoining regions—or alternatively, such that a discontinuity in some high-order derivative can exist across Γ? In finding such a *characteristic*,† it is useful, as before, to introduce a local (ξ, η) coordinate system such that Γ is one of the curves $\xi(x, y) = $ const.

In terms of ξ and η, Eq. (12.7) becomes

$$F\big(x(\xi, \eta),\, y(\xi, \eta),\, u(\xi, \eta),\, u_\xi \xi_x + u_\eta \eta_x,\, u_\xi \xi_y + u_\eta \eta_y\big) = 0 \quad (12.8)$$

Differentiation with respect to ξ shows that $u_{\xi\xi}$, for example, cannot be determined from the given data on Γ (or alternatively, may be discontinuous across Γ) if

$$F_p \xi_x + F_q \xi_y = 0 \qquad (12.9)$$

on Γ.‡

The ratio $-\xi_x / \xi_y$, equal from Eq. (12.9) to F_q / F_p, is the slope of Γ; as in Chapter 6, it is convenient to introduce a parameter s such that the equation of the characteristic curve Γ becomes

$$\frac{dx}{ds} = F_p, \qquad \frac{dy}{ds} = F_q \qquad (12.10)$$

---

† More strictly, perhaps, a *characteristic base curve*, if we want to use the term *characteristic* to denote a curve in (x, y, u) space. The nomenclature in the literature is not consistent.

‡ Another way of looking at Eq. (12.9) is that this is the condition that the left-hand side of Eq. (12.8) not really depend on u_ξ—i.e., $\partial F / \partial u_\xi = 0$.

Also as in Chapter 6, we can obtain an equation for du/ds:

$$\frac{du}{ds} = u_x \frac{dx}{ds} + u_y \frac{dy}{ds} = pF_p + qF_q \tag{12.11}$$

by use of Eqs. (12.10). There is now, however, a difference from the quasi-linear case, for Eqs. (12.10) and (12.11) will generally involve p and q on the right-hand sides, and these quantities are not known a priori as functions of s. We therefore also need equations for dp/ds and dq/ds. To get these, we differentiate Eq. (12.7) with respect to x to obtain

$$F_x + F_u p + F_p p_x + F_q q_x = 0$$

or using $q_x = p_y$ and Eqs. (12.10),

$$F_x + F_u p + (dx/ds)p_x + (dy/ds)p_y = 0$$

Thus

$$dp/ds = -F_x - F_u p \tag{12.12}$$

and similarly,

$$dq/ds = -F_y - F_u q \tag{12.13}$$

Equations (12.10), (12.11), (12.12), and (12.13) provide a set of five ordinary differential equations for $x(s)$, $y(s)$, $u(s)$, $p(s)$, and $q(s)$, respectively. If we start from some initial point, say $x(0) = x_0$, $y(0) = y_0$, $u(0) = u_0$, $p(0) = p_0$, $q(0) = q_0$, then this set of equations generates a curve in (x, y, u) space and also provides values of p and q along this curve.

A typical initial value problem requires that one find a solution surface $u = u(x, y)$ for Eq. (12.7) containing some initial curve defined parametrically by $x = x(t)$, $y = y(t)$, $u = u(t)$, say. Then at a point on this curve corresponding to $t = t_0$, the values of x_0, y_0, u_0 are given by $x_0 = x(t_0)$, $y_0 = y(t_0)$, $u_0 = u(t_0)$. Moreover, in accordance with the discussion earlier in this section, we assume that $p(t)$ and $q(t)$ are also specified, so that p_0 and q_0 can be determined; these quantities must of course satisfy the equations

$$u'(t_0) = p_0 x'(t_0) + q_0 y'(t_0)$$
$$F(x_0, y_0, u_0, p_0, q_0) = 0 \tag{12.14}$$

Thus we have the information required to construct the characteristic curve emanating from any point on the initial curve, and as in Chapter 6, the totality of such characteristic curves should sweep out the desired solution surface. We actually get more than this from the solution of the differential equations for the characteristic, for these equations also provide values of p and q along each characteristic. The question of consistency immediately arises—are the values of p and q so provided the same as u_x and u_y, respectively, for the solution surface as obtained by direct partial differentiation of $u(x, y)$? In Problem 12.4.6 the reader is asked to verify this consistency, and we henceforward accept this result.

The fact that Eqs. (12.10) through (12.13) yield $p(s)$ and $q(s)$, as well as $x(s)$, $y(s)$, $u(s)$, means that at each point we also know a tangent plane orientation; consequently, a solution of these equations is often said to provide a *characteristic strip*. As explained above, a solution surface containing some initial curve, $x = x(t), y = y(t), u = u(t)$, can be obtained by use of these characteristic strips, provided we also know $p(t)$ and $q(t)$ along the initial curve. It may be that $p(t)$ and $q(t)$ are prescribed as part of the initial data (if so, Eqs. (12.14) must of course be satisfied); if not, we may be able to solve Eqs. (12.14) for $p(t)$ and $q(t)$. This process may break down if the initial strip satisfies some of the characteristic strip equations. In fact, if the given $u(t)$, $p(t)$, $q(t)$ data is such that $x'(t)$: $y'(t)::F_p:F_q$, then there will be no solution unless all of the strip equations are satisfied, i.e., unless

$$x'(t):y'(t):u'(t):p'(t):q'(t)::$$

$$F_p:F_q:pF_p + qF_q:-F_x - F_up:-F_y - F_uq$$

In the event this consistency condition is indeed met, there will be an infinite number of solutions—in fact, we can then think of the given data as representing a characteristic strip emanating from some other, reasonably arbitrary, initial curve.

A simple example of the generation of a solution by means of characteristic strips may be worthwhile. Let $u(x, y)$ satisfy

$$u - px - \tfrac{1}{2}q^2 + x^2 = 0 \tag{12.15}$$

with $u(x, 0) = x^2 - \tfrac{1}{6}x^4$ for $0 < x < 1$. Since $p(x, 0) = 2x - \tfrac{2}{3}x^3$, we find from Eq. (12.15) that $q(x, 0) = \pm x^2$. Thus there will be two possible solutions for the problem as stated; for definiteness, we choose that for which $q(x, 0) = +x^2$. The characteristic strip equations become

$$\frac{dx}{ds} = -x, \qquad \frac{dy}{ds} = -q, \qquad \frac{dp}{ds} = -2x, \qquad \frac{dq}{ds} = -q \tag{12.16}$$

where we omit Eq. (12.11) as unnecessary here, since the right-hand sides of Eqs. (12.16) do not involve u explicitly. We find easily that the characteristic that cuts the initial curve at the point $x = t$ is given by

$$y = tx - t^2, \qquad q = tx, \qquad p = 2x - \tfrac{2}{3}t^3 \tag{12.17}$$

The characteristic base curves are sketched in Fig. 12.1. They possess

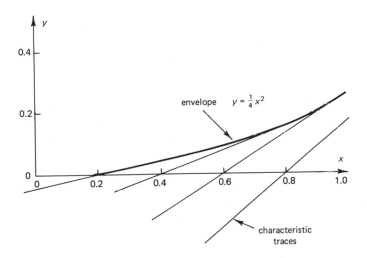

Fig. 12.1 Characteristics for Eq. (12.15).

the envelope $y = \frac{1}{4}x^2$. In that region below this parabola which is traversed by the characteristics, the value of u from Eq. (12.15) is

$$u = x^2 - \tfrac{2}{3}t^3x + \tfrac{1}{2}t^2x^2$$

and we can eliminate t by use of $y = tx - t^2$ (choosing the root of the quadratic that yields $t = x$ at $y = 0$) to obtain

$$u = x^2 - \frac{x^4}{12} + \frac{x^2y}{2} + \left(\frac{xy}{3} - \frac{x^3}{12}\right)(x^2 - 4y)^{1/2} \qquad (12.18)$$

The reader may verify that the values of p and q, as obtained from Eqs. (12.17), are consistent with those obtained by direct partial differentiation of Eq. (12.18). It is clear that Eq. (12.18) ceases to be valid at the envelope curve $y = \frac{1}{4}x^2$.

Although we were able in this example to obtain an explicit expression for u, we would generally have to content ourselves with an implicit solution, with each of x, y, and u expressed as functions of the two parameters s and t.

12.3 COMPLETE INTEGRAL

Suppose that $u = \phi(x, y, \tau)$ is a solution of Eq. (12.7), in some region of the (x, y) plane, for all values (in some range) of the parameter τ. Suppose also that this family has an envelope; then this envelope surface is tangent at each point to a solution surface and so must also satisfy Eq. (12.7). Incidentally, a geometric interpretation of Eq. (12.7) is sometimes useful. Equation (12.7) provides, at each point of (x, y, u) space, a relationship between p and q, and thus also provides some information concerning the tangent plane to any solution surface passing through this (x, y, u) point. If in fact we choose some value of p, then a corresponding value of q may be determined from Eq. (12.7) [holding our (x, y, u) point fixed]; considering all such (p, q) pairs, we obtain a family of possible tangent plane elements at the (x, y, u) point. This family will in general have an envelope, called a *Monge cone*, and any solution surface of Eq. (12.7) must, at each point (x, y, u), be tangent to the Monge cone at that point. The envelope surface will of course share this property of the solution surfaces.

As an example, $u = x^2 + \frac{1}{2}(y - \tau)^2$ is a solution of Eq. (12.15) for all τ, and so the envelope surface $u = x^2$ must also be a solution.

Our main interest in this section is in an attempt to obtain a general solution of Eq. (12.7). It would appear that the knowledge of a family of solutions, such as the family $u = \phi(x, y, \tau)$ discussed above, is not adequate, since this family involves only an arbitrary constant, rather than the arbitrary function we might expect to appear in the general solution of a PDE. [For example, the general solution of the equation $p = 0$ is given by $u = f(y)$, where f is arbitrary.] Surprisingly, however, it turns out that one small additional step—the introduction of just one more parameter—results in much more generality of solution structure than might be anticipated. The process of envelope generation will turn out to play an essential role.

Consider then $u = \phi(x, y, \tau, w)$, which is to be solution of Eq. (12.7) for all values of the parameters τ and w in some region. The crucial observation is that this function will continue to be a solution of Eq. (12.7) if w is made a suitable function of τ, via $w = w(\tau)$; moreover, the envelope defined by eliminating τ from

$$u = \phi\big(x, y, \tau, w(\tau)\big)$$
$$0 = \phi_\tau\big(x, y, \tau, w(\tau)\big) + w'(\tau) \cdot \phi_w\big(x, y, \tau, w(\tau)\big)$$

(12.19)

is also a solution of Eq. (12.7). But since $w(\tau)$ is arbitrary, the function defined by Eqs. (12.19) is a solution of Eq. (12.7), which contains (implicitly, of course) an arbitrary function of x and y, and so in a sense—to use conventional nomenclature—is a *general solution* of Eq. (12.7). Because a "general solution" of Eq. (12.7) may be constructed from a two-parameter solution in this way, such a solution involving two arbitrary parameters is called a *complete integral*. An example of a complete integral for Eq. (12.15) is $u = x^2 + \tau x + \frac{1}{2}(y - w)^2$. We can set $w = \tau$, or $w = \tau^2$, or $w = \sin \tau$, or almost anything we please, and in each case use Eqs. (12.19) to generate a new solution for Eq. (12.7) [e.g., $w = \tau$ leads to $u = \frac{1}{2}x^2 + xy$]; we consider the totality of such solutions as a set of solutions characterized by the arbitrary function $w(\tau)$.

The terms "general solution" and "complete integral" are somewhat misleading, since they need not provide all solutions of a given equation. A simple example to illustrate this point is given by the equation $(u - p) \times (u - q) = 0$. By the above definition, the function $u = f(y)e^x$, where $f(y)$ is arbitrary, is a "general solution"; however it does not include the solution $u = e^y$.

A complete integral of Eq. (12.7), $u = \phi(x, y, \tau, w)$, may have another interesting property. Suppose that by holding w fixed and varying τ we get an envelope surface; there will be one such envelope surface for each choice of w, and if we now vary w, these envelope surfaces may themselves have an envelope. The previous geometric interpretation shows that this "super envelope" will still represent a solution (sometimes called a "singular solution") of Eq. (12.7); we obtain it analytically by eliminating τ and w from the set

$$u = \phi(x, y, \tau, w), \qquad 0 = \phi_\tau(x, y, \tau, w), \qquad 0 = \phi_w(x, y, \tau, w) \qquad (12.20)$$

As an example, a complete integral of Eq. (12.15) is given by

$$u = x^2 + w\tau x + \tfrac{1}{2}(y - w - \tau)^2$$

and Eqs. (12.20) lead to the new solution

$$u = x^2 + xy^2/2(x + 2) \qquad (12.21)$$

We turn next to the relationship between a complete integral and characteristics. If $u = \phi(x, y, \tau, w)$ is a complete integral of Eq. (12.7), then by making $w = w(\tau)$ we obtain an envelope surface defined by $u = \phi$ and $0 = \phi_\tau + \phi_w w'$, as before. For a particular choice of τ, say $\tau = a$, these two equations define a curve of tangency, and since the envelope surface and the $u = \phi$ surface need not have the same curvature along this curve, this curve must be a characteristic—for we have seen in Section 12.2 that only across such curves can discontinuities in higher derivatives occur. Moreover, since $w(\tau)$ is arbitrary, we can choose it so that $w(a) = b$, $w'(a) = c$, where b and c are arbitrary constants. Thus one way of obtaining at least some of the characteristics of Eq. (12.7) in integrated form, if a complete integral $u = \phi$ is known, is via

$$u = \phi(x, y, a, b), \qquad 0 = \phi_\tau(x, y, a, b) + c\phi_w(x, y, a, b)$$

For example, the complete integral $u = x^2 + \tau x + \frac{1}{2}(y - w)^2$ for Eq.

(12.15) leads to the set of characteristics

$$u = x^2 + ax + \tfrac{1}{2}(y - b)^2, \qquad 0 = x - c(y - b)$$

This family includes the set (12.17), and if we borrow the condition $dy/dx = q/x = x$ when $y = 0$ from Eqs. (12.16) and their sequel, we can easily determine a, b, and c so as to recover Eq. (12.18).

12.4 PROBLEMS

12.4.1 (a) Use the method of characteristic strips to solve the equation

$$p^2 + q^2 - x = 0 \tag{12.22}$$

where $u(x, 0) = f(x)$ for x in $(1, 2)$. Discuss the nature of the solution; consider also the special cases $f(x) = \tfrac{1}{2}$, $f(x) = x$, $f(x) = \tfrac{2}{3}x^{3/2}$.

(b) It is sometimes feasible to obtain a solution of an equation like (12.7) by separation of variables. In Eq. (12.22), set $u = F(x) + G(y)$ to obtain a complete integral for u in the form

$$u = \tfrac{2}{3}(x - t^2)^{3/2} + ty + w$$

and use this expression to generate a general solution in the sense of Section 12.3. Is there a singular solution?

12.4.2 *Clairaut's equation* is given by

$$u = xp + yq + f(p, q) \tag{12.23}$$

where $f(p, q)$ is a prescribed function of p and q. Discuss solutions of this equation, considering both a characteristic strip and a complete integral approach [the latter, for example, via the family of planes $u = tx + wy + f(t, w)$]. Generate a general solution and a singular solution.

As another approach to Clairaut's equation, differentiate with respect to each of x and y to obtain a pair of equations from which it may be deduced that either $u_{xx}u_{yy} - u_{xy}^2 = 0$ or $x + f_p = y + f_q = 0$. [*Note:* The implication of the equation $u_{xx}u_{yy} = u_{xy}^2$ is discussed in Problem 12.4.8).]

12.4.3 Let $u = \phi(x, y, \tau, w)$ be a solution of Eq. (12.7) for all values of the parameters τ and w in some range. Then a singular solution, if it exists, can be obtained by replacing τ and w by whatever functions of x and y satisfy the equations $\phi_\tau = 0$ and $\phi_w = 0$. Since Eq. (12.7) is satisfied by $u = \phi$ for all values of τ and w, we may differentiate Eq. (12.7) with respect to either τ or w and set the results equal to zero. Use this approach to show that the singular solution normally satisfies $F = 0$, $F_p = 0$, $F_q = 0$, and so can be obtained directly from the differential equation without first finding a complete integral. Discuss Problem 12.4.2 in this light. Why is Eq. (12.21) not obtainable in this manner?

12.4.4 Discuss the possibility of a perturbation approach to the problem of using characteristic strips to solve for $u(x, y)$ in $F(x, y, u, p, q, \varepsilon) = 0$, where $|\varepsilon| \ll 1$, and where we assume that the strips are known for $\varepsilon = 0$. Can a similar approach be used if ε occurs in the initial data rather than in the equation? Illustrate your discussion by inventing nontrivial examples.

12.4.5 Discuss characteristic strips, Monge cones, general solutions, complete integrals, and singular solutions for linear and quasi-linear equations of the kinds already considered in Chapter 6. Relate your results to those previously obtained.

12.4.6 Suppose that u is prescribed along some curve $x = x(t)$, $y = y(t)$, and that $p(t)$, $q(t)$ have been determined along this curve. Let a surface $u = u(s, t)$ then be constructed by use of the characteristic strip equations, as in Section 12.2. The solution of these equations also provides functions $x(s, t)$, $y(s, t)$, $p(s, t)$, and $q(s, t)$. If the determinant $x_s y_t - x_t y_s$ is assumed nonzero in a region near the initial curve (i.e., if this initial curve is not a characteristic), then implicit function theory implies that s and t are determinable in terms of x and y, so that u may be considered to be a function of x and y. We can therefore compute u_x and y_y; show that these quantities will coincide with p and q as determined by the characteristic strip equations.

[*Hint:* we have $u_s = p x_s + q y_s$ from Eq. (12.11), where p and q are determined from the characteristic strip equations. If we can also show that $u_t = p x_t + q y_t$, then since these two equations will have a unique solution for p and q, and since the same equations with p and q replaced by u_x and u_y must hold—where u_x and u_y are calculated from the surface $u = u(x, y)$—it will follow that $p \equiv u_x$ and $q \equiv u_y$. To verify this second equation, obtain and solve a differential equation for $(\partial/\partial s)[u_t - p x_t - q y_t]$, using the fact that $F \equiv 0$ (why?) in our region.] Comment on uniqueness.

12.4.7 Show that the method of characteristic strips can be extended to treat a first-order equation for $u(x_1, x_2, \ldots, x_n)$. As a particular example, solve for $u(x, y, z)$ if $u_x^2 + u_y^2 + u_z^2 = 1$ in a region.

12.4.8 We define a *developable surface* to be the envelope of a one-parameter family of planes. Excluding cylinders whose generators are parallel to the u-axis, such a family of planes in (x, y, u) space may be written $u = \alpha x + f(\alpha)y + g(\alpha)$, where f and g are functions of the parameter α. The envelope equation is then obtained by eliminating α between this equation and $0 = x + f'(\alpha)y + g'(\alpha)$, to give $\alpha = \alpha(x, y)$ and thus $u = u(x, y)$. Show that $u_x = \alpha$ and $u_y = f(\alpha)$, $u_{xy} = f'(\alpha)\alpha_x$, etc., and deduce that a developable surface satisfies the equation $u_{xx}u_{yy} = u_{xy}^2$.

Second, prove that if u satisfies the equation $u_{xx}u_{yy} - u_{xy}^2 = 0$, then $u = u(x, y)$ is a developable surface. [*Hint:* define $\alpha = u_x$, $f = u_y$, and note that the Jacobian $J(\alpha, f) = 0$, so that f must be a function of α.]

12.5   LEGENDRE TRANSFORMATION

If a function $u = u(x, y)$ satisfies a certain partial differential equation, it may happen that the equation takes a simpler form if x and y are replaced by other independent variables. One possibility is to choose p and q, where $p = u_x$ and $q = u_y$, as such a set; it may then also be convenient to replace u by a new dependent variable. A particularly simple choice is suggested by the equation

$$du = p \, dx + q \, dy$$

If we define

$$v = px + qy - u \tag{12.24}$$

then

$$dv = x \, dp + y \, dq$$

so that if v is considered to be a function of p and q, we have

$$x = v_p, \qquad y = v_q \tag{12.25}$$

[Of course if p and q are to be legitimate independent variables, we must have the Jacobian $p_x q_y - p_y q_x$ nonzero, i.e., $u_{xx}u_{yy} - u_{xy}^2 \neq 0$. This is a reasonable condition, since the vanishing of this expression would imply that $u = u(x, y)$ is a developable surface (cf. Problem 12.4.8), which necessarily contains straight lines along which p and q do not alter.]

The process that led to Eqs. (12.24) and (12.25) is termed a *Legendre transformation*. As an example of its use, consider the equation

$$p^2 + qx = 0 \tag{12.26}$$

The function $v(p, q)$ defined by Eq. (12.24) thus satisfies the equation

$$p^2 + qv_p = 0$$

which may be integrated at once to yield

$$v = -(p^3/3q) + f(q) \tag{12.27}$$

where $f(q)$ is arbitrary. We also have

$$x = v_p = -(p^2/q), \qquad y = v_q = (p^3/3q^2) + f'(q) \qquad (12.28)$$

and finally, from Eq. (12.24),

$$u = px + qy - v = -(p^3/3q) + qf'(q) - f(q) \qquad (12.29)$$

Equations (12.28) and (12.29) give each of u, x, and y in terms of the parameters p and q, and in terms of an arbitrary function $f(q)$. The solution so obtained is the most general nondevelopable one.

It is clear that a Legendre transformation may be useful in the problem of Eq. (12.7) if p and q occur in a complicated manner, but if x and y occur in a simple manner. It may even happen, as in Eq. (12.26), that a nonlinear equation becomes linear as a result of a Legendre transformation. This device has been much used in the literature; cf. Problem 12.6.3 and Problem 12.6.5.

12.6 PROBLEMS

12.6.1 (a) Let $p = u_x$, $q = u_y$ as above. Define $w = px - u$, and show that $dw = x\,dp - q\,dy$, so that if p and y are new independent variables, and $w(p, y)$ the new dependent variable, we have

$$x = \partial w/\partial p, \qquad -q = \partial w/\partial y$$

Devise a simple partial differential equation in which this form of the Legendre transformation is useful
  (b) Repeat with $\zeta = yq - u$.

12.6.2 Discuss the use of Legendre transformations for each of the equations

  (a) $pq = 1$,
  (b) Clairaut's equation of Problem 12.4.2,
  (c) $u + x \sin(p + q) = 0$.

In which cases do developable surfaces play a role?

12.6.3 A differentiation with respect to x, holding y fixed, of the equation $x = v_p$ yields $1 = v_{pp}p_x + v_{pq}q_x = v_{pp}u_{xx} + v_{pq}u_{xy}$. Obtain a number of equations of this kind, and deduce that:

$$v_{pp} = \frac{1}{D}u_{yy}, \qquad v_{pq} = -\frac{1}{D}u_{xy}, \qquad v_{qq} = \frac{1}{D}u_{xx}$$

where $D = u_{xx}u_{yy} - u_{xy}^2 = 1/(v_{pp}v_{qq} - v_{pq}^2)$. Since we exclude developable surfaces, $D \neq 0$.

Using these results, apply the Legendre transformation to the equation of minimal surfaces (Problem 10.2.1c) to obtain a linear equation. Examine similarly the compressible flow equation of Problem 7.2.9.

12.6.4   In thermodynamics, the state of a simple system is characterized by specifying the values of any two of the following: pressure P, absolute temperature T, volume V, internal energy U, or entropy S; moreover, the equation defining the change in entropy is given by $T\,dS = dU + P\,dV$. Use Legendre transformations to obtain Maxwell's relations:

$$\left(\frac{\partial T}{\partial V}\right)_S = -\left(\frac{\partial P}{\partial S}\right)_V, \qquad \left(\frac{\partial T}{\partial P}\right)_S = \left(\frac{\partial V}{\partial S}\right)_P$$

$$\left(\frac{\partial S}{\partial V}\right)_T = \left(\frac{\partial P}{\partial T}\right)_V, \qquad \left(\frac{\partial S}{\partial P}\right)_T = -\left(\frac{\partial V}{\partial T}\right)_P$$

Show also that

$$\left(\frac{\partial P}{\partial V}\right)_T \left(\frac{\partial V}{\partial T}\right)_P \left(\frac{\partial T}{\partial P}\right)_V = -1$$

12.6.5   In changing the independent variables as in a Legendre transformation, it may be convenient to leave the dependent variable unaltered. As an illustration of this, consider the problem of incompressible flow from an opening in an infinite reservoir, as in Fig. 12.2. If the stream function is $\psi(x, y)$, then the velocity components in the x and y directions are given by $u = \psi_y$, $v = -\psi_x$, respectively; we postulate irrotationality, which requires $\Delta\psi = 0$ (cf. Problem 7.2.9). On the curve AB of Fig. 12.2, the pressure (atmospheric) is constant, so that Bernoulli's theorem requires that $u^2 + v^2 = $ const. Show that if ψ is considered to be a function of (u, v),† then ψ is harmonic in a portion of a circular region in the (u, v) plane, with $\psi = 0$ on part of the boundary, and a constant on the remainder. Show also that if ψ is known in terms of u and v, we can obtain x and y via equations such as

$$x_u = -\frac{v\psi_u + u\psi_v}{u^2 + v^2}$$

---

† In fluid mechanics, this is known as a *hodograph transformation*.

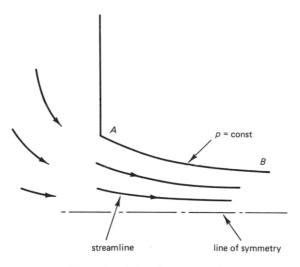

Fig. 12.2   Flow from reservoir.

12.7   PROPAGATION OF A DISTURBANCE

In Problem 12.4.7, the reader is asked to extend the method of characteristic strips to problems involving more than two independent variables; we now consider one such example in detail.

Let a disturbance be propagating through an isotropic medium, initially at rest, occupying some portion of (x, y, z) space. At any instant, some surface in space will separate the disturbed from the undisturbed region; this surface will move in time. Denote the time at which the surface passes through a point (x, y, z) by u; this correspondence constructs a functional relationship $u = u(x, y, z)$. Denote the local disturbance speed by c, also a function of (x, y, z); by this we mean that each point on the disturbance surface is acting as an instantaneous source for a local disturbance spreading out from that point with velocity c.

Thus if (ξ, η, ζ) is some point on the disturbance surface at time u, the new disturbance surface after a time du will be the envelope of all spheres of the form

$$(x - \xi)^2 + (y - \eta)^2 + (z - \zeta)^2 = [c(\xi, \eta, \zeta) \cdot du]^2$$

A straightforward envelope calculation (using the constraint that (ξ, η, ζ) lies on the original surface),[†] shows that as $du \to 0$, the vector whose

---

† The basic result is that, for any $(d\xi, d\eta, d\zeta)$ satisfying $u_\xi\, d\xi + u_\eta\, d\eta + u_\zeta\, d\zeta = 0$, we must have $(x - \xi)\, d\xi + (y - \eta)\, d\eta + (z - \zeta)\, d\zeta = 0$, where a higher order term in du^2 has been omitted from the second equation.

components are $(x - \xi, y - \eta, z - \zeta)$ must approach perpendicularity to the disturbance surface. Thus the disturbance surface tends to propagate orthogonally to itself, and the local velocity of this propagation is of course c. These orthogonal trajectories to the sequence of disturbance surfaces (obtained by considering a sequence of different values for u) are termed *rays*.

The direction of the orthogonal trajectory at any point is parallel to that of the gradient vector of the function $u(x, y, z)$. As we move along a particular ray from one position of the disturbance surface to a neighboring position, corresponding to a time change from u to $u + du$ and to a distance increment ds, we therefore have

$$| \mathbf{grad}\ u\ | = |\ du/ds\ | = \frac{1}{c}$$

so that

$$u_x{}^2 + u_y{}^2 + u_z{}^2 = \frac{1}{c^2} \tag{12.30}$$

Equation (12.30) is known as the *eikonal equation*.

Denoting u_x by p, u_y by q, u_z by r, the characteristic strip equations are

$$dx/d\alpha = p, \qquad dy/d\alpha = q, \qquad dz/d\alpha = r$$

$$dp/d\alpha = (-1/c^3)c_x, \qquad dq/d\alpha = (-1/c^3)c_y, \tag{12.31}$$

$$dr/d\alpha = (-1/c^3)c_z \qquad du/d\alpha = p^2 + q^2 + r^2 = 1/c^2$$

where α is a parameter (into which we have absorbed a factor 2). The first three of these equations imply that the characteristic directions are those of the rays; i.e., they are again the orthogonal trajectories to the family of surfaces $u(x, y, z) = $ const.

It is convenient to use s, the distance along a ray, as the parameter instead of α. Since

$$(ds/d\alpha)^2 = (dx/d\alpha)^2 + (dy/d\alpha)^2 + (dz/d\alpha)^2 = p^2 + q^2 + r^2 = 1/c^2$$

we have

$$dx/ds = cp, \qquad dy/ds = cq, \qquad dz/ds = cr$$

so that

$$d^2x/ds^2 = (dc/ds)p + c\ dp/ds = (c_x cp + c_y cq + c_z cr)p - (1/c)c_x$$

and similarly for d^2y/ds^2 and d^2z/ds^2. In index notation, with $x = x_1$, $y = x_2$, $z = x_3$, the resulting ray equations may be written compactly as

$$c\frac{d^2x_i}{ds^2} = \sum_{j=1}^{3} \frac{\partial c}{\partial x_j}\frac{dx_j}{ds}\frac{dx_i}{ds} - \frac{\partial c}{\partial x_i} \tag{12.32}$$

for $i = 1$, 2, or 3. The three equations (12.32) permit the whole path of a ray to be traced out if its initial point and direction are specified, or alternatively, if two points on the ray are given, provided in this latter case that a ray joining the two points does indeed exist.

Consider next two points between which a ray path exists, and let v denote the time required for a disturbance to propagate along this ray path. If we introduce a path parameter τ such that the initial and final points correspond to $\tau = 0$ and $\tau = 1$, then the ray path may be conveninently represented via $x_i = x_i(\tau)$, $i = 1, 2, 3$, and we have

$$v = \int_0^1 \frac{1}{c} A \, d\tau \tag{12.33}$$

where

$$A = \left[\left(\frac{dx_1}{d\tau} \right)^2 + \left(\frac{dx_2}{d\tau} \right)^2 + \left(\frac{dx_3}{d\tau} \right)^2 \right]^{1/2}$$

We now want to compute δv, the change in v, resulting from a slight alteration in the path $\delta x_i(\tau)$, and also, perhaps, a slight alteration in the end point positions $\delta x_i(0)$ and $\delta x_i(1)$. The usual variational calculation (cf. Chapter 10) gives

$$\delta v = \int_0^1 \sum_{i=1}^3 \left[-\frac{A}{c^2} \frac{\partial c}{\partial x_i} - \frac{d}{d\tau} \left\{ \frac{dx_i/d\tau}{cA} \right\} \right] \delta x_i \, d\tau + \left[\sum_{i=1}^3 \frac{dx_i/d\tau}{cA} \delta x_i \right]_0^1 \tag{12.34}$$

If s denotes arc length along the ray path, then $A = ds/d\tau$, and use of Eqs. (12.32) shows at once that the first term on the right-hand side of Eq. (12.34) vanishes. Thus, using again $ds/d\tau = A$ for the ray,

$$\delta v = \left[\sum_{i=1}^3 \frac{1}{c} \frac{dx_i}{ds} \delta x_i \right]_0^1 \tag{12.35}$$

If as a special case we keep the end points fixed, so that $\delta x_i(0) = \delta x_i(1) = 0$ for $i = 1, 2, 3$, then Eq. (12.35) shows that $\delta v = 0$. This is *Fermat's principle*, which characterizes a ray path linking two fixed points as having the property that the traversal time is stationary with respect to small changes from that path.

The function v defined by Eq. (12.33) is a function of the positions of the two points; denote the initial point by (ξ_1, ξ_2, ξ_3) and the final point by (x_1, x_2, x_3). Then $v = v(x_1, x_2, x_3; \xi_1, \xi_2, \xi_3)$, and from Eq. (12.35) we have

$$\frac{\partial v}{\partial x_i} = \frac{1}{c} \frac{dx_i}{ds}, \qquad \frac{\partial v}{\partial \xi_i} = -\frac{1}{c} \frac{dx_i}{ds} \tag{12.36}$$

for $i = 1, 2, 3$, where dx_i/ds is the ith component of the direction vector for the ray path at the point (x_1, x_2, x_3) for the first equation and at the point (ξ_1, ξ_2, ξ_3) for the second equation. From Eqs. (12.36) it follows that

$$(\partial v/\partial x_1)^2 + (\partial v/\partial x_2)^2 + (\partial v/\partial x_3)^2 = [1/c(x_1, x_2, x_3)]^2$$
$$(\partial v/\partial \xi_1)^2 + (\partial v/\partial \xi_2)^2 + (\partial v/\partial \xi_3)^2 = [1/c(\xi_1, \xi_2, \xi_3)]^2$$
(12.37)

so that v satisfies the eikonal equation (12.30). The function v is termed the *eikonal function*, or the *characteristic function*. Again, since it is a complete integral [the three parameters being (ξ_1, ξ_2, ξ_3)], a knowledge of it can lead to a description of the ray path in integrated form. We now turn to this calculation.

12.8   COMPLETE INTEGRAL AND EIKONAL FUNCTION

Equations (12.32) are nonlinear, and to solve them for a given $c(x_1, x_2, x_3)$ function one may well have to resort to numerical methods. There are, however, some cases in which these ray paths are available in analytical form; a particularly interesting situation is that in which a complete integral of Eq. (12.30) can be found (e.g., by separation of variables), for then curves of contact with certain envelope solutions give ray paths, as we shall now see. Our discussion provides a generalization to a higher number of dimensions for the considerations of Section 12.3.

Let $u = \phi(x, y, z, \alpha, \beta, \gamma)$, where α, β, γ are parameters, be a complete integral of Eq. (12.30). Replace one of these parameters by an arbitrarily chosen function of the other two—say $\gamma = \gamma(\alpha, \beta)$. Next, let α and β be determined in terms of x, y, and z from the pair of simultaneous equations

$$0 = \phi_\alpha + \phi_\gamma \gamma_\alpha, \qquad 0 = \phi_\beta + \phi_\gamma \gamma_\beta \qquad (12.38)$$

and substitute (at least in principle) the results into $u = \phi(x, y, z, \alpha, \beta, \gamma(\alpha, \beta))$ so as to obtain $u = \psi(x, y, z)$. [Although the geometrical interpretation is not so easy, this function ψ is the analog of the envelope function obtained in Section 12.3 for the two-dimensional case.] We now claim that $\psi(x, y, z)$ is another solution of Eq. (12.30). To see this, consider any set of values for (x, y, z), and via Eqs. (12.38), the corresponding values for α, β and $\gamma(\alpha, \beta)$. Then at this point,

$$\psi_x = \phi_x + \phi_\alpha \alpha_x + \phi_\beta \beta_x + \phi_\gamma \{\gamma_\alpha \alpha_x + \gamma_\beta \beta_x\}$$

$$= \phi_x + \alpha_x(\phi_\alpha + \phi_\gamma \gamma_\alpha) + \beta_x(\phi_\beta + \phi_\gamma \gamma_\beta)$$

$$= \phi_x$$

by use of Eqs. (12.38). Similarly, $\psi_y = \phi_y$ and $\psi_z = \phi_z$ at this point. Thus, if ϕ satisfies Eq. (12.30) at this point, so does ψ.

We next consider the contact curves between the envelope function $u = \psi(x, y, z)$ and any member of the family of solutions $u = \phi(x, y, z, \alpha, \beta, \gamma(\alpha, \beta))$. Let (x, y, z) be any point of contact; then, as before, the corresponding values of α and β may be found from Eqs. (12.38), and this procedure picks out the particular member of the ϕ family that touches the envelope at that point. Holding α and β [and so $\gamma = \gamma(\alpha, \beta)$] fixed, other contact points between the envelope and our chosen member of the ϕ family are obtained by considering all values of (x, y, z) that satisfy the two simultaneous equations (12.38); the result is to define some curve in (x, y, z) space. Corresponding values of u along this curve are obtainable from $u = \psi(x, y, z)$, or equivalently, from $u = \phi(x, y, z, \alpha, \beta, \gamma(\alpha, \beta))$. Thus the set of contact points is indeed a space curve; it must be a characteristic, since in general ψ will not have the same second derivatives along the curve as the particular member in contact with ψ along this curve.[†]

Moreover, for any particular choices for α, β, and for function $\gamma(\alpha, \beta)$, we can presumably find some set of values of (x, y, z) satisfying Eqs. (12.38); thus α, β, γ, γ_α, γ_β should be reasonably arbitrary parameters for the resulting characteristic curves. Five parameters is an appropriate number, because any characteristic will be defined by its initial point and direction—and this requires the specification of five quantities. Finally, we remark that although we have considered here a special equation—viz., Eq. (12.30)—and only three independent variables, it is clear that the technique may readily be extended to arbitrary first-order equations in any number of variables.

We now illustrate the above considerations by considering the example $c(x, y, z) = (1 + x)^{-1/2}$, where we consider only the region $x > 0$, say. Then Eq. (12.30) becomes

$$u_x^2 + u_y^2 + u_z^2 = 1 + x$$

and separation of variables, [i.e., $u = X(x) + Y(y) + Z(z)$] leads to the complete integral

$$u = \tfrac{2}{3}(x + \alpha)^{3/2} + \beta y + (1 - \alpha - \beta^2)^{1/2}z + \gamma$$

Replacing γ_α and γ_β by the parameters δ and ε, respectively, Eqs. (12.38)

[†] A direct formal verification of the fact that the contact curve satisfies the characteristic differential equations can also be given. See for example Courant and Hilbert (1962, p. 104).

then provide the ray paths

$$(x + \alpha)^{1/2} - z/2(1 - \alpha - \beta^2)^{1/2} + \delta = 0, \quad y - \beta z/(1 - \alpha - \beta^2)^{1/2} + \varepsilon = 0$$

$$(12.39)$$

12.9 HAMILTON–JACOBI EQUATION

In Section 12.8 a set of ordinary differential equations was solved by the use of a complete integral for an associated partial differential equation. We now follow a similar approach for the differential equations of classical mechanics.

Consider a mechanical system with holonomic constraints (i.e., certain functional equations relate position variables to one another or to time) and with external applied forces possessing a potential. Denote the n generalized coordinates by q_i, $i = 1, 2, \ldots, n$, and the Lagrangian (i.e., kinetic minus potential energy) by L. Using a dot to indicate a time derivative, so that $L = L(q_1, \ldots, q_n, \dot{q}_1, \dot{q}_2, \ldots, \dot{q}_n, t)$, Lagrange's equations† are

$$\frac{d}{dt}\left(\frac{\partial L}{\partial \dot{q}_i}\right) - \frac{\partial L}{\partial q_i} = 0, \quad i = 1, 2, \ldots, n \qquad (12.40)$$

This is a set of n second-order equations. An alternative form, in which we deal with $2n$ first-order equations, can be obtained by means of a Legendre transformation. Define the generalized momentum p_i by $p_i = \partial L/\partial \dot{q}_i$, $i = 1, 2, \ldots, n$; conversely, this set of n equations defines each of the $\dot{q}_i$ as a function of the p_j and q_j.

Define now the Hamiltonian $H(p_1, p_2, \ldots, p_n, q_1, q_2, \ldots, q_n, t)$ by the Legendre transformation

$$H = \sum_{j=1}^{n} p_j \dot{q}_j - L(q_1, \ldots, q_n, \dot{q}_1, \ldots, \dot{q}_n, t) \qquad (12.41)$$

where all $\dot{q}_j$ on the right-hand side are considered functions of the p_j and q_j. Then

$$dH = \sum_{j=1}^{n}\left(p_j\, d\dot{q}_j + dp_j \dot{q}_j - \frac{\partial L}{\partial q_j}\, dq_j - \frac{\partial L}{\partial \dot{q}_j}\, d\dot{q}_j\right) - \frac{\partial L}{\partial t}\, dt$$

† An excellent discussion of the Lagrangian–Hamiltonian formulation of mechanics will be found in Goldstein (1950).

so that, using $p_j = \partial L/\partial \dot{q}_j$ and Eqs. (12.40),

$$dH = \sum_{j=1}^{n} \left(\dot{q}_j \, dp_j - \dot{p}_j \, dq_j \right) - \frac{\partial L}{\partial t} \, dt$$

It follows that

$$\dot{q}_j = \partial H/\partial p_j, \qquad \dot{p}_j = -\partial H/\partial q_j, \qquad j = 1, 2, \ldots, n \qquad (12.42)$$

which is a set of $2n$ first-order equations.

We now ask if there is a simple first-order partial differential equation for which Eqs. (12.42) represent characteristics? If so, we can hope to solve Eqs. (12.42) by means of the complete integral approach. The answer is yes; a suitable equation, for an unknown function $J(q_1, q_2, \ldots, q_n, t)$, is given by

$$\frac{\partial J}{\partial t} + H \left(\frac{\partial J}{\partial q_1}, \frac{\partial J}{\partial q_2}, \ldots, \frac{\partial J}{\partial q_n}, q_1, q_2, \ldots, q_n, t \right) = 0 \qquad (12.43)$$

known as the *Hamilton–Jacobi equation*. The first of the characteristic equations, in terms of a parameter s, is simply $dt/ds = 1$, so that we can equivalently use t as a parameter, and the other equations become (denoting $\partial J/\partial q_i$ by p_i)

$$dq_i/dt = \partial H/\partial p_i, \qquad dp_i/dt = -\partial H/\partial q_i \qquad (12.44)$$

for $i = 1, 2, \ldots, n$. We also have, of course, an equation for dJ/dt, viz.,

$$\frac{dJ}{dt} = -H + \sum_{j=1}^{n} p_j \frac{\partial H}{\partial p_j} \qquad (12.45)$$

but the equations of primary interest are Eqs. (12.44), which are indeed identical in form to Eqs. (12.42).

A standard example of the complete integral approach is that of central body motion. Let a mass M be fixed at the origin, and let a mass m move in the (x, y) plane under the gravitational field of the first mass. Within a multiplicative constant, the Hamiltonian has the form

$$H = \tfrac{1}{2}(p^2 + q^2) - k^2/(x^2 + y^2)^{1/2}$$

where the constant k^2 involves M and the gravitational constant. The equations of motion are $\dot{x} = H_p, \dot{y} = H_q, \dot{p} = -H_x, \dot{q} = -H_y$; to integrate them, we seek a complete integral of

$$\frac{\partial J}{\partial t} + \tfrac{1}{2}(J_x{}^2 + J_y{}^2) - k^2/(x^2 + y^2)^{1/2} = 0$$

If we change to polar coordinates $x = r \cos \theta$ and $y = r \sin \theta$, then separation of variables is effective. We obtain

$$J_t + \tfrac{1}{2}[J_r^2 + (1/r^2)J_\theta^2] - (k^2/r) = 0$$

which has the complete integral

$$J = \alpha t + \beta \theta + \int_{r_0}^r \left(2\frac{k^2}{\rho} - \frac{\beta^2}{\rho^2} - 2\alpha\right)^{1/2} d\rho + \gamma$$

By envelope formation, as in Section 12.8, the trajectories are easily seen to be the usual conic sections.

12.10   PROBLEMS

12.10.1   Let $u = \phi(x_1, x_2, \ldots, x_n, \alpha_1, \ldots, \alpha_n)$, where the α_j are parameters, be a complete integral of

$$F(x_1, x_2, \ldots, x_n, u, p_1, p_2, \ldots, p_n) = 0$$

where $p_j = \partial u/\partial x_j$, $j = 1, 2, \ldots, n$. Obtain the equations of envelope formation analogous to Eqs. (12.38), with $\alpha_n = \gamma(\alpha_1, \alpha_2, \ldots, \alpha_{n-1})$. Show that the contact curves between the envelope and the members of the generating family of solutions have $(2n - 1)$ parameters, and explain why this is an appropriate number. [*Hint:* any characteristic can be characterized by the values of $x_2, \ldots, x_n$ at which it cuts the $x_1 = 0$ plane and by the values of $u, p_1, \ldots, p_n$ there; moreover, one always has $F = 0$).

12.10.2   Consider a hypothetical motion of a mechanical system, corresponding to $q_i = q_i(\tau)$, $i = 1, 2, \ldots, n$, and $t = t(\tau)$, where τ is a parameter ranging from zero to one. Define

$$K = \int_0^1 L\left(q_i(\tau), \frac{q_i'(\tau)}{t'(\tau)}, t(\tau)\right) \cdot t'(\tau)\, d\tau$$

where a prime indicates differentiation, and where $L(q_i, \dot{q}_i, t)$ is the Lagrangian. Compute δK, corresponding to $\delta q_i(\tau)$, $\delta t(\tau)$, where the end points $q_i(0)$ and $q_i(1)$, as well as the end times $t(0)$ and $t(1)$, may vary.

Show that the true motion of a mechanical system between two specified configurations, where the initial and final times are also specified, makes the time integral of the Lagrangian stationary with respect to small alterations in the path.

Show also that if an initial configuration and time are specified, and if we define I as the time integral of the Lagrangian from this initial state to any accessible final state (i.e., final values of the q_i, and final time t specified), then $I(q_i, t)$ satisfies Eq. (12.43). Compare with the results of Section 12.8.

12.10.3 Let $F(x, y, u, p, q) = 0$ be a first-order partial differential equation, with $p = u_x$ and $q = u_y$. Show that if a function $v(x, y, u)$ is a solution of

$$F(x, y, u, -(v_x/v_u), -(v_y/v_u)) = 0$$

then the function $u(x, y)$ defined by $\{v(x, y, u) = \text{const}\}$ is a solution of $F(x, y, u, p, q) = 0$. Explain how this idea permits us to replace any first-order equation with a new equation involving one more independent variable, in which the dependent variable does not occur explicitly; explain also why this new equation takes the form of a Hamilton–Jacobi equation when we think of one of the partial derivatives as being expressed in terms of the remaining quantities.

13

MORE ON CHARACTERISTICS

13.1 DISCONTINUITIES—A PRELIMINARY EXAMPLE

In Chapter 12, the method of characteristics was used to obtain a solution for a first-order equation in some region adjoining an initial curve. The method will provide contradictory information at points at which characteristics intersect one another, so we cannot extend a continuous solution beyond such points. However, a variety of contexts lead to problems in which discontinuous solutions are possible, provided certain constraints are imposed.

To illustrate, let $u(x, y)$ satisfy

$$u_t + uu_x = 0 \tag{13.1}$$

for $t > 0$, $-\infty < x < \infty$, where $u(x, 0) = \phi(x)$, with

$$\phi(x) = \begin{cases} \frac{1}{2}, & \text{for } x < 0 \\ 5/4 - [(5/4)^2 - (1 - x)^2]^{1/2}, & \text{for } 0 < x < 1 \\ 0, & \text{for } x > 1 \end{cases} \tag{13.2}$$

The characteristics are sketched in Fig. 13.1. They have an envelope (also sketched), whose equation in the (x, t) plane is given by

$$x = 1 + (5/4)t - (5/4)(1 + t^2)^{1/2}, \qquad t > \tfrac{3}{4}$$

Thus for $t < \tfrac{3}{4}$, there is no difficulty, and the solution is given implicitly by $u = \phi(x - ut)$, where $\phi(x)$ is defined in Eq. (13.2). For $t > \tfrac{3}{4}$, however, no continuous solution exists. For example, the point $x = \tfrac{1}{2}$, $t = 1$ lies

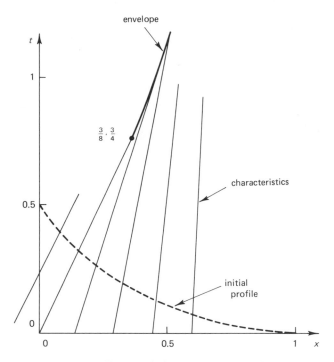

Fig. 13.1   Characteristics for Eqs. 13.1 and 13.2.

on each of the characteristics $x = \frac{1}{2}t$, $x = \frac{1}{4} + \frac{1}{4}t$; along these charac-teristics $u = \frac{1}{2}$ and $u = \frac{1}{4}$, respectively, so that inconsistent values of u would be obtained for this intersection point.

Let us suppose, however, that the physical context that led to Eq. (13.1) permits discontinuities, or "shocks." For uniqueness, some additional in-formation is usually necessary, and here we will postulate that if $x = X(t)$ is a curve across which $u(x, t)$ is discontinuous, then†

$$(u_1 - u_2) \, dX/dt + \tfrac{1}{2}(u_2{}^2 - u_1{}^2) = 0 \qquad (13.3)$$

where u_1 and u_2 are the limiting values of u on the two sides of the dis-

---

† We note that Eq. (13.1) may be written in the "conservation form" $u_t + \frac{1}{2}(u^2)_x = 0$. Integration between the two fixed points $x = A$ and $x = B$ yields

$$\frac{\partial}{\partial t} \int_A^B u \, dx + \tfrac{1}{2}[u(B, t)]^2 - \tfrac{1}{2}[u(A, t)]^2 = 0$$

and Eq. (13.3) is equivalent to the assumption that this integrated form of the con-servation law continues to hold even if there is a moving discontinuity (cf. Fig. 13.2) between the points A and B. Of course whether or not this is a reasonable requirement will depend on the actual physical problem.

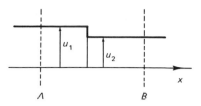

Fig. 13.2 Moving discontinuity.

continuity curve, defined by (for $\varepsilon > 0$)

$$u_1 = \lim_{\varepsilon \to 0} u[X(t) - \varepsilon, t], \qquad u_2 = \lim_{\varepsilon \to 0} u[X(t) + \varepsilon, t]$$

Returning now to the characteristics diagram of Fig. 13.1, we try to extend the solution into the $t > \frac{3}{4}$ region by permiting a curve of discontinuity to emanate from the point $x = \frac{3}{8}$, $t = \frac{3}{4}$. From Eq. (13.3), the slope $V = dX/dt$ of this curve of discontinuity will satisfy

$$V = \tfrac{1}{2}(u_1 + u_2)$$

and therefore will equal the average of the local values of u on the two sides of this curve. To find the curve itself, let each point (x, t) on it be the intersection point for two characteristics—one emanating from the point $\xi_1 < 0$, and the other emanating from $\xi_2 > 0$. Then, for $t > \frac{3}{4}$, we have

$$x = \xi_1 + \tfrac{1}{2}t$$

$$x = \begin{cases} \xi_2 + \{5/4 - [(5/4)^2 - (1 - \xi_2)^2]^{1/2}\}t, & \text{for } 0 < \xi_2 < 1 \\ \xi_2, & \text{for } \xi_2 > 1 \end{cases}$$

(13.4)

$$\frac{dx}{dt} = \frac{1}{2}\left[\frac{1}{2} + \begin{cases} 5/4 - [(5/4)^2 - (1 - \xi_2)^2]^{1/2}, & \text{for } 0 < \xi_2 < 1 \\ 0, & \text{for } \xi_2 > 1 \end{cases}\right]$$

from which the discontinuity curve may be determined.† This curve lies to the right of the envelope curve in Fig. 13.1 and for $x > 1$ has the slope $dx/dt = \frac{1}{4}$, so that the whole $x > 0$ region eventually experiences a nonzero value of u. On either side of the discontinuity curve the solution is found by the previous characteristics approach, where of course no characteristic is permitted to cross the curve (13.4).

As an alternative to the discontinuity condition (13.3), it might happen

---

† The last of Eqs. (13.4) gives dx/dt in terms of the parameter ξ_2, and the second gives ξ_2 in terms of x and t. The combination of the two equations provides a differential equation for dx/dt in terms of x and t.

that Eq. (13.1) is in reality an approximation to a more accurate equation, such as

$$u_t + uu_x = \varepsilon u_{xx} \tag{13.5}$$

where $\varepsilon > 0$ is a small parameter.† Equation (13.1) would then result from the neglect of the term εu_{xx}, but this neglect would no longer be justified if u_{xx} were large—as would happen in the neighborhood of intersecting characteristics. We will see in Section 13.3 that the effect of the term εu_{xx} is to make the solution of Eq. (13.5) unique and to obviate discontinuities; as $\varepsilon \to 0$, the solution turns out (in this special case) to approach that described in connection with Eqs. (13.4).

In general, different constraints associated with possible discontinuities will yield different results, and it is essential in the formulation of a scientific problem that the appropriate choice be made.

13.2 WEAK SOLUTIONS

The solution we have just obtained for $u(x, t)$, which is discontinuous across the curve defined by Eqs. (13.4) but which satisfies a conservation law everywhere, is termed a *weak solution* of Eq. (13.1). Wherever it is adequately differentiable—i.e., everywhere except on the discontinuity curve—it satisfies Eq. (13.1) itself.

It is conventional to express the condition of Eq. (13.3) in somewhat different form. Returning to Eq. (13.1), let $w(x, t)$ be any continuously differentiable function, and let R be any region in the (x, t) plane in which Eq. (13.1) is valid. Then

$$\int_R w(u_t + uu_x) \, dx \, dt = 0$$

i.e.,

$$\int_R w[u_t + (\tfrac{1}{2}u^2)_x] \, dx \, dt = 0$$

Impose the further constraint on w that it vanish on the contour Γ of the region R. Then an application of the divergence theorem to the last integral

---

† Because of the interpretation in an analogous equation in fluid mechanics, the term εu_{xx} is often said to represent a "viscosity" term. Sometimes such a term is introduced quite arbitrarily in order to smooth out possible discontinuities.

yields

$$\int_R \left[uw_t + \tfrac{1}{2}u^2 w_x \right] dA = 0 \tag{13.6}$$

where $dA = dx\, dt$ is the element of area. In this form, derivatives of u no longer appear. We now define, formally, a weak solution of Eq. (13.1) to be one that is piecewise continuous and for which Eq. (13.6) is valid for any continuously differentiable function w vanishing on Γ. If as a special case u is differentiable, then the above calculation may be reversed to show that u must satisfy Eq. (13.1). We can think of Eq. (13.6) as being an equivalent form of Eq. (13.1) that remains valid for situations in which u becomes discontinuous.

More generally, if a given equation has the "conservation form"

$$u_t + (\partial/\partial x)\, f(x, t, u) + g(x, t, u) = 0 \tag{13.7}$$

where f and g are prescribed functions, then a weak solution of this equation is defined to be one for which (in any region R in the domain of interest)

$$\int_R \left[w_t u + w_x f - gw \right] dA = 0 \tag{13.8}$$

for any continuously differentiable function w vanishing on the boundary of R. Still more generally, u, f, and g could be vector functions (cf. Problem 13.4.2).

Suppose now that u has a discontinuity across some curve Γ in the (x, t) plane, but that on either side of that curve Eq. (13.7) is satisfied. Moreover, let u be a weak solution, so that the condition of Eq. (13.8) is satisfied for any suitable function w. If we choose our region R so as to contain a portion of the discontinuity curve, as shown in Fig. 13.3, then we may remove from the region of integration a subregion of infinitesimal

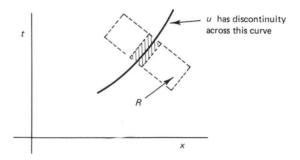

Fig. 13.3 Discontinuity curve.

size that includes the discontinuity curve (shaded in the figure), so as to replace R by two new regions of integration within each of which u is differentiable. A straightforward use of the divergence theorem [using the fact that u satisfies Eq. (13.7) except on Γ] shows that Eq. (13.8) reduces to the statement that the integral along that portion of the discontinuity curve contained within R satisfies

$$\int w[(u_1 - u_2)\, dx - (f_1 - f_2)\, dt] = 0$$

where the subscripts 1 and 2 refer to values on the two sides of the discontinuity curve. Since this result must hold for any suitable function w, we obtain the discontinuity condition

$$(dx/dt)(u_1 - u_2) = f_1 - f_2 \qquad (13.9)$$

where dx/dt is the slope of the discontinuity curve. Equation (13.3) is now seen to be compatible with this result for the special case $f = \frac{1}{2}u^2$.

In generalizing the idea of a solution of Eq. (13.7) in this manner we may expect to lose something, and one such loss is that we can no longer necessarily expect the solution to be unique. This possible loss of uniqueness can show up in several ways. First of all, there may be more than one conservation form into which the given equation can be put. For example, the change is variable $u = v^2$ in Eq. (13.1) yields the new conservation law $v_t + (\frac{1}{3}v^3)_x = 0$, and the reader may verify that the resulting discontinuity condition for v, when rewritten in terms of u, does not agree with Eq. (13.3). [How would Eqs. (13.4) alter?] The choice as to which conservation law is the right one to use must be made on physical grounds. Second, there may be more than one solution even if the form of the conservation law is fixed. For example, either of the functions

$$u^{(1)} = \begin{cases} 0, & \text{for } x/t < 1/2 \\ 1, & \text{for } x/t > 1/2 \end{cases}$$

$$u^{(2)} = \begin{cases} 0, & \text{for } x < 0 \\ x/t, & \text{for } 0 < x/t < 1 \\ 1, & \text{for } x/t > 1 \end{cases}$$

is a weak solution of Eq. (13.1), satisfying Eq. (13.3) with the initial condition $u(x, 0) = 0$ for $x < 0$, and $u(x, 0) = 1$ for $x > 0$. Again this kind of ambiguity can be resolved only by the use of more physical information. A well-known example is afforded by the Rankine–Hugoniot jump conditions across a shock wave in compressible flow, where one of two possible solutions must be discarded because it would correspond to a decrease in entropy.

A discussion of conservation laws, and uniqueness of shock conditions, in general hyperbolic systems will be found in Lax (1973).

13.3 BURGERS' EQUATION

Equation (13.5), studied by Burgers as a one-dimensional analog of viscous compressible flow, is of sufficient interest that we will discuss it further. An exact solution of it, valid for any ε, was found by Hopf and Cole. Rewrite it as

$$u_t = (\varepsilon u_x - \tfrac{1}{2}u^2)_x$$

which implies the existence of a function $Q(x, t)$ such that $Q_x = u$, $Q_t = \varepsilon u_x - \tfrac{1}{2}u^2$. Using the first of these equations in the second, we obtain

$$Q_t = \varepsilon Q_{xx} - \tfrac{1}{2}Q_x^2$$

In terms of

$$\psi(x, t) = \exp\left[-\frac{1}{2\varepsilon} Q(x, t)\right] \qquad (13.10)$$

this equation becomes

$$\psi_t = \varepsilon\psi_{xx} \qquad (13.11)$$

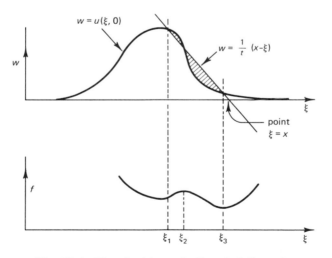

Fig. 13.4 Plot of $w(\xi) = u(\xi, 0)$ and of $f(x, t, \xi)$.

which is the familiar diffusion equation. Note that the definition of ψ as an exponential function means that it is always nonzero; once ψ is known, we obtain u from $u = -2\varepsilon\psi_x/\psi$.

As an example of the use of this result, let $-\infty < x < \infty$, and let $u(x, 0)$ be prescribed, with $u(\pm\infty, 0) = 0$. One such function $u(x, 0)$ is depicted in Fig. 13.4. We can write

$$\psi(x, 0) = \exp\left[-\frac{1}{2\varepsilon}\int_{-\infty}^{x} u(\eta, 0)\, d\eta\right] \tag{13.12}$$

From Eq. (2.15), the solution of Eq. (13.11) now has the form

$$\psi(x, t) = \frac{1}{2(\pi t\varepsilon)^{1/2}}\int_{-\infty}^{\infty} \exp\left[-\frac{1}{2\varepsilon}\int_{-\infty}^{\xi} u(\eta, 0)\, d\eta - \frac{1}{4t\varepsilon}(x - \xi)^2\right] d\xi$$

$$= \frac{1}{2(\pi t\varepsilon)^{1/2}}\int_{-\infty}^{\infty} \exp\left[-\frac{1}{\varepsilon}f(x, t, \xi)\right] d\xi \tag{13.13}$$

where this last equation defines the function $f(x, t, \xi)$. Thus

$$u(x, t) = \frac{\displaystyle\int_{-\infty}^{\infty} \{(x - \xi)/t\} \exp[-(1/\varepsilon) f(x, t, \xi)]\, d\xi}{\displaystyle\int_{-\infty}^{\infty} \exp[-(1/\varepsilon) f(x, t, \xi)]\, d\xi}$$

$$= \frac{\displaystyle\int_{-\infty}^{\infty} u(\xi, 0) \exp[-(1/\varepsilon) f(x, t, \xi)]\, d\xi}{\displaystyle\int_{-\infty}^{\infty} \exp[-(1/\varepsilon) f(x, t, \xi)]\, d\xi} \tag{13.14}$$

following an integration by parts.

An easily interpreted approximation to this expression for small positive values of ε can be obtained as follows. Note that

$$\frac{\partial f}{\partial \xi} = \frac{1}{2} u(\xi, 0) - \frac{1}{2t}(x - \xi)$$

so that a plot of the function f would look something like Fig. 13.4. As $\varepsilon \to 0$, only that part of the integration region near ξ_3 (or near ξ_1, whichever corresponds to the least positive value of f) is of importance,[†] and we obtain

$$u(x, t) \sim u(\xi_3, 0) \qquad \text{for} \quad t < T$$

$$u(x, t) \sim u(\xi_1, 0) \qquad \text{for} \quad t > T$$

where the time T is that value of t for which the two shaded regions in Fig. 13.4 have the same area. At this time, there is a rapid transition in the value of u from $u(\xi_3, 0)$ to $u(\xi_1, 0)$ corresponding to the passage of a shock wave. It is clear that the maximum value of $u(x, t)$ never exceeds the maximum height of the $u(\xi, 0)$ curve, and that the amplitude of the shock wave gradually decreases. The reader may also show that the asymptotic speed of the shock wave, obtained by differentiating[‡] the condition

$$\int_{\xi_1}^{\xi_3} \left[u(\xi, 0) - \frac{1}{t} (x - \xi) \right] d\xi = 0$$

with respect to t, is given by $v = \frac{1}{2}[u(\xi_1, 0) + u(\xi_3, 0)]$, which justifies the remark made in the sequel to Eq. (13.5). In Problem 13.4.3 the reader is asked to find an asymptotic thickness for the shock wave.

Thus as $\varepsilon \to 0$ in Burgers' equation, we obtain again the "shockfitting" condition of Section 13.1. However, we emphasize once more that there are many equations for which Eq. (13.1) is a limiting form, and the discontinuity conditions obtainable from these in the limit would not necessarily agree with that obtained in this section. The appropriate choice of equation has to be made on physical grounds.

[†] If a function $g(\xi)$ has a minimum at ξ_0, then as $\varepsilon \to 0$, only the immediate neighborhood of the point $\xi = \xi_0$ can be of importance in the integration, so

$$\int_{-\infty}^{\infty} \exp\left[-\frac{1}{\varepsilon} g(\xi) \right] h(\xi) \, d\xi \sim \int_{-\infty}^{\infty} \exp\left(-\frac{1}{\varepsilon} \left[g(\xi_0) + \frac{1}{2} g''(\xi_0) \cdot (\xi - \xi_0)^2 \right] \right) h(\xi_0) \, d\xi$$

$$\sim h(\xi_0) \cdot \exp\left[-\frac{1}{\varepsilon} g(\xi_0) \right] \left[\frac{2\pi\varepsilon}{g''(\xi_0)} \right]^{1/2}$$

This calculation represents an application of the *saddle point method* to the special case of real variables; the name arises from the geometry of the more general case of a complex integration path. Higher-order terms may also be obtained. See Carrier *et al.* (1966, p. 257).

[‡] Note that the integrand vanishes at each end of the interval of integration, so that the derivative of the left-hand side is simply

$$\int_{\xi_1}^{\xi_3} \left[\frac{1}{t^2} (x - \xi) - \frac{V}{t} \right] d\xi, \qquad \text{where } V = \frac{dx}{dt}$$

Burgers' equation is one of the few authentically nonlinear PDEs for which an exact solution is possible. Moreover, it models certain fluid mechanics equations, possesses solutions representing propagating (and deforming) waves, and provides a good test case for the interaction of characteristics and shocks; as a result, it has been much studied. Another rather similar equation that has received considerable attention is that of Kortweg and de Vries (cf. Problems 13.4.6–13.4.8). A detailed discussion of these and similar equations will be found in Whitham (1974); Leibovich and Seebass (1974); and Lick (1974).

13.4 PROBLEMS

13.4.1 With reference to Eq. (13.1), let

$$
u(x, 0) = \begin{cases} 1, & \text{for } x < 0 \\ \cos^2(\pi x/2), & \text{for } 0 < x < 1 \\ 0, & \text{for } x > 1 \end{cases}
$$

Draw the characteristics and obtain their envelope (both branches). Find a weak solution valid for all time. Find the shock path in the (x, t) plane in as simple a form as you can.

13.4.2 Let p, ρ, T, u denote the pressure, density, temperature, and velocity, respectively, for an ideal compressible gas in one-dimensional flow in the x direction. From Section 7.1, we have

$$
\rho_t + (\rho u)_x = 0, \qquad \rho(u_t + u_x u) = -p_x
$$

In a similar way, the formulation of energy conservation for a moving volume leads to

$$
c_V \rho (T_t + u T_x) + p u_x = 0
$$

where c_V is the specific heat at constant volume, assumed constant (we take the internal energy as $c_V T$ per unit mass). The gas law requires $p = \rho R T$, where R is the gas constant. Show that the first three of these equations can be rewritten in conservation form as

$$
\rho_t + (\rho u)_x = 0
$$

$$
(\rho u)_t + (\rho u^2 + p)_x = 0
$$

$$
(c_V \rho T + \tfrac{1}{2}\rho u^2)_t + (c_V \rho u T + \tfrac{1}{2}\rho u^3 + p u)_x = 0
$$

and obtain the conditions that a weak solution must satisfy across a discontinuity. If the slope dx/dt of the discontinuity curve in the (x, t) plane is V, define $u_1' = u_1 - V$ and $u_2' = u_2 - V$, where u_1 and u_2 are the values of u on the two sides of the discontinuity curve, and show that the discontinuity conditions become particularly simple when written in terms of u_1' and u_2'.

13.4.3 Show that the asymptotic thickness (as $\varepsilon \to 0$) of the shock wave discussed in connection with Eq. (13.14) is of order $2\varepsilon/[u(\xi_1, 0) - u(\xi_3, 0)]$. [*Hint:* consider that value of x for which $f(x, t, \xi_1) = f(x, t, \xi_3)$, and write the asymptotic expression for $u(x, t)$. Observe next that a change in x, such that the change in $(f_3 - f_1)$ is of $O(2\varepsilon)$, will correspond to the shock wave thickness.]

13.4.4 Let $u(x, t)$ satisfy Eq. (13.1), with

$$u(x, 0) = \begin{cases} 0 & \text{for} \quad x < -1 \\ U & \text{for} \quad -1 < x < 0 \\ 0 & \text{for} \quad x > 0 \end{cases}$$

where $U > 0$ is a given constant. Sketch the characteristics, noting that a fan emanates from the point $(-1, 0)$ in the (x, t) plane. Use the condition of Eq. (13.3) to find the shock path. Next, use Eq. (13.5) and the steepest-descent method of Section 13.3 to evaluate u near the shock path for small $\varepsilon > 0$.

13.4.5 Multiply the numerator and denominator of the right-hand side of Eq. (13.14) by $\exp[(1/2\varepsilon) \int_{-\infty}^{\infty} u(\eta, 0) \, d\eta]$ so as to alter the form of the equation, and show that for the case

$$u(\eta, 0) = \begin{cases} U = \text{const}, & \text{for} \quad \eta < 0 \\ 0, & \text{for} \quad \eta > 0 \end{cases}$$

the solution of Eq. (13.5) is given by

$$\frac{U}{u} = 1 + \exp\left[\frac{U}{2\varepsilon}\left(x - \frac{1}{2} Ut\right)\right] \frac{\text{erfc}[-x/(2\sqrt{\varepsilon t})]}{\text{erfc}[(x - Ut)/(2\sqrt{\varepsilon t})]}$$

Discuss the nature of this solution for small ε.

13.4.6 A nonlinear equation that arises in shallow-water theory is the *Korteweg–de Vries equation:*

$$u_t + (\alpha + \varepsilon u) u_x + \beta u_{xxx} = 0 \tag{13.15}$$

where $\alpha, \beta, \varepsilon$ are constants.

If we are interested in a traveling-wave solution of this equation, we can write $u = f(\xi)$, where $\xi = x - Vt$, and where V is some appropriate constant; substitution into Eq. (13.15) shows that f must satisfy

$$(\alpha - V) f + \tfrac{1}{2}\varepsilon f^2 + \beta f'' = C$$

where C is a constant of integration. Suppose now that ε is small, and consider a perturbation approach to this equation in which the lowest-order contribution to f is a sinusoidal wave. Specifically, write

$$f = A \sin k\xi + \varepsilon f_1 + \varepsilon^2 f_2 + \cdots$$
$$V = V_0 + \varepsilon V_1 + \varepsilon^2 V_2 + \cdots$$
$$C = C_0 + \varepsilon C_1 + \varepsilon^2 C_2 + \cdots$$

where $A, k, V_0, V_1, \ldots, C_0, C_1, \ldots$ are constants, and where $f_1, f_2, \ldots$ are functions of ξ. Show that the zeroth-order calculation leads to $V_0 = \alpha - \beta k^2$ and to $C_0 = 0$; show next that the first-order calculation gives

$$\beta k^2 f_1 + \beta f_1'' = C_1 - (A^2/4) + (A^2/4) \cos 2k\xi + V_1 A \sin k\xi$$

whence we are led to choose $C_1 = A^2/4$ (we postulate that the average value of each f_i is to be zero), and also $V_1 = 0$ (to avoid a term in f_1 whose magnitude is proportional to t).

Compute f_1 and continue the calculation to show that

$$f = A \sin k\xi - (A^2\varepsilon/12\beta k^2) \cos 2k\xi + \cdots$$
$$V = (\alpha - \beta k^2) + \varepsilon^2(A^2/24\beta k^2) + \cdots$$

Thus the velocity with which the wave form moves has a correction term proportional to the square of the amplitude A of the wave.

Finally, suppose the velocity V had not been expanded in powers of ε as above, but simply fixed at its value V_0. Obtain the resulting new expansion for f, and explain why it is not as useful as that obtained above.

The idea of expanding both the dependent variable and the wave velocity in a perturbation series seems to have been first used by Stokes.† This technique is a forerunner of the method of multiple time scales; compare Section (15.9) and Problem 15.10.6.

13.4.7   In Problem 13.4.6 the equation $(\alpha - V)f + \tfrac{1}{2}\varepsilon f^2 + \beta f'' = C$ was obtained for a traveling-wave solution of Eq. (13.15), where $u = f(\xi)$, $\xi = x - Vt$. This equation for f can be integrated exactly to obtain (at least implicitly) f in terms of ξ; do so, and show in particular that one solution (the solitary wave, or *soliton*) is given by

$$u = A \operatorname{sech}^2[(\varepsilon A/12\beta)^{1/2}(x - Vt)], \qquad V = \alpha + \tfrac{1}{3}\varepsilon A$$

† Stokes (1847, p. 441).

Despite the nonlinear form of Eq. (13.15), it turns out that two such solitary waves can propagate through one another and emerge unaltered in shape.†

13.4.8 The equations of shallow-water theory derived in Problem 3.2.4 read

$$u_t + uu_x + gw_x = 0, \qquad w_t + [u(w + d)]_x = 0$$

where u is the horizontal velocity (taken as independent of vertical position), w is the wave height, d is the depth, and g is the acceleration of gravity.

The second of these equations is already in the form of a conservation law. Obtain another conservation law by multiplying the first equation by $(w + d)$, the second by u, and adding. Interpret these two laws in terms of conservation of matter and momentum, respectively. Follow the procedure of Section 13.1 to obtain the conditions governing a discontinuity in u and w (called a *hydraulic jump*). [*Note:* In considering the jump, one can set $d_x = 0$.]

13.5 A COMPRESSIBLE FLOW PROBLEM

As an instructive application of the method of characteristics, we turn now to a situation involving a pair of coupled first-order equations in which the independent variables are both spacelike. Let $u(x, y)$ and $v(x, y)$ denote the velocity components in the x and y directions, respectively, of a two-dimensional steady flow of an ideal compressible fluid of density $\rho(x, y)$. Denote the stream velocity by $q(x, y)$, given by $q^2 = u^2 + v^2$.

From Section 7.1, the continuity equation reads

$$(\rho u)_x + (\rho v)_y = 0 \tag{13.16}$$

We consider the case in which the flow has emanated from a reservoir in which the fluid is either at rest or in a state of uniform motion, so that by Problem 7.2.9 the flow is always irrotational; thus

$$u_y - v_x = 0 \tag{13.17}$$

Moreover, ρ will be some function of q. Specifically, we have (cf. Problem 7.2.9)

$$\tfrac{1}{2}q^2 + c^2/(\gamma - 1) = B, \qquad c^2 = A\gamma\rho^{\gamma-1} \tag{13.18}$$

---

† A discussion of this and other properties of solutions of the Kortweg de Vries equation may be found in Jeffrey and Kakutani (1972, p. 582).

where c is the local velocity of sound, γ the (constant) ratio of specific heats, and A and B constants [the constant A is the same as that in Eq. (7.5)]. A consequence of Eqs. (13.18) is that $d\rho/dq = -q\rho/c^2$.

Our problem now is to analyze the properties of Eqs. (13.16) and (13.17), where ρ is that function of q defined by Eqs. (13.18). It is reasonable to ask if the method of characteristics, as exploited for single nonlinear equations in Chapter 12, can be applied to a coupled nonlinear pair such as this. We might begin by asking again whether there are curves in the (x, y) plane across which there could be discontinuities in slope, and the reader is in fact asked to follow this direct approach in Problem 13.7.1. It turns out, however, that the discussion can be simplified by means of a preliminary transformation based on a physical visualization of the problem.

At any chosen point P in the flow, let the stream velocity q make an angle θ with the x axis (Fig. 13.5). Locally, let s and n denote distances from P as measured in the streamline and normal directions, respectively. Then Eqs. (13.16) and (13.17) become, in these intrinsic coordinates,[†]

$$q_s + q\theta_n + (q/\rho)\rho_s = 0, \qquad q_n - q\theta_s = 0$$

Using the relation $d\rho/dq = -q\rho/c^2$ mentioned above, we obtain

$$(1 - M^2)q_s + q\theta_n = 0, \qquad q_n - q\theta_s = 0 \qquad (13.19)$$

where the local *Mach number* M is defined by $M = q/c$. (The flow is termed locally supersonic if $M > 1$, and subsonic if $M < 1$.)

If the curve Γ defined by $\xi(x, y) = 0$ is one across which there can be a discontinuity in slope, then in terms of a new (ξ, η) coordinate system [formed from the curves $\xi = $ const, $\eta = $ const, where $\eta(x, y)$ is some function whose contour curves are not parallel to those of $\xi(x, y)$], Eqs.

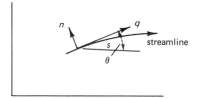

Fig. 13.5 Intrinsic coordinates.

[†] An easy way in which to derive these equations is to rewrite Eqs. (13.16) and (13.17) in terms of a local (x', y') coordinate system with origin at P and with the x'- and y'-axes tangent to the s and n directions at P, respectively. Then, for example, $u' = q \cos \theta'$, and $\partial u'/\partial y' = \partial(q \cos \theta')/\partial y' = \partial(q \cos \theta')/\partial n = q_n \cos \theta' - q \theta_n' \sin \theta' = q_n$, since $\theta' = 0$ at P.

(13.19) become

$$(1 - M^2)q_\xi\xi_s + q\theta_\xi\xi_n + \text{other terms} = 0$$
$$q_\xi\xi_n - q\theta_\xi\xi_s + \text{other terms} = 0$$

(13.20)

If either or both of q_ξ, θ_ξ can be discontinuous across Γ, with all other quantities continuous, then the coefficient determinant must vanish, so that

$$(1 - M^2)\xi_s{}^2 + \xi_n{}^2 = 0$$

(13.21)

Thus real characteristics exist only for $M > 1$, and they make an angle $(+\alpha)$ or $(-\alpha)$ with the streamline direction, where $\sin \alpha = 1/M$. There are two families of characteristics. We will designate generic curves belonging to the two families by C_+ and C_-, as sketched in Fig. 13.6. Thus a C_+ characteristic direction is obtained by rotating the vector $\mathbf{q}$ counterclockwise through the acute angle $(+\alpha)$. Because these characteristics are curves along which different solutions of Eqs. (13.16) and (13.17) may be melded together (with q and θ continuous across the curves), they sometimes appear in photographs of supersonic flow patterns, where optical techniques are used to make density gradients visible. In such cases, the characteristics—called Mach lines—often emanate from minute irregularities on wind tunnel surfaces. [Of course, stronger discontinuities, or shocks, in which there are discontinuities in q or θ, need not (and will not) coincide with characteristic curves.]

We might expect that Eqs. (13.19) will lead to particularly simple formulas for rates of change along a characteristic direction. In the direction of a C_+ characteristic, we have in fact

$$dq = q_s \cos \alpha + q_n \sin \alpha, \qquad d\theta = \theta_s \cos \alpha + \theta_n \sin \alpha$$

and by use of Eqs. (13.19) we find that

$$d\theta = (M^2 - 1)^{1/2} \frac{dq}{q} \qquad \text{in the } C_+ \text{ direction} \qquad (13.22)$$

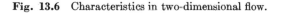

Fig. 13.6 Characteristics in two-dimensional flow.

Similarly,

$$d\theta = -(M^2 - 1)^{1/2}\frac{dq}{q} \quad \text{in the } C_- \text{ direction} \quad (13.23)$$

We could, alternatively, derive these results by noticing that the vanishing of the coefficient determinant in Eqs. (13.20) implies the existence of a certain linear dependence. In any event, we see that θ and q cannot be prescribed arbitrarily along a C_+ or C_- curve; the appropriate equation, either (13.22) or (13.23), must be satisfied.

Equations (13.22) and (13.23) can be put in somewhat simpler form. Since $M = q/c$, and since c is a function of q via Eq. (13.18), it follows that we can define

$$\mu(q) = \int_{q_0}^{q} (M^2 - 1)^{1/2}\frac{dq}{q}$$

where q_0 is any convenient constant. Thus

$$\theta - \mu(q) = \text{constant along a } C_+ \text{ curve}$$
$$\theta + \mu(q) = \text{constant along a } C_- \text{ curve}$$

$$(13.24)$$

13.6 A NUMERICAL APPROACH

Suppose now, for the problem discussed in the preceding section, that q and θ are known along some noncharacteristic curve C_0 in a region in which the flow is supersonic (see Fig. 13.7). Choose a number of points along this

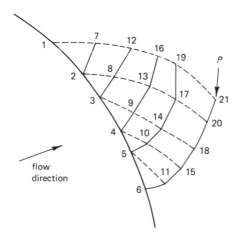

Fig. 13.7 Domain of dependence.

curve, numbered as 1, 2, ... , 6 in the figure. At each such point, q and θ are known; from Eqs. (13.18) we can compute c (we assume the constant B is known from upstream flow conditions) and so M; thus α is known, and we can sketch the local C_+ and C_- directions at each point. In the figure, C_+ directions are solid lines, C_- directions are dashed lines. Locally, we approximate characteristics by straight lines. At point 7, we can now use Eqs. (13.24) to write

$$\theta_2 - \mu_2 = \theta_7 - \mu_7, \qquad \theta_1 + \mu_1 = \theta_7 + \mu_7$$

so that θ_7 and μ_7 (and so q_7 and M_7) can be determined. Similarly, we can find θ and q at points 8, 9, 10, 11, and working our way outwards in the same way, we eventually determine θ and q at points 12, 13, ... , 21. We can of course make our calculation more accurate by using more points. However, we observe that if P is the point at which the characteristics through points 1 and 6 intersect, then the values of q and θ at P depend only on the specified data on that part of C_0 (the "domain of dependence" of P) lying between points 1 and 6. Conversely, the data specified at point 3, say, can affect values of q and θ only in that part of the plane (the "domain of influence" of point 3) contained between the two characteristics emanating from point 3.

As an extension of this idea, let q and θ be specified across a two-dimensional supersonic nozzle, i.e., on the curve C_0 in Fig. 13.8. Here C_1 and C_2 are rigid, frictionless walls. Using the above step-by-step approach, we can find q and θ in the shaded triangular region enclosed by C_0 and by the characteristics emanating from the points A and B where C_0 meets C_1 and C_2. Consider now a point D lying on the C_- characteristic emanating from A, with D close to A. A C_+ characteristic through D (at which point we now know q and θ) would strike the wall C_1 at point E, say; we necessarily have $\theta_D - \mu_D = \theta_E - \mu_E$, so that $\theta_E - \mu_E$ is known. However, the flow direction at E must be parallel to the wall, so θ_E is known; consequently we can also calculate μ_E. We next draw characteristic elements from points E and G to meet at F, compute μ_F and θ_F, and then find μ_H

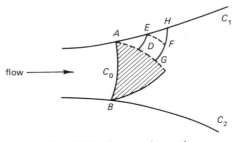

Fig. 13.8 Supersonic nozzle.

and θ_H just as before; in this way we can fill up the region above the shaded triangle, and similarly, the region below it. In other situations, q, rather than θ, might be known along some boundary (e.g., a free-stream situation, where the pressure, and so c, and in turn q, is prescribed). We may not be able to extend the characteristics approach beyond a certain region; if so, it may be necessary to introduce shocks, just as in previous sections.

There is another relation, beyond Eqs. (13.22) or (13.23), that must hold along a characteristic. Let a characteristic C_+ make a slope angle ω_+ with the x-axis. Then as we move along the characteristic,

$$d\theta = d(\alpha + \theta) - d\alpha = d\omega_+ - d\alpha \qquad (13.25)$$

Replacing α by $\sin^{-1}(1/M)$, and using Eq. (13.18), we can rewrite Eq. (13.25) as

$$d\theta = d\omega_+ + \frac{1 + \frac{1}{2}(\gamma - 1)M^2}{(M^2 - 1)^{1/2}} \frac{dq}{q} \qquad (13.26)$$

Similarly, for a C_- characteristic, we would obtain

$$d\theta = d\omega_- - \frac{1 + \frac{1}{2}(\gamma - 1)M^2}{(M^2 - 1)^{1/2}} \frac{dq}{q} \qquad (13.27)$$

A special case of interest is that in which the characteristic is a straight line. By eliminating $d\theta$ between Eqs. (13.22) and (13.26) [or (13.23) and (13.27)], it follows *that q and θ are constant along a straight-line characteristic.* Another interesting property[†] of a straight-line characteristic is that *neighboring members of the same family of characteristics are also straight.* To prove this statement, consider for example a C_+ characteristic, say Γ_0 in Fig. 13.9, which is straight, and let Γ_1 be a neighboring C_+ characteristic.

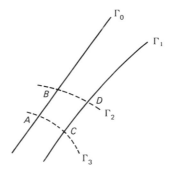

Fig. 13.9 Straight-line characteristic.

[†] These properties of straight-line characteristics are, of course, not necessarily true in problems leading to different equation sets.

Let Γ_2 and Γ_3 be a pair of C_- characteristics, intersecting Γ_0 and Γ_1 at the points A, B, C, D as shown. Then $\mu_A = \mu_B$, and $\theta_A = \theta_B$, since Γ_0 is straight. The two Eqs. (13.24) then lead to the conclusion that $\mu_C = \mu_D$ and $\theta_C = \theta_D$. Thus μ and θ are constant along Γ_1, which consequently must be straight.

There are some problems in two-dimensional supersonic flow in which families of straight-line characteristics play a useful role. A popular technique is to guess that such a family (said to constitute a "simple wave") is appropriate in some region of the field, and to justify this guess on an a posteriori basis; we will shortly consider an example. First, however, let us obtain a formula governing changes in q and θ as we move from one straight-line characteristic to a neighboring one.

Let $C_-{}^{(1)}$ and $C_-{}^{(2)}$ be a pair of neighboring straight-line characteristics, as shown in Fig. 13.10, and let the angle between them be $d\beta$, measured as shown. Then if the values of q, α, and θ on $C_-{}^{(1)}$ change to $(q + dq)$, $(\alpha + d\alpha)$, $(\theta + d\theta)$ on $C_-{}^{(2)}$, we must have

$$d\theta = -d\beta + d\alpha, \qquad d\theta = (M^2 - 1)^{1/2}\, dq/q \qquad (13.28)$$

With the help of $\sin \alpha = 1/M$ and Eq. (13.18), Eqs. (13.28) provide formulas for $d\theta$, $d\alpha$, dq, and dM in terms of $d\beta$. In particular, after some algebra, we obtain

$$\frac{(\gamma + 1)\, d\alpha}{\gamma - \cos 2\alpha} = d\beta \qquad (13.29)$$

which may be integrated to give

$$\tan \alpha = \left(\frac{\gamma - 1}{\gamma + 1}\right)^{1/2} \tan \left\{\left(\frac{\gamma - 1}{\gamma + 1}\right)^{1/2} \beta + \text{const}\right\}$$

As an example of the way in which a supersonic flow pattern can be constructed using a family of straight-line C_- characteristics, consider the families of characteristics sketched in Fig. 13.11. To the left of Γ_1 and to

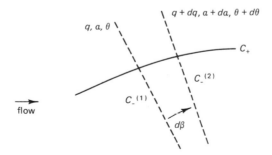

Fig. 13.10 Neighboring C_- characteristics.

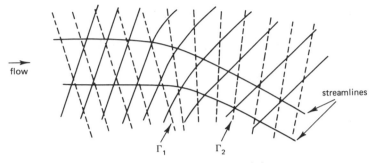

Fig. 13.11 Supersonic channel flow.

the right of Γ_2 (both of which are C_- characteristics), the C_+ and C_- characteristics are straight, so that the flow conditions are there uniform. Transition between these two regions is accomplished by means of a straight-line family of C_- characteristics, drawn in any way in which the slopes progress uniformly between Γ_1 and Γ_2. Once these C_- characteristics have been chosen, Eq. (13.29) may be used to find α on each such characteristic, and (13.28) to find θ. The C_+ characteristics may therefore be drawn, as can the streamlines. Two of these streamlines may then be designated as channel walls within which this kind of supersonic flow is possible. If one encounters a region in which characteristics of the same family intersect one another, one would presumably have to fit a shock wave discontinuity into that region, much as in Section 13.1.

13.7 PROBLEMS

13.7.1 Find the characteristics for Eqs. (13.16) and (13.17) by a direct transformation from (x, y) coordinates to (ξ, η) coordinates, and show that the results lead again to Eq. (13.21). Discuss similarly the two-dimensional form of the last equation in Problem 7.2.9, which provides an alternative formulation for this problem.

13.7.2 Show that Eqs. (13.28) imply that the vector increment in velocity is perpendicular to the C_- line. Find a compact expression for its magnitude.

13.7.3 Obtain results, similar to those given by Eqs. (13.28), for the case in which the C_+ characteristics form a straight-line family. Construct an example for this case analogous to that of Fig. 13.11. Discuss, as a special case, flow around a corner with the C_+ characteristics all emanating from the corner point.

13.7.4 Consider the pair of equations for $u(x, y)$ and $v(x, y)$ given by

$$L_1(u, v) = A_1 u_x + B_1 u_y + C_1 v_x + D_1 v_y + E_1 = 0$$
$$L_2(u, v) = A_2 u_x + B_2 u_y + C_2 v_x + D_2 v_y + E_2 = 0$$

(13.30)

where the coefficients are given functions of u, v, x, and y. Given that the system is hyperbolic, obtain the characteristics by the requirement that first derivatives may be discontinuous across certain curves. As an alternative approach, let λ_1 and λ_2 be functions of position chosen so that the derivatives of u and v occurring in $\lambda_1 L_1 + \lambda_2 L_2 = 0$ may be interpreted as directional derivatives in the same directions—i.e., so that

$$\lambda_1 A_1 + \lambda_2 A_2 : \lambda_1 B_1 + \lambda_2 B_2 :: \lambda_1 C_1 + \lambda_2 C_2 : \lambda_1 D_1 + \lambda_2 D_2$$

$$:: x_\tau : y_\tau$$

where the curve $x = x(\tau)$, $y = y(\tau)$ is chosen tangent to this common direction. Show that the existence of such λ_1 and λ_2 requires that

$$(A_1 C_2 - A_2 C_1) y_\tau{}^2 + [(B_2 C_1 - B_1 C_2) + (A_2 D_1 - A_1 D_2)] x_\tau y_\tau$$
$$+ (B_1 D_2 - B_2 D_1) x_\tau{}^2 = 0$$

and deduce that the resulting directions are again those of characteristics. Show that the combination $\lambda_1 L_1 + \lambda_2 L_2 = 0$ leads to Eqs. (13.22) and (13.23) for the special case of Eqs. (13.30) considered in Section 13.5.

13.7.5 From Section 7.1 the equations of one-dimensional time-dependent gas flow are

$$\rho_t + (\rho u)_x = 0, \qquad u_t + u u_x + (c^2/\rho) \rho_x = 0, \qquad c^2 = A \gamma \rho^{\gamma-1} \quad (13.31)$$

where we have assumed $p = A\rho^\gamma$,† A and γ constant, throughout the flow. Here ρ is density, u is velocity, p is pressure, and c is the local velocity of sound. Show that the characteristics in the (x, t) plane are given by

$$\frac{dx}{dt} = \begin{cases} u + c & \text{on} \quad C_+ \\ u - c & \text{on} \quad C_- \end{cases}$$

that $du + (c/\rho) \, d\rho = 0$ along a C_+ curve, and that $du - (c/\rho) \, d\rho = 0$ along a C_- curve. [The quantities $u \pm \int (c \, d\rho/\rho)$ are sometimes termed *Riemann invariants*.] Show that a characteristic is a straight line if and only if both u and ρ are constant along it. Show that if a characteristic is

---

† The condition $p = A\rho^\gamma$ is equivalent to the requirement that the flow remain isentropic; it is not valid across a shock wave, since dissipative mechanisms, which alter the entropy, are important inside the narrow transition region corresponding to a shock. Thus in Problem 13.7.5, we are considering shock-free flow only.

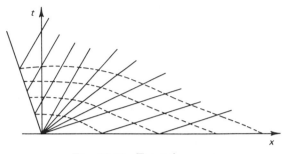

Fig. 13.12   Expansion wave.

a straight line, so are neighboring members of the same family. Devise a numerical method, based on characteristics, analogous to that discussed in Section 13.6, and show that domains of dependence and regions of influence may be defined similarly. Show that the "centered simple wave" sketched in Fig. 13.12 is an appropriate characteristics approach to a problem in which a compressible gas, initially at rest in the region $x > 0$, is displaced to the left as a result of the barrier at $x = 0$ moving to the left with velocity V. What happens if V is variable, with (a) $dV/dt > 0$, (b) dV/dt arbitrary?†

13.7.6   In the wave equation $\phi_{tt} = c^2\phi_{xx}$, with $c = c(x)$, define $u = \phi_t$ and $v = \phi_x$ to obtain $u_x = v_t$, $u_t = c^2v_x$. Find the characteristics for this pair of equations. Show that $du \pm c\,dv = 0$ along appropriate characteristics, and find that part of the (x, t) plane in which a solution is determined by a prescription of u and v (or equivalently ϕ and its normal derivative) along a portion of a noncharacteristic curve. Consider similarly the more general hyperbolic equation

$$\phi_{xy} + D\phi_x + E\phi_y + F\phi = G$$

where D, E, F, G are functions of (x, y). [*Hint:* write $\phi_y = \psi - A\phi$, etc.]

13.7.7   (a)   Let $\phi(x, y)$ satisfy the equation

$$L(\phi) = \phi_{xy} + D\phi_x + E\phi_y + F\phi = G$$

where D, E, F, G are functions of x and y (from Chapter 5, any linear hyperbolic equation can be brought into this form).

---

† Many problems of this character are discussed in Courant and Friedrichs (1948). The equations of shallow-water theory are essentially similar; see Stoker (1957, Chap. 10).

The operator M defined by

$$M(\psi) = \psi_{xy} - (D\psi)_x - (E\psi)_y + F\psi$$

is said to be *adjoint* to L; the motivation for this definition is based on the fact that the usual "multiply and subtract" process leads to a divergence expression. Show in fact that, if A is a region in the (x, y) plane, with boundary Γ, and if u, v are a pair of adequately differentiable functions,

$$\int_A [vL(u) - uM(v)]\, dA = \int_\Gamma [(1 - k)vu_y - kv_yu + Duv]n_1\, ds$$

$$+ \int_\Gamma [kvu_x - (1 - k)v_xu + Euv]n_2\, ds$$

where n_1 and n_2 are the components of the outward unit normal vector erected at the line element ds of Γ. Here k is any constant.

(b)   For symmetry, set $k = \frac{1}{2}$, and apply this identity to the shaded region A of Fig. 13.13, with $u = \phi$, $L(\phi) = G$, $v = \psi$, $M(\psi) = 0$, where P_1P and P_2P are characteristics, and where C is a curve along which ϕ and its normal derivative are specified, to obtain

$$\int_A \psi G\, dA = \int_C \{[\tfrac{1}{2}\psi\phi_y - \tfrac{1}{2}\psi_y\phi + D\phi\psi]n_1 + [\tfrac{1}{2}\psi\phi_x - \tfrac{1}{2}\psi_x\phi + E\phi\psi]n_2\}\, ds$$

$$+ \int_{P_2}^{P} [\tfrac{1}{2}(\psi\phi)_y - \psi_y\phi + D\phi\psi]\, dy + \int_{P_1}^{P} [(\tfrac{1}{2}\psi\phi)_x - \psi_x\phi + E\phi\psi]\, dx$$

(c)   Next, specialize ψ so that, in addition to $M(\psi) = 0$, it satisfies $\psi_y = D\psi$ along P_2P, $\psi_x = E\psi$ along P_1P, and $\psi(P) = 1$. Show that the preceding identity now gives $\phi(P)$ in terms of ϕ and $\partial\phi/\partial n$ as specified along that part of C lying between P_1 and P_2. Explain how this *Riemann function* ψ could be constructed numerically.

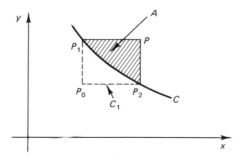

Fig. 13.13 Riemann function.

(d)   Since the Riemann function ψ depends not only on the coefficients in $M(\psi) = 0$ but also on the location (ξ, η) of the point P, it is conventional to write it as $\psi = R(x, y; \xi, \eta)$. As a special case of the curve C in Fig. 13.13, let C consist of the other two sides of the rectangle of which P_1P and P_2P are two sides (i.e., C_1 as shown dotted in the figure). Let the coordinates of P_0 be (x_0, y_0). Choose ϕ to be the Riemann function for the adjoint equation $M(v) = 0$ [so that $L(\phi) = 0$], with respect to the point P_0; denote this Riemann function by $R^*(x, y; x_0, y_0)$. Deduce the symmetry condition

$$R^*(\xi, \eta; x_0, y_0) = R(x_0, y_0; \xi, \eta)$$

and explain why $L_{(\xi,\eta)}R(x, y; \xi, \eta) = 0$, where $L_{(\xi,\eta)}$ denotes the L operation with respect to the variables (ξ, η).

(e)   Let $u(x, y)$ satisfy the equation

$$u_{tt} - u_{xx} + Du_x + Eu_t + Fu = \delta(x - \alpha)\cdot\delta(t - \beta)$$

with u and $\partial u/\partial n$ both equal to zero along some initial curve. Obtain a relationship between u and R, and so relate Riemann's function to the Green's functions of Chapter 9.

(f)   Find R for the case in which $L = u_{xy} + cu$, where c is a constant.

13.8 MORE DEPENDENT VARIABLES

From Section 7.1, two of the equations governing the one-dimensional time-dependent flow of an ideal compressible gas are

$$\rho_t + (\rho u)_x = 0, \qquad u_t + uu_x + (1/\rho)p_x = 0 \qquad (13.32)$$

where p, ρ, u represent pressure, density, and velocity, respectively; x is position and t is time. We now relax the thermodynamic requirement of Section 7.1 that the specific entropy $s = c_v \ln(p/\rho^\gamma)$ (where c_v is the constant specific heat and γ the constant ratio of specific heats) be constant throughout the gas, and require only that the flow be isentropic, with the specific entropy permitted to vary from particle to particle. Thus we can say only that $(\partial/\partial t + u\,\partial/\partial x)s = 0$, and by the first of Eqs. (13.32) this leads to

$$p_t + up_x + \gamma pu_x = 0 \qquad (13.33)$$

The local velocity of sound, c, is still given by $c^2 = \gamma p/\rho$.

In Eqs. (13.32) and (13.33) we have a set of three equations in two independent and three dependent variables. It is natural to ask whether characteristic curves, $\xi(x, t) = $ const, exist in the (x, t) plane. In terms of our usual (ξ, η) coordinate system, the above equations become

$$\rho_\xi(\xi_t + u\xi_x) + u_\xi(\rho\xi_x) + \text{other terms} = 0$$

$$u_\xi(\xi_t + u\xi_x) + p_\xi[(1/\rho)\xi_x] + \text{other terms} = 0 \qquad (13.34)$$

$$u_\xi(\gamma p\xi_x) + p_\xi(\xi_t + u\xi_x) + \text{other terms} = 0$$

where the "other terms" involve derivatives of ρ, u, or p with respect to η. If values of ρ, u, and p are prescribed on a certain curve $\xi = $ const, then we can solve Eqs. (13.34) for p_ξ, u_ξ, and ρ_ξ only if the coefficient determinant is nonzero; the condition defining a characteristic curve is then

$$\begin{vmatrix} \xi_t + u\xi_x & \rho\xi_x & 0 \\ 0 & \xi_t + u\xi_x & \dfrac{1}{\rho}\xi_x \\ 0 & \gamma p\xi_x & \xi_t + u\xi_x \end{vmatrix} = 0 \qquad (13.35)$$

Denoting the slope of the curve $\xi = $ const by $\lambda = dx/dt = -\xi_t/\xi_x$, this condition leads to

$$\lambda = u \quad \text{or} \quad \lambda = u + c \quad \text{or} \quad \lambda = u - c \qquad (13.36)$$

as the three possibilities for a characteristic curve. We thus have three families of characteristics that could be drawn if the flow field were known in (x, t) space, and that, conversely, could be used to sketch the flow field if suitable initial data were prescribed.

Let us label typical characteristic curves belonging to the above three families by C_0, C_+, and C_-. Along these curves, we have dx/dt equal to u, $u + c$, and $u - c$, respectively. Along a C_0 curve,

$$d\rho = (\rho_t + u\rho_x)\, dt, \qquad du = (u_t + uu_x)\, dt, \qquad dp = (p_t + up_x)\, dt$$

and by use of Eqs. (13.32) and (13.33), we obtain

$$dp - c^2\, d\rho = 0 \qquad \text{along a } C_0 \text{ curve}$$

Similarly,

$$dp + \rho c\, du = 0 \qquad \text{along a } C_+ \text{ curve}$$

$$dp - \rho c\, du = 0 \qquad \text{along a } C_- \text{ curve} \qquad (13.37)$$

Suppose now that Cauchy data—values of ρ, u, p—are given along some curve Γ in the (x, t) plane, which is not a characteristic. Then a numerical procedure very similar to that described in Section 13.6 can be used; the only difference is that we must use three characteristics emanating from Γ to determine the values of ρ, u, and p at neighboring points. In Fig. 13.14, we extend our knowledge of flow variables to points 6 and 7 in this way; to obtain information at point 9, we need data at point 8, which can be

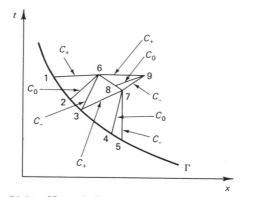

Fig. 13.14 Numerical method for three characteristics.

found by a rather obvious interpolation or iteration procedure. In any event, turning to Fig. 13.15, we see that values of u, ρ, and p at some point P will depend only on the data prescribed on that (darker) part of Γ intercepted by the extreme pair of characteristics (C_- and C_+) passing through P.

We note also that if Γ were itself a characteristic, then the prescribed data on Γ would have to satisfy whichever of Eqs. (13.37) was the appropriate consistency condition, and that if this condition were satisfied, there would be an infinity of solutions.

More generally, one might encounter a number n of first-order equations (for n unknowns $u_1, u_2, \ldots, u_n$) having the quasi-linear form

$$\frac{\partial}{\partial t}\begin{bmatrix} u_1 \\ u_2 \\ \cdot \\ \cdot \\ \cdot \\ u_n \end{bmatrix} + A \frac{\partial}{\partial x}\begin{bmatrix} u_1 \\ u_2 \\ \cdot \\ \cdot \\ \cdot \\ u_n \end{bmatrix} = F \qquad (13.38)$$

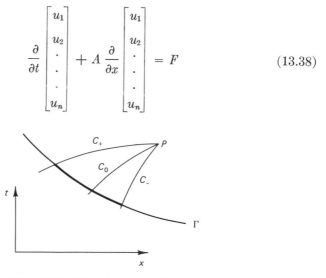

Fig. 13.15 Domain of dependence.

where the $n \times 1$ matrix F and the $n \times n$ matrix A involve only the u_i, x, and t. Changing to (ξ, η) variables, the slopes λ of the characteristic curves are the roots of the determinantal equation $|A - \lambda I| = 0$. Only if all of these roots are real is the system said to be hyperbolic. Along a characteristic, since $|A - \lambda I| = 0$, it follows that a certain linear combination of the equations must exist such that all net ξ derivatives vanish; thus only η derivatives will occur in this linear combination, so that a differential relationship analogous to those of Eq. (13.37) must exist along each characteristic.† For Cauchy data given along a characteristic, this differential relationship would be the consistency condition.

In Problem 13.10.1, the reader is asked to show that the general nonlinear second-order equation problem

$$F(x, t, u, u_x, u_t, u_{xx}, u_{xt}, u_{tt}) = 0$$

may be transformed to the form of Eq. (13.38). Alternatively, this problem can be treated directly by a generalization of the characteristic-strip idea of Chapter 12, as in Courant and Hilbert (1962, pp. 408ff).

13.9 MORE INDEPENDENT VARIABLES

As an example of a situation in which there are more than two independent variables, consider the wave equation in x, y and t:

$$\phi_{tt} - c^2(\phi_{xx} + \phi_{yy}) = 0 \tag{13.39}$$

where $c = c(x, y)$ is a given function of position. In (x, y, t) space, a surface $t = \psi(x, y)$ is *characteristic* if one of the usual kinds of conditions holds—i.e., if a knowledge of Cauchy data (ϕ and its normal derivative) is insufficient to determine higher derivatives, or if there may be a second- or higher-order derivative discontinuity across the surface, or if the differential equation reduces to a consistency condition for Cauchy data on the surface. Using any one of these definitions, and introducing a suitable new coordinate system (ξ, η, ζ) in which $\xi = t - \psi(x, y)$, we find easily that our surface is characteristic if

$$1/c^2 = \psi_x{}^2 + \psi_y{}^2 \tag{13.40}$$

which is again the eikonal equation [cf., Eq. (12.30)].

---

† In fact, the characteristics may be found by posing this requirement in reverse—i.e., find a direction in (x, t) space such that some linear combination of Eqs. (13.38) reduces to a statement involving derivatives only in this direction.

This result should not be surprising, since we know from Section 12.7 that if a disturbance wave front is defined by $t = \psi(x, y)$, then $\psi(x, y)$ satisfies the eikonal equation, and our characteristic surface in (x, y, t) space is by definition a possible trajectory of a certain kind of discontinuity (or disturbance).

If s denotes distance along a ray in the (x, y) plane, then the ray equations of Section 12.7 become, for the case of Eq. (13.40),

$$dx/ds = cp, \qquad dy/ds = cq, \qquad d\psi/ds = 1/c,$$
$$dp/ds = -c_x/c^2, \qquad dq/ds = -c_y/c^2 \qquad (13.41)$$

where we have denoted ψ_x by p and ψ_y by q. Identifying t with ψ, the (x, y, ψ) space curves represented by Eqs. (13.41) lie on the characteristic surface, and can be thought of as generating that surface. Because of the close relationship between the characteristic surface of Eq. (13.39) and the rays of Eq. (13.40), these rays are often termed *bicharacteristics* of Eq. (13.39).

Choose now any point (x_0, y_0, t_0) in (x, y, t) space. Construct the (x, y) plane rays emanating from (x_0, y_0) and with each point on such a ray associate a value of t, as above, to obtain a set of space curves emanating from (x_0, y_0, t_0). This set of space curves forms a conical surface†; if we permit $t < t_0$ as well as $t > t_0$, the conical surface will have two sheets. This special characteristic surface, associated with the point (x_0, y_0, t_0), is termed a *characteristic conoid*.

A natural problem involving Cauchy data is that in which ϕ and ϕ_t are specified on the plane $t = 0$. We can think of this Cauchy data as being impulsively produced in a quiescent system (e.g., a membrane) at $t = (0-)$; the disturbances will spread with velocity c, and it follows that the value of ϕ at a point (x_0, y_0, t_0), with $t_0 > 0$ say, can depend only on the initial data given on the part of the plane $t = 0$ that is contained within the characteristic conoid centered on (x_0, y_0, t_0). More generally, if Cauchy data is prescribed along some surface S in (x, y, t) space, then the value of ϕ at an adjoining point (x_0, y_0, t_0) will—for similar reasons— depend only on the data specified on that part of S contained within the characteristic conoid; if S "slopes" so steeply that a part of the conoid misses S entirely, then the above thought process is not applicable, and we cannot so easily delineate zones of dependence. In this latter case (in which S is sometimes termed *timelike*, as contrasted to *spacelike*) the Cauchy problem may in fact no longer be reasonable.

† Since $c = c(x, y)$, the cone need not, of course, have a circular cross section.

13.10 PROBLEMS

13.10.1   Consider the general second-order equation

$$F(x, t, u, p, q, r, s, v) = 0$$

for a function $u(x, y)$, where $p = u_x$, $q = u_t$, $r = u_{xx}$, $s = u_{xt}$, $v = u_{tt}$. Let u and u_t be prescribed along some portion of the line $t = 0$, via $u(x, 0) = f(x)$, $u_t(x, 0) = g(x)$.

Differentiate the equation $F = 0$ with respect to t to obtain an equation involving u_t, p_t, ..., and write equations like $p_t = s$, $r_t = s_x$, etc., to obtain a quasi-linear set of six equations, of the form (13.38), for the quantities u, p, q, r, s, v. Find, in terms of f and g, the initial condition satisfied by each of these six quantities. Write the determinantal equation defining the characteristic directions (because of hyperbolicity, we assume all roots to be real—for the special case of a linear equation, does this criterion correspond to that of Chapter 5?). In principle (via the use of characteristics, for example), a solution of the quasi-linear system, satisfying these initial conditions, may now be obtained; show that this solution is also a solution for the original problem. [*Hint:* $r_t = s_x = p_{tx} \Rightarrow (r - p_x)_t = 0 \Rightarrow r \equiv p_x$, etc.]

13.10.2   In steady rotational two-dimensional flow of an ideal gas, the two velocity components u, v and the density ρ satisfy the equations (cf. Section 7.1)

$$(\rho u)_x + (\rho v)_y = 0, \qquad \rho(u_x u + u_y v) = -c^2 \rho_x, \qquad \rho(v_x u + v_y v) = -c^2 \rho_y$$

where c is the local sound velocity. Show that real characteristic directions exist only if $u^2 + v^2 > c^2$ and find the directions for that case. Is it physically reasonable for one of these directions to be that of a streamline? What are the consistency conditions along characteristics?

13.10.3   The equations of two-dimensional time-dependent isentropic compressible flow (cf. Problem 13.10.2) are

$$\rho_t + (\rho u)_x + (\rho v)_y = 0$$

$$\rho(u_t + u_x u + u_y v) = -c^2 \rho_x$$

$$\rho(v_t + v_x u + v_y v) = -c^2 \rho_y$$

where $c = c(\rho)$. Discuss the characteristic surfaces in (x, y, t) space.

13.10.4   Modify the discussion of Section 13.9 so as to apply to the three-dimensional wave equation

$$\phi_{tt} = c^2(\phi_{xx} + \phi_{yy} + \phi_{zz})$$

where $c = c(x, y, z)$.

13.10.5   Let $\phi(x_1, x_2, \ldots, x_n)$ satisfy

$$\left(\sum_{i,j=1}^{n} a_{ij} \frac{\partial^2 \phi}{\partial x_i \, \partial x_j} \right) + F = 0$$

where F is a function of ϕ, its first derivatives, and the (x_i) variables, and where each a_{ij} coefficient is a function of the (x_i) variables.

Suppose now that, at a certain point $(x_1^0, x_2^0, \ldots, x_n^0)$, where the value of a_{ij} is denoted by a_{ij}^0, a local linear transformation of variables [to new variables $(x_1', x_2', \ldots, x_n')$] exists such that

$$\sum_{i,j} a_{ij}^0 \frac{\partial^2 \phi}{\partial x_i \, \partial x_j} = b_{11} \frac{\partial^2 \phi}{\partial x_i' \, \partial x_1'} + b_{22} \frac{\partial^2 \phi}{\partial x_2' \, \partial x_2'} + \cdots + b_{nn} \frac{\partial^2 \phi}{\partial x_n' \, \partial x_n'}$$

where all of $(b_{11}, b_{22}, \ldots, b_{nn})$ are positive, except for one that is negative. Is it reasonable to say that the original equation is hyperbolic at the point $(x_1^0, x_2^0, \ldots, x_n^0)$? Discuss.

13.10.6   Carry out an analysis, similar to that of Section 13.8, for the problem of shallow-water waves described in Problem 3.2.4.

14

FINITE-DIFFERENCE EQUATIONS AND NUMERICAL METHODS

We have already discussed a number of approximation methods suitable for numerical use on a computer; these include the Rayleigh–Ritz and Galerkin methods of Sections 10.6 and 11.5, the finite-element method of Section 10.8, the integral equation method of Section 9.5, and the method of characteristics of Section 13.6. In this chapter we deal with numerical methods based on finite difference approximations to the governing equations.

14.1 ACCURACY AND STABILITY; A DIFFUSION EQUATION EXAMPLE

Let $u(x, t)$ satisfy the diffusion equation

$$u_t = a^2 u_{xx} \tag{14.1}$$

over the interval $0 < x < L$, for $t > 0$, where $a(x) > 0$ is a given function, and where $u(0, t) = u(L, t) = 0$ for all t. The initial condition is $u(x, 0) = f(x)$, prescribed.

We choose uniformly spaced mesh points along the x-axis, at $x_j = jh$, $j = 0, 1, 2, \ldots, M$, where $Mh = L$, and we seek to determine (at least approximately) the values of u at these mesh points, at times $t_n = nk$, where $n = 1, 2, 3, \ldots$, and where k is a chosen time interval. Let $u_j^{(n)}$ denote the approximation, as obtained by our numerical process, to the exact solution $u(jh, nk)$. Denoting the value of $a(x_j)$ by a_j, we require the quantities $u_j^{(n)}$ to satisfy a finite-difference equivalent of Eq. (14.1), and a consideration of the basic definition of a derivative suggests that a reasonable choice† is

$$\frac{u_j^{(n+1)} - u_j^{(n)}}{k} - (a_j^2) \frac{u_{j+1}^{(n)} - 2u_j^{(n)} + u_{j-1}^{(n)}}{h^2} = 0 \qquad (14.2)$$

for $n = 0, 1, 2, \ldots$, and for $j = 1, 2, \ldots, (M-1)$. The values of $u_j^{(0)}$ are obtained from the initial conditions, via $u_j^{(0)} = f(jh)$, and for endpoint conditions we choose of course $u_0^{(n)} = u_M^{(n)} = 0$, for all n.

With $n = 0$, Eq. (14.2) permits an explicit calculation of $u_j^{(1)}$ for all j in terms of the known values of $u_j^{(0)}$; with $n = 1$, the equation can then be used to obtain the $u_j^{(2)}$ values, and so on. A question of immediate interest is the extent to which $u_j^{(n)}$ will differ from the exact solution $u(jh, nk)$. We certainly anticipate that there will be some difference, in general, since the solution of Eq. (14.1), at any chosen value of t, must depend upon $f(x)$ for all values of x in the interval $(0, L)$, whereas the solution of Eq. (14.2) will depend only upon the mesh point values $f(jh)$. [Moreover, $u_j^{(n)}$ will depend only on $f(rh)$ for $r = j, j \pm 1, j \pm 2, \ldots, j \pm n$, a situation that is in contrast to the infinite-signal-speed property of the diffusion equation.]

Useful guidance concerning the error made in using Eq. (14.2) rather than Eq. (14.1) can be obtained by considering special cases in which corresponding exact solutions of these two equations are available. One such case is that in which $a(x) = $ const, $f(x) = A \sin(P\pi x/L)$, where A is a constant and where P is some integer [we can think of this choice for $f(x)$ as representing one component in a Fourier representation of a more general function]. The exact solutions are

$$u(x, t) = A \exp\left(-\frac{a^2 P^2 \pi^2 t}{L^2}\right) \sin\left(\frac{P\pi x}{L}\right) \qquad (14.3)$$

$$u_j^{(n)} = A \left[1 - 4\frac{a^2 k}{h^2} \sin^2\left(\frac{P\pi h}{2L}\right)\right]^n \sin\left(\frac{P\pi jh}{L}\right) \qquad (14.4)$$

No matter what value P has, Eq. (14.3) shows that $u(x, t)$ will decay as t increases. In remarkable contrast, we see that if $(a^2k/h^2) > \frac{1}{2}$, then there can be values of P for which

$$1 - 4\frac{a^2 k}{h^2} \sin^2\left(\frac{P\pi h}{2L}\right) < -1 \qquad (14.5)$$

---

† We will subsequently examine in detail the discrepancy between Eqs. (14.2) and (14.1); compare Eq. (14.10).

so that $|u_j^{(n)}|$ increases exponentially with increasing n. Moreover, even if P is not such as to make the inequality (14.5) hold, the condition $(a^2k/h^2) > \frac{1}{2}$ will still lead to computational disaster. For a computer introduces some roundoff error at each step, so that the solution for $u_j^{(n)}$ as obtained on a computer is repeatedly distorted slightly, and this distortion pattern can easily possess a sinusoidal component whose frequency fulfills condition (14.5). We conclude that we cannot expect *computational stability* unless $(a^2k/h^2) < \frac{1}{2}$. For the more general case in which $a = a(x)$, we would reasonably require

$$a^2_{\max}k/h^2 \leq \tfrac{1}{2} \tag{14.6}$$

as a stability criterion, where $a_{\max}$ denotes the maximum over j of $a(jh)$ (cf. Problem 14.3.1).

Returning now to the special case that led to Eqs. (14.3) and (14.4), we want to examine the discrepancy between the two results, under the assumption $a^2k/h^2 = \omega \leq \frac{1}{2}$. In particular, we will compare

$$u(jh, nk) = A \exp(-a^2\lambda^2 nk) \sin \lambda jh \tag{14.7}$$

with

$$u_j^{(n)} = A \exp[n \ln(1 - 4a^2(k/h^2) \sin^2 \tfrac{1}{2}\lambda h)] \sin \lambda jh \tag{14.8}$$

where we have replaced $P\pi/L$ by λ, for brevity.

If h is such that $\lambda h \ll 1$, then the right-hand side of Eq. (14.8) is closely approximated by

$$u_j^{(n)} \cong A \exp[-a^2\lambda^2 nk(1 + \lambda^2 h^2[(\omega/2) - (1/12)])] \sin \lambda jh \tag{14.9}$$

A comparison between Eqs. (14.7) and (14.9) indicates that (still for $\lambda h \ll 1$) the accuracy of Eq. (14.2) will be excellent, until the number of time steps, n, becomes large enough that $\exp\{(-a^2\lambda^2 nk)\lambda^2 h^2[(\omega/2) - (1/12)]\}$ is significantly different from unity. We note in passing that the choice $\omega = \frac{1}{6}$ [in which case of course we would need the next correction term in Eq. (14.9)] should lead to particularly accurate results.

14.2 ERROR ANALYSIS

Exact solutions for suitable special cases of the original equation and its finite-difference analog may not be easy to find. An alternative procedure is to examine the discrepancy between the two equations "in the small"— i.e., to ask, for example, to what extent a solution of the partial differential equation fails to satisfy the finite difference equation. In the present example, let the exact solution $u(x, t)$ of Eq. (14.1) be supposed to be continuously differentiable to as high an order as required in the following calculation. The first term of the finite-difference expression (14.2), if applied to $u(x, t)$ itself, would yield

$$\frac{u(jh, [n + 1]k) - u(jh, nk)}{k}$$

and Taylor's theorem permits us to replace this by

$$u_t(jh, nk) + \tfrac{1}{2}ku_{tt}(jh, [n + \alpha]k)$$

where $0 \leq \alpha \leq 1$. Treating the second term similarly, the result of applying the finite-difference operator of Eq. (14.2) to $u(x, t)$ is found to be

$$u_t - a^2u_{xx} + R$$

where

$$R = \tfrac{1}{2}k\bar{u}_{tt} - (a^2h^2/12)\bar{u}_{xxxx} \tag{14.10}$$

Here a, u_t, and u_{xx} are evaluated at (jh, nk), and $\bar{u}_{tt}$, $\bar{u}_{xxxx}$ represent values of u_{tt} and u_{xxxx} at nearby points. The quantity R, which as we have said measures the extent to which $u(x, t)$ fails to satisfy Eq. (14.2), is termed the *discretization error* or the *truncation error*. We observe that as $k, h \to 0$ in any manner, $R \to 0$; standard nomenclature is that Eq. (14.2) is *consistent* with Eq. (14.1). We observe also that for the case $a = $ const so that $u_{tt} = a^4u_{xxxx}$, the choice $(a^2k/h^2) = \tfrac{1}{6}$ should be particularly effective in keeping R small, a result we have already noted.

The functional dependence of R on k and h provides some indication of the rapidity with which the solution of the finite-difference equation, computed at some chosen (x, t) value, approaches $u(x, t)$ as h and $k \to 0$. In the present example we can derive a quantitative result. Define

$$w_j^{(n)} = u(jh, nk) - u_j^{(n)}$$

and apply the finite difference operator of Eq. (14.2) to $w_j^{(n)}$ so as to obtain [using Eq. (14.10)]

$$w_j^{(n+1)} = w_j^{(n)}\left(1 - \frac{2a_j^2k}{h^2}\right) + \frac{a_j^2k}{h^2}\left(w_{j+1}^{(n)} + w_{j-1}^{(n)}\right) + Rk \tag{14.11}$$

Suppose now that the inequality $(a^2k/h^2) \leq \tfrac{1}{2}$ is satisfied and that we have estimated upper bounds for u_{tt} and u_{xxxx}, so that

$$|u_{tt}| < A, \qquad |u_{xxxx}| < B$$

for all x and t, where A and B are some constants. Then if $\varepsilon^{(n)}$ denotes the maximum over j of $|w_j^{(n)}|$, we have from Eq. (14.11)

$$\varepsilon^{(n+1)} \le \varepsilon^{(n)} + k\,|R| \le \varepsilon^{(n)} + \tfrac{1}{2}k^2A + \tfrac{1}{12}a_{\max}^2kh^2B$$

Since $\varepsilon^{(0)} = 0$, it follows that

$$\varepsilon^{(n)} \le nk[\tfrac{1}{2}kA + \tfrac{1}{12}a_{\max}^2h^2B]$$

where $nk = t$, of course. Thus we have an upper bound for the error, and we find that it does indeed approach zero as k, $h \to 0$. In practice, computer round-off error must be included in addition.

The diffusion equation we have considered in this section is one example of an "equation of evolution" representing a situation in which an initial configuration alters with time. The results we obtained for the diffusion equation are typical of those that hold for more general equations of evolution. In particular, a finite-difference approximation to such an equation must be consistent with the differential equation, and it must be stable, if we are to expect convergence to the exact solution as the mesh size in space and time is refined.†

A remark on stability may be worthwhile. The reader may have noticed that the stability constraint $\omega \le \tfrac{1}{2}$ was not essential in the derivation of Eq. (14.9). In fact, if we set $k = Ch^2$, where C is any positive constant, and let $h \to 0$, then Eq. (14.9) is valid and shows that the solution of the difference equation approaches that of the differential equation, no matter what choice has been made for P. The point is, however, that this assumes a fixed choice for P has been made once and for all; what we have to be concerned about is the possibility that, for any particular choice of h, there is *some* signal component whose wavelength is such as to lead to the computational disaster associated with Eq. (14.5), and this will indeed be the case if the constraint $\omega \le \tfrac{1}{2}$ is violated. Notice, incidentally, that our discussion of accuracy is concerned with a fixed value of $t = nk$; as k decreases, n increases so as to keep the value of the product nk constant. In some situations one might be concerned with what happens as k is held fixed and as $n \to \infty$, but we do not consider this matter here.

---

† For problems in which the initial and boundary data is appropriate to the nature of the differential equation ("properly posed problems"), it has been proved by Lax that, for a consistent finite-difference approximation, stability is the necessary and sufficient condition for convergence. See Richtmyer and Morton (1967, p. 45). A more general discussion is given by Thomée (1969, p. 152).

14.3   PROBLEMS

14.3.1   Show that if condition (14.6) is satisfied, and if $f(jh) < K$ for all j, where K is some constant, then the solution of Eq. (14.2) requires

$$| u_j^{(n)} | < K \qquad \text{for all } j, \quad \text{and for all } n \geq 0$$

14.3.2   Since the first term in Eq. (14.2) is a better approximation to u_t as evaluated at some time between nk and $(n + 1)k$ than it is at nk itself, an improvement on Eq. (14.2) might read

$$u_j^{(n+1)} = u_j^{(n)} + \left(a_j^2 \frac{k}{h^2} \right) \{ A \, (u_{j+1}^{(n)} - 2u_j^{(n)} + u_{j-1}^{(n)})$$

$$+ (1 - A) \, (u_{j+1}^{(n+1)} - 2u_j^{(n+1)} + u_{j-1}^{(n+1)}) \} \qquad (14.12)$$

where A is a constant (presumably, $A \cong \frac{1}{2}$). In Eq. (14.12) we have formed a weighted average of the finite-difference expressions for u_{xx}, as evaluated at $t = nk$ and at $t = (n + 1)k$.

Compare the discretization error with that of Eq. (14.2), and find that value of A for which this error is least. Discuss comparative accuracy. For constant a, find that range of A values such that Eq. (14.12) is stable for all h and k. Is Eq. (14.12), with optimal choice for A, the most accurate finite-difference equation possible using six mesh points in time and space?

Equation (14.12) requires the solution of a set of algebraic equations at each time step, and so is *implicit* rather than *explicit*—as Eq. (14.2) was. Show nevertheless that if the problem involves a finite interval, $0 < x < L$, with boundary conditions given at the ends of this interval, then because this set of algebraic equations involves a tridiagonal coefficient matrix, it may be very efficiently solved by means of Gaussian elimination.

With $A = \frac{1}{2}$, the method of Eq. (14.12) is termed the *Crank–Nicholson* method.

14.3.3   A popular technique for an empirical accuracy test is to rerun a problem with smaller values of h and k, and see if the answer changes "significantly." Discuss. List all the practical criteria you can think of that should affect the choice of h and k in a diffusion-type problem. In particular, how would the boundary conditions enter into this choice? Is nondimensionalization useful? What is the practical effect, as k and h are made smaller, of a requirement such as that of Eq. (14.6)?

14.3.4   As an alternative to Eq. (14.12), one might hope for improved accuracy by replacing Eq. (14.2) by

$$\frac{u_j^{(n+1)} - u_j^{(n-1)}}{2k} = a_j^2 \frac{u_{j+1}^{(n)} - 2u_j^{(n)} - u_{j-1}^{(n)}}{h^2} \tag{14.13}$$

If $u_j^{(0)}$ were prescribed, then one would have to first find $u_j^{(1)}$ by some other means (perhaps by use of a local Taylor series—how?), and then use Eq. (14.13) with $n = 1, 2, 3, \ldots$, in succession. This difference equation is second-order in n, however, so that we might expect trouble in using it to approximate Eq. (14.1), which is first-order in time. Show† in fact that Eq. (14.13) is generally unstable (take $a =$ const), although it (a) is consistent, and (b) has smaller discretization error than Eq. (14.2).

14.3.5 Since Eq. (14.13) is unstable, the following equation was suggested by Du Fort and Frankel:

$$\frac{u_j^{(n+1)} - u_j^{(n-1)}}{2k} = a_j^2 \frac{u_{j+1}^{(n)} - u_j^{(n+1)} - u_j^{(n-1)} + u_{j-1}^{(n)}}{h^2} \tag{14.14}$$

Show that this equation is always stable (at least for constant a), but that it is not consistent with Eq. (14.1) unless $(k/h) \to 0$ as k and $h \to 0$. Evaluate the discretization error.

14.3.6 One numerical approximation to Eq. (14.1) is via a set of mesh functions $u_j(t)$ satisfying

$$(d/dt)u_j = (a^2/h^2)[u_{j+1} - 2u_j + u_{j-1}]$$

where $h = \Delta x$, and where we take a as constant. This set of equations is a set of ordinary differential equations, which could then be solved by standard numerical methods. Assuming that the solution of the differential equation set is essentially exact, investigate consistency, accuracy, and stability.

14.3.7 Let $u_t = a^2 u_{xx} + bu$, where a and b are positive constants. Modify Eq. (14.2) so as to apply to this equation, and discuss accuracy and stability. Note that, for certain choices of a^2 and b, solutions of both the exact and the stable difference equations may grow in magnitude as time passes; discuss stability in this context.

---

† A discussion of second-order difference equations will be found in Carrier and Pearson (1968, Chap. 16).

14.4 MORE DIMENSIONS, OR OTHER COMPLICATIONS

The two-dimensional form of Eq. (14.1) is

$$u_t = a^2(u_{xx} + u_{yy}) \tag{14.15}$$

where $a = a(x, y)$ is a given function of position. The point (x, y) lies within a region V, on the boundary of which some information involving u and/or $\partial u/\partial n$ is given, and $u(x, y, 0)$ is prescribed; our problem is to determine $u(x, y, t)$ for $t > 0$.

If we use a square space mesh, with $\Delta x = \Delta y = h$ and a time mesh $\Delta t = k$, and if we denote the approximation to u at the mesh point (x_i, y_j, t_n) by $u_{i,j}^{(n)}$, then the analog to the explicit equation (14.2) becomes

$$\frac{u_{i,j}^{(n+1)} - u_{i,j}^{(n)}}{k} = a_{i,j}^2 \, \Delta_{i,j}^{(n)} u \tag{14.16}$$

where we denote the "finite Laplacian" by

$$\Delta_{i,j}^{(n)} u = (1/h^2) \left[u_{i+1,j}^{(n)} + u_{i-1,j}^{(n)} + u_{i,j+1}^{(n)} + u_{i,j-1}^{(n)} - 4u_{i,j}^{(n)} \right] \tag{14.17}$$

Similarly, the analog to the implicit equation (14.12) reads

$$\frac{u_{i,j}^{(n+1)} - u_{i,j}^{(n)}}{k} = a_{i,j}^2 \{ A \, \Delta_{i,j}^{(n)} u + (1 - A) \, \Delta_{i,j}^{(n+1)} u \} \tag{14.18}$$

where A is a constant.

To investigate stability, we can consider the special case $a = \text{const}$, and let V be the region $-\infty < x < \infty$, $-\infty < y < \infty$, with $u(x, y, 0) = B \sin \lambda x \sin \gamma y$, where B, λ, and γ are constants. Then the solution of Eq. (14.16) is easily found to be

$$u_{i,j}^{(n)} = B[1 - 4(ka^2/h^2)(\sin^2(\lambda h/2) + \sin^2(\gamma h/2)]^n \sin \lambda i h \sin \gamma j h$$

Thus the process is unstable unless $(ka^2/h^2) < \frac{1}{4}$. We note that the most rapid growth occurs for values of λ and γ such that the period of the oscillation is about two mesh intervals; since the behavior of such a high-frequency component would presumably not depend too much on boundary data, we expect this criterion to apply reasonably well to finite regions also. If $a = a(x, y)$, then as in Section 14.1 we would "play safe" by replacing a in the criterion by $a_{\max}$. The stability criterion for the explicit equation becomes, then,

$$k(a_{\max})^2/h^2 \leq 1/4 \tag{14.19}$$

for any region V. If $0 \leq A \leq \frac{1}{2}$, then a similar calculation indicates that the implicit equation (14.18) is stable irrespective of the sizes of h and k. Both equations are consistent; with $A = \frac{1}{2}$, Eq. (14.18) is more accurate than Eq. (14.16). Although the implicit equation has advantages of accuracy and stability (thus permitting the use of larger time steps), it is no longer possible to use Gaussian elimination to solve in as simple a manner as for the one-dimensional case (cf. Problem 14.3.2) the set of linear equations that occur at each time step. One possibility is to use a relaxation process to solve these equations, with Eq. (14.16) being used to provide a first guess for the solution (for a discussion of relaxation schemes, see Section 14.11).

There is an interesting alternative to Eq. (14.18), due to Peaceman and Rachford, and to Douglas, that provides computational stability without the necessity of a relaxation process. To illustrate the idea, let V be a rectangular region, $0 < x < L_1$, $0 < y < L_2$, with time-dependent boundary values of u prescribed on the four edges of V, and with $u(x, y, 0)$ prescribed. For the 1st, 3rd, 5th, ... time steps we use

$$\frac{u_{i,j}^{(n+1)} - u_{i,j}^{(n)}}{k} = a_{i,j}^2 \left[\frac{u_{i+1,j}^{(n+1)} - 2u_{i,j}^{(n+1)} + u_{i-1,j}^{(n+1)}}{h^2} + \frac{u_{i,j+1}^{(n)} - 2u_{i,j}^{(n)} + u_{i,j-1}^{(n)}}{h^2} \right]$$

(14.20)

and for the 2nd, 4th, 6th, ... time steps we use

$$\frac{u_{i,j}^{(n+1)} - u_{i,j}^{(n)}}{k} = a_{i,j}^2 \left[\frac{u_{i+1,j}^{(n)} - 2u_{i,j}^{(n)} + u_{i-1,j}^{(n)}}{h^2} + \frac{u_{i,j+1}^{(n+1)} - 2u_{i,j}^{(n+1)} + u_{i,j-1}^{(n+1)}}{h^2} \right]$$

(14.21)

where each of these is solved by simple Gaussian elimination. The reader may show that this process is stable for all h, k, and that the discretization error is comparable to that of Eq. (14.18) with $A = \frac{1}{2}$. The method is usually referred to as the *alternating-direction-implicit* or *ADI method*.

An alternative method (due to Bagrinovskii and Godunov) is to divide each time step $\Delta t = k$ into two halves, with $u_{xx} + u_{yy}$ replaced by $2u_{xx}$ in the first half step and by $2u_{yy}$ in the second half step. An implicit technique is used for each half step. Thus

$$u_{i,j}^{(n+\frac{1}{2})} - u_{i,j}^{(n)} = 2a_{i,j}^2 (\tfrac{1}{2}k/h^2) [\tfrac{1}{2}(u_{i+1,j}^{(n+\frac{1}{2})} - 2u_{i,j}^{(n+\frac{1}{2})} + u_{i-1,j}^{(n+\frac{1}{2})})$$
$$+ \tfrac{1}{2}(u_{i+1,j}^{(n)} - 2u_{i,j}^{(n)} + u_{i-1,j}^{(n)})]$$
$$u_{i,j}^{(n+1)} - u_{i,j}^{(n+\frac{1}{2})} = 2a_{i,j}^2 (\tfrac{1}{2}k/h^2) [\tfrac{1}{2}(u_{i,j+1}^{(n+1)} - 2u_{i,j}^{(n+1)} + u_{i,j-1}^{(n+1)})$$
$$+ \tfrac{1}{2}(u_{i,j+1}^{(n+\frac{1}{2})} - 2u_{i,j}^{(n+\frac{1}{2})} + u_{i,j-1}^{(n+\frac{1}{2})})]$$

(14.22)

Since each half step is of the Crank–Nicholson form (cf. Problem 14.3.2), this method is unconditionally stable. If we write these equations in the shorthand form

$$(1 - \tfrac{1}{2}ka^2\,\delta_i^2)u^{(n+\frac{1}{2})} = (1 + \tfrac{1}{2}ka^2\,\delta_i^2)u^{(n)}$$

$$(1 - \tfrac{1}{2}ka^2\,\delta_j^2)u^{(n+1)} = (1 + \tfrac{1}{2}ka^2\,\delta_j^2)u^{(n+\frac{1}{2})}$$

(where δ_i^2 is the Laplacian difference operator in the x direction, etc.), then since the operators commute we have

$$(1 - \tfrac{1}{2}ka^2\,\delta_i^2)(1 - \tfrac{1}{2}ka^2\,\delta_j^2)u^{(n+1)} = (1 + \tfrac{1}{2}ka^2\,\delta_i^2)(1 + \tfrac{1}{2}ka^2\,\delta_j^2)u^{(n)}$$

Expanding, we obtain

$$u^{(n+1)} - u^{(n)} = ka^2(\delta_i^2 + \delta_j^2)(\tfrac{1}{2}u^{(n+1)} + \tfrac{1}{2}u^{(n)}) - \tfrac{1}{4}k^2a^4\,\delta_i^2\,\delta_j^2(u^{(n+1)} - u^{(n)})$$

In the last term on the right-hand side, $(u^{(n+1)} - u^{(n)})$ is $O(k)$; thus the discretization error is of the same order as the Crank–Nicholson scheme.

The most natural extension of the ADI method to three space dimensions would use three kinds of time steps in cyclic sequence, each implicit in a different space direction. The reader may verify that this scheme is no longer unconditionally stable (however, modified schemes possessing this desirable feature are feasible[†]). On the other hand, the reader may show that the natural extension of the method of Eq. (14.22) to three space dimensions is unconditionally stable.

For a more general diffusion-type equation, say

$$u_t = au_{xx} + bu_{yy} + cu_{xy} + du_x + eu_y + fu + g$$

where the coefficients are functions of position and perhaps also of time, and where $4ab - c^2 > 0$, with $a > 0$, $b > 0$, the study of simple solutions for special cases indicates that stability of the finite-difference analog generally requires $(a + b)k/h^2 < \tfrac{1}{2}$ for the explicit case, and that there is no restriction for the implicit case if A (defined as in Problem 14.3.2) is less than or equal to one half. If the coefficients depend on u as well as on x, y, t, then one may want to adjust the time step for the explicit equation, as the computation proceeds, to maintain the inequality just stated.

It frequently happens that the boundary of the region does not coincide with mesh line elements, as in Fig. 14.1. One procedure is to approximate the boundary by such mesh lines, as in the figure; the boundary data on the boundary nodes can be obtained from the given boundary data by an interpolation process. Accuracy requires that the fineness of the mesh (at least near the boundary) be appropriate to the irregularity of the boundary. It may also be useful to use a mesh other than rectangular. For example,

† See Richtmeyer and Morton (1967, p. 212).

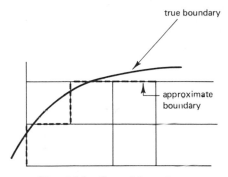

Fig. 14.1   Curved boundary.

equilateral mesh triangles could be used; a finite-difference Laplacian, for example, for an interior point would then involve the difference between the value of a function at a nodal point and the average of its values on the six surrounding nodes. Irregular triangles could also be used, in which case it might be more efficient to proceed via the finite-element method of Section 10.8 so as to obtain appropriate time derivatives of nodal values.

14.5 SERIES EXPANSIONS

In preceding sections we have alluded to Fourier-type components of a function representing an initial condition. We now ask whether series expansions of the kinds discussed in Chapter 1 can be applied to finite-difference equations.

Consider the problem of Eq. (14.2), with $a^2 = $ const, for a finite interval $0 < x < L$, with $h = L/M$, where M is some integer. For boundary conditions we choose $u_0^{(n)} = u_M^{(n)} = 0$, for all n. The initial condition is $u_j^{(0)} = f(jh)$, where $f(x)$ is a given function; we avoid for the present a minor complication, and require $f(0) = f(L) = 0$ so as to match the boundary conditions.

In analogy with Chapter 1, we look for a series solution of the form

$$u_j^{(n)} = \sum_q F_q(n) \cdot G_q(j) \qquad (14.23)$$

where F_q and G_q are functions of the integer variables n and j, respectively. Suitable functions are obtainable by separation of variables; set $u_j^{(n)} = F(n) \cdot G(j)$ in Eq. (14.2) to obtain

$$\frac{F(n+1) - F(n)}{F(n)} = \left(a^2 \frac{k}{h^2}\right) \frac{G(j+1) - 2G(j) + G(j-1)}{G(j)} \qquad (14.24)$$

Equation (14.24) can be valid for all choices of j and n only if each side is a constant. After some experimentation, we choose a convenient form for this constant so that the $G(j)$ equation becomes

$$\frac{G(j+1) - 2G(j) + G(j-1)}{G(j)} = -4 \sin^2 \frac{\omega}{2}$$

where ω is a constant. The solution of this difference equation requires $G(j)$ to be a linear combination of $\cos \omega j$ and $\sin \omega j$. Choosing ω so as to satisfy the boundary conditions $G(0) = G(M) = 0$, we obtain the final result $G(j) = \sin(\pi q j / M)$ where q is some integer (we omit an arbitrary multiplicative constant).

It is reasonable then to write Eq. (14.23) as

$$u_j^{(n)} = \sum_{q=1}^{M-1} F_q(n) \sin \frac{\pi q j}{M} \tag{14.25}$$

(In the summation, we include only those values of q that involve essentially distinct functions.) Substitution into Eq. (14.2) now yields an equation for $F_q(n)$, which is easily solved to give

$$F_q(n) = c_q \left(1 - 4a^2 \frac{k}{h^2} \sin^2 \frac{\pi q}{2M} \right)^n \tag{14.26}$$

where c_q is a constant. The initial conditions require

$$\sum_{q=1}^{M-1} c_q \sin \frac{\pi q j}{M} = f(jh)$$

and by use of the easily proved orthogonality relation

$$\sum_{j=1}^{M-1} \sin \frac{\pi q j}{M} \sin \frac{\pi r j}{M} = \begin{cases} 0, & \text{if } q \neq r \\ \frac{1}{2}M, & \text{if } q = r \neq 0 \end{cases} \tag{14.27}$$

we find

$$c_q = \frac{2}{M} \sum_{j=1}^{M-1} f(jh) \sin \frac{\pi q j}{h} \tag{14.28}$$

so that the solution (14.25) is complete.

If the boundary conditions were not homogeneous, then we could reformulate the problem, as in Section 1.7, to obtain a new problem involving homogeneous conditions—or alternatively, we could use a version of the finite transform method of Problem 1.8.3.

A generating function method, analogous to the Laplace transform method of Chapter 2, is also sometimes feasible. In the present example, let z be a complex variable and define the function

$$U_j(z) = \sum_{n=0}^{\infty} u_j^{(n)} z^n \qquad (14.29)$$

(The change in variable $z = e^{-s}$ brings out the Laplace transform analogy.) If we multiply Eq. (14.2) by z^n, and sum the equations so obtained over all values of n, we obtain

$$(1/z)(U_j - u_j^{(0)}) - U_j = a^2(k/h^2)[U_{j+1} - 2U_j + U_{j-1}] \qquad (14.30)$$

Since $u_j^{(0)} = f(jh)$ is known, the formal solution of this finite difference equation, subject to the boundary conditions $U_0 = U_M = 0$, may be obtained (note that the coefficients are constant, in the sense that they do not involve j). Each $u_j^{(n)}$ is then the coefficient of z^n in the power series expansion of the result. Contour integration techniques are frequently useful in the last process.

More generally, almost all of the techniques for the solution of partial *differential* equations, as discussed in previous chapters, may be adapted to apply to partial *difference* equations. We will not give further examples here; however, it is worthwhile to comment briefly on the matter of possible singularities.

Consider again the model problem of this section, with the single change that $f(0)$ no longer vanish. Since $u(0, t) = 0$, however, we have a choice in computing $u_1^{(1)}$ via Eq. (14.2)—we can set $u_0^{(0)}$, which then appears in this equation, equal to either zero or to $f(0)$. In fact, we could probably convince ourselves that we should set $u_0^{(0)}$ equal to some average value of these two quantities. An analysis of this kind of situation shows that this last suggestion is indeed worthwhile; a practical alternative is to begin with a number of very small time steps.

14.6 PROBLEMS

14.6.1 (a) Obtain a series expansion, similar to that of Eqs. (14.25)–(14.28), for the same model problem, but using Eq. (14.12) instead of Eq. (14.2).

(b) Extend the solution (14.25) to include the effect of a nonhomogeneous term in Eq. (14.2) corresponding to the equation $u_t = a^2 u_{xx} + p(x, t)$.

14.6.2 Discuss the appropriate modification for the analysis of Section 14.5 that would result from a change in the right-hand boundary condition to one requiring zero flux.

14.6.3 Consider $u_t = a^2 u_{xx} + B u_x$, with a^2 and B constant. In an (explicit) numerical approximation, one can replace u_x by the left-hand difference $(u_j^{(n)} - u_{j-1}^{(n)})/h$, by the right-hand difference $(u_{j+1}^{(n)} - u_j^{(n)})/h$, or by the centered difference $(u_{j+1}^{(n)} - u_{j-1}^{(n)})/2h$. With $\alpha = a^2 k/h^2$, $\xi = \frac{1}{2} Bk/h$, find those regions in the (α, ξ) plane in which the explicit procedure is stable.

14.7 WAVE EQUATION

As a prototype wave equation, consider

$$u_{tt} = c^2 u_{xx} \tag{14.31}$$

where c is constant, for $-\infty < x < \infty$, $0 < t$. Values of $u(x, 0) = f(x)$ and $u_t(x, 0) = g(x)$ are prescribed. Choosing time and space intervals $h = \delta x$ and $k = \delta t$, and denoting an approximation to $u(jh, nk)$ by $u_j^{(n)}$ as before, a simple finite-difference equivalent to Eq. (14.31) would be

$$\frac{u_j^{(n+1)} - 2u_j^{(n)} + u_j^{(n-1)}}{k^2} = c^2 \frac{u_{j+1}^{(n)} - 2u_j^{(n)} + u_{j-1}^{(n)}}{h^2} \tag{14.32}$$

From the initial conditions, we know $u_j^{(0)} = f(jh)$, and we could determine $u_j^{(1)}$ by use of

$$u_j^{(1)} - u_j^{(0)} = kg(jh) \tag{14.33}$$

or, more accurately, by

$$u_j^{(1)} - u_j^{(0)} = kg(jh)$$
$$+ \tfrac{1}{2}(k^2 c^2/h^2)\{f([j+1]h) - 2f(jh) + f([j-1]h)\} \tag{14.34}$$

The choices $j = 1, 2, 3, \ldots$ in Eq. (14.32) then determine $u_j^{(2)}, u_j^{(3)}, \ldots$, in sequence.

Stability can be investigated in terms of a disturbance of the form

$$u_j^{(n)} = P(n) \sin \lambda jh$$

where λ is some constant. Substitution into Eq. (14.32) gives a difference equation for $P(n)$:

$$P(n+1) + 2\left[2\alpha^2 \sin^2 \frac{\lambda h}{2} - 1\right]P(n) + P(n-1) = 0$$

where $\alpha = ck/h$; to avoid exponential growth of $P(n)$ with n, we require $\alpha^2 < 1/\sin^2(\lambda h/2)$, and considering all values of λ, this condition becomes $\alpha < 1$ for stability.

We know via Eq. (14.31) that the domain of dependence for a point (x, t) is that part of the x-axis intercepted by the two characteristics through the chosen point. A corresponding point (jh, nk) in the (x, t) plane will also have a domain of dependence on the x-axis, via Eq. (14.32). It is clear that if $\alpha > 1$, then the latter domain of dependence will be contained within the former, so that a change in values near the endpoints of the domain of dependence for Eq. (14.31) will affect $u(jh, nk)$ but not $u_j^{(n)}$. Consequently, $u_j^{(n)}$ cannot be correct if $\alpha > 1$, so from this second point of view we are again led to the requirement $\alpha < 1$.

Even if the condition $\alpha < 1$ is satisfied, however, there are some possible difficulties. First of all, the speed with which signals propagate under Eq. (14.32) is $h/k = c/\alpha$, and for $\alpha < 1$ this speed will be greater than c. Second, since the domain of dependence of a point (x, t) via Eq. (14.32) is greater than the corresponding domain via Eq. (14.31) if $\alpha < 1$, it should be possible to alter the initial condition so as to affect the solution of Eq. (14.32) but not that of Eq. (14.31). Fortunately, a study of the properties of Eq. (14.32) shows† that the spurious contributions due to these effects are exponentially small.

The convergence of a solution of Eq. (14.32) to the corresponding solution of Eq. (14.31), as h and $k \to 0$, with α always less than one, may be shown in much the same way as in Section 14.1.‡

To ease the stability constraint, it is possible to replace Eq. (14.32) by one in implicit form. There are a number of choices; two of these are

$$u_j^{(n+1)} - 2u_j^{(n)} + u_j^{(n-1)} = \tfrac{1}{2}\alpha^2 \{ [u_{j+1}^{(n+1)} - 2u_j^{(n+1)} + u_{j-1}^{(n+1)}]$$
$$+ [u_{j+1}^{(n-1)} - 2u_j^{(n-1)} + u_{j-1}^{(n-1)}] \} \qquad (14.35)$$

and

$$u_j^{(n+1)} - 2u_j^{(n)} + u_j^{(n-1)} = \tfrac{1}{4}\alpha^2 \{ [u_{j+1}^{(n+1)} - 2u_j^{(n+1)} + u_{j-1}^{(n+1)}]$$
$$+ 2[u_{j+1}^{(n)} - 2u_j^{(n)} + u_{j-1}^{(n)}]$$
$$+ [u_{j+1}^{(n-1)} - 2u_j^{(n-1)} + u_{j-1}^{(n-1)}] \} \qquad (14.36)$$

The reader should compare the discretization errors with that of Eq. (14.32), and show that each of Eqs. (14.35), (14.36) is stable for any choice of h and k. Extensions to two or more space dimensions, to the inclusion of more terms, or to the case in which coefficients are position dependent, may be carried out along the same lines as in preceding sections and problems.

† Pearson (1969).
‡ A fundamental investigation of the convergence problem for hyperbolic equations was given in a classical paper by R. Courant *et al.* (1928, p. 32).

14.8   A NONLINEAR EQUATION

We begin with the simple first-order equation discussed in Section 13.1:

$$u_t + uu_x = 0 \tag{14.37}$$

for $-\infty < x < \infty$, $t > 0$, with $u(x, 0) = \phi(x)$, a given function. Except for discontinuity situations, u is constant along each characteristic defined by $dx/dt = u$ in the (x, t) plane. In replacing Eq. (14.37) by a finite-difference equation, there are many choices; two particularly simple ones are

$$u_j^{(n+1)} - u_j^{(n)} + u_j^{(n)}(k/h)(u_{j+1}^{(n)} - u_j^{(n)}) = 0 \tag{14.38}$$

$$u_j^{(n+1)} - u_j^{(n)} + u_j^{(n)}(k/h)(u_j^{(n)} - u_{j-1}^{(n)}) = 0 \tag{14.39}$$

where $h = \delta x$, $k = \delta t$ as before. If $u_j^{(n)} > 0$, then the special choice $h = ku_j^{(n)}$ (which corresponds to $dx/dt = u$) reduces these equations to

$$u_j^{(n+1)} = 2u_j^{(n)} - u_{j+1}^{(n)} \qquad \text{and} \qquad u_j^{(n+1)} = u_{j-1}^{(n)}$$

respectively; only the second of these agrees directly with the corresponding result for Eq. (14.37). If on the other hand $u_j^{(n)} < 0$, the first equation would yield the desired result for $h = k \mid u_j^{(n)} \mid$. This fact suggests that it might be desirable to use Eq. (14.38) or Eq. (14.39) wherever $u_j^{(n)}$, locally, is less or greater than zero, respectively. A stability analysis [set $u_j^{(n)} = \text{Re}\{P_n e^{i\lambda jh}\}$ and treat the coefficient $(u_j^{(n)}k/h)$ as locally constant] bears this out; the stability analysis also indicates that we must choose $k \leq h/\mid u_j^{(n)} \mid$.

Actually, one would probably not use either of Eqs. (14.38) or (14.39) in any event, because each is only first-order accurate. Moreover, neither would be accurate in the neighborhood of a discontinuity of the kind that arises where characteristics of Eq. (14.37) intersect. If Eq. (14.37) is appropriately replaced by a conservation law valid across such a discontinuity (cf. Sections 13.1 and 13.2), say

$$u_t + \tfrac{1}{2}(u^2)_x = 0 \tag{14.40}$$

then it might be useful to base the finite-difference equation on Eq. (14.40). One such equation is that of Lax and Wendroff, based on the idea

$$u_t = -\tfrac{1}{2}(u^2)_x, \qquad u_{tt} = (-uu_t)_x = [\tfrac{1}{2}u(u^2)_x]_x$$

so that

$$
\begin{aligned}
u_j^{(n+1)} - u_j^{(n)} = {}& -(k/4h)[(u_{j+1}^{(n)})^2 - (u_{j-1}^{(n)})^2] \\
& + (k^2/4h)[\{\tfrac{1}{2}[u_{j+1}^{(n)} + u_j^{(n)}]\}\{(1/h)[(u_{j+1}^{(n)})^2 - (u_j^{(n)})^2]\} \\
& - \{\tfrac{1}{2}[u_j^{(n)} + u_{j-1}^{(n)}]\}\{(1/h)[(u_j^{(n)})^2 - (u_{j-1}^{(n)})^2]\}] \tag{14.41}
\end{aligned}
$$

On the other hand, it may be that Eq. (14.37) is in actuality an approximation to

$$u_t + uu_x = \varepsilon u_{xx} \qquad (14.42)$$

where $0 < \varepsilon \ll 1$. In Section 13.3 we saw that Eq. (14.42) will not generate discontinuities as long as $\varepsilon > 0$; there may, however, be regions of rapid transition (shocks). For very small ε, the term εu_{xx} will be effective only where $|u_{xx}|$ is large, and with a reasonable mesh spacing such regions may not be detected, so that a finite-difference approximation to Eq. (14.42) may well reduce to one of the preceding approximations to Eq. (14.37). This may lead to a difficulty near shock transitions. One way around this is to make ε, artificially, large enough that the finite-difference equivalent of εu_{xx} is of significant magnitude near a shock wave; the effect of this *artificial viscosity method* will be to smooth out a shock over a number of mesh points.

Similar considerations may arise in obtaining numerical solutions for more general nonlinear equations of evolution:

$$u_t = F(x, t, u, u_x, u_{xx}, \ldots)$$

Here x is a space variable and t is time; of course, more than one space dimension might occur. Also, u and F could be vector functions.

Stability analyses are usually carried out with the help of a local linearization, as in the above example. If it is anticipated that shock waves or other rapid transitions may be encountered, a nonuniform space mesh could be used; the spacing would be adjustable during the computation process so as to have high mesh point density in regions in which u_x or u_{xx} is large.

14.9 PROBLEMS

14.9.1 Discuss the discretization error and stability of each of Eqs. (14.38), (14.39), (14.41). Invent implicit and more accurate versions of Eqs. (14.38), (14.39), and analyze them also. Examine the behavior of some of these equations near a discontinuity.

14.9.2 If one is interested in u_t or u_x, rather than in the solution u itself of the wave equation (14.31), one might construct finite-difference equivalents to

$$v_t = c^2 w_x, \qquad v_x = w_t$$

where $v = u_t$, $w = u_x$. Analyze the properties of such an approach.

14.9.3 Discuss the stability and accuracy for your choice of a finite-difference approximation to the equation

$$u_{tt} = c^2(u_{xx} + u_{yy}) - \alpha u_t$$

where $c = c(x, y)$, $\alpha = \alpha(x, y) > 0$ are given functions; initial values of u and u_t are prescribed, as well as time-dependent values for u on the finite boundary of the region of interest.

14.9.4 Devise, and discuss, a numerical procedure for analyzing the one-dimensional flow of a gas governed by the equations of Problem 13.4.2.

14.9.5 Discuss from the numerical point of view the equations (of beam vibration)

$$u_t = -v_{xx}, \qquad v_t = u_{xx}$$

for the region $-\infty < x < \infty$.

14.9.6 For Eq. (14.37), Lax has suggested the numerical scheme

$$u_j^{(n+1)} - \tfrac{1}{2}[u_{j+1}^{(n)} + u_{j-1}^{(n)}] + (k/4h)[(u_{j+1}^{(n)})^2 - (u_{j-1}^{(n)})^2] = 0$$

Relate this equation to the conservation law $u_t + \tfrac{1}{2}(u^2)_x = 0$, investigate its behavior near a discontinuity, and show that if k and h are related to one another by $k = \lambda h^{3/2}$, where $\lambda = $ const, then as $h \to 0$ this equation approaches

$$u_t + u u_x = (\sqrt{h}/2\lambda) u_{xx}$$

which has the form of Eq. (14.42).

14.9.7 Consider $u_t + A u_x = 0$, where A is a constant. Since u must be constant along any line of the form $x = At + $ const, it follows that $u(x_j, t_{n+1}) = u(x_j - A \cdot \delta t, t_n)$, and the right-hand side can be obtained by linear or quadratic interpolation involving $u_j^{(n)}$, $u_{j-1}^{(n)}$, $u_{j+1}^{(n)}$. Follow this idea to obtain a suitable difference scheme, and discuss its properties. Generalize to the case in which A is a function of position or of u.

14.10 BOUNDARY VALUE PROBLEMS

As a prototype boundary value problem, let it be required that a function $u(x, y)$ in a region R be found, where

$$u_{xx} + u_{yy} = g(x, y) \quad \text{in } R, \qquad u = f \quad \text{on } \Gamma \qquad (14.43)$$

Here $g(x, y)$ is a given function, and f a function of position on the boundary Γ of R. In a finite-difference approach we content ourselves with the approximate determination of R at a number of mesh points in R, and we replace Eq. (14.43) with a relationship involving these mesh point values.

The simplest case for theoretical study is that in which R is a square and the mesh points are lattice points of a square grid. Let R be defined by $0 < x < A$, $0 < y < A$, with $A = Mh$, where M is an integer. Let $u_{i,j}$ denote the desired approximation to $u(ih, jh)$; values of $u_{0,j}$, $u_{M,j}$, $u_{i,0}$, $u_{i,M}$ are known from the boundary condition. A suitable replacement for Eq. (14.43) is

$$u_{i,j+1} + u_{i,j-1} + u_{i+1,j} + u_{i-1,j} - 4u_{i,j} = h^2 g_{i,j} \qquad (14.44)$$

where $g_{i,j} = g(ih, jh)$. [Note that if u is to be harmonic so that $g_{i,j} = 0$, then Eq. (14.44) requires each $u_{i,j}$ to be the average of the mesh point values in its immediate neighborhood; this is a finite difference analog of the mean value theorem.]

Eqs. (14.44) are a set of linear algebraic equations for the unknown $u_{i,j}$ values; there is one such equation for each interior mesh point, so the number of equations is just right. The number of equations is $(M - 1)^2$; in three dimensions, the corresponding number would be $(M - 1)^3$. For large values of M (say $M = 100$ or more), a direct solution of this set of linear equations—by elimination, say—becomes computationally laborious and error-sensitive, so that an alternative procedure is desirable.

A well-known procedure of this type is *relaxation*. One starts by guessing solution values for the $u_{i,j}$, and then iteratively improves these solution values. A natural iterative technique, and the one we consider first, is to proceed through the mesh on a point-by-point basis, adjusting the value of $u_{i,j}$ at each point in turn so as to satisfy Eq. (14.44) locally. Thus, designating the values of $u_{i,j}$ prior to a relaxation sweep by $u_{i,j}^{(\text{old})}$, one computes at each point

$$u_{i,j}^{(\text{new})} = \tfrac{1}{4}[u_{i,j+1}^{(\text{old})} + u_{i,j-1}^{(\text{old})} + u_{i+1,j}^{(\text{old})} + u_{i-1,j}^{(\text{old})} - h^2 g_{i,j}] \qquad (14.45)$$

Once all such $u_{i,j}^{(\text{new})}$ have been computed—and only then—does one replace the $u_{i,j}^{(\text{old})}$ values with the $u_{i,j}^{(\text{new})}$ values and begin a new relaxation sweep. We will refer to this process as *Jacobi relaxation*. Experience with such relaxation techniques indicates that it is often more effective to *overrelax* or *underrelax*—i.e., to correct more or less, respectively, than the amount called for by Eq. (14.45). Let α denote the relaxation parameter; we then replace Eq. (14.45) by

$$u_{i,j}^{(\text{new})} = u_{i,j}^{(\text{old})} + (\alpha/4)[u_{i,j+1}^{(\text{old})} + u_{i,j-1}^{(\text{old})} + u_{i+1,j}^{(\text{old})} + u_{i-1,j}^{(\text{old})} - 4u_{i,j}^{(\text{old})} - h^2 g_{i,j}]$$

$$(14.46)$$

For $\alpha = 1$, Eq. (14.45) is recovered. We examine now the speed with which this relaxation process, for a particular choice of α, converges to the desired exact solution of Eq. (14.44). If we denote the difference between the solution of Eq. (14.44) and the value of $u_{i,j}^{(\text{old})}$ at any stage of the relaxation process by $w_{i,j}^{(\text{old})}$, (and similarly for "new" values) then we have

$$w_{i,j}^{(\text{new})} = w_{i,j}^{(\text{old})} + (\alpha/4)[w_{i,j+1}^{(\text{old})} + w_{i,j-1}^{(\text{old})} + w_{i+1,j}^{(\text{old})} + w_{i-1,j}^{(\text{old})} - 4w_{i,j}^{(\text{old})}]$$

(14.47)

with $w_{i,j}^{(\text{new})} = w_{i,j}^{(\text{old})} = 0$ at each boundary point. Hopefully, each $w_{i,j}$ will approach zero as the iterations proceed. Now if we think of the $(M-1)^2$ quantities $w_{i,j}^{(\text{old})}$ as forming a column vector, then Eq. (14.47) can be thought of as representing the multiplication of this column vector by a certain square matrix K so as to produce a column vector whose $(M-1)^2$ rows are the quantities $w_{i,j}^{(\text{new})}$. The iterative process thus consists in the repeated application of K to the column vectors as produced, and the condition that the column vectors tend to the null vector is that all eigenvalues of K be less than one in absolute magnitude. But an eigenvalue λ of K must satisfy the set of equations

$$\lambda w_{i,j} = w_{i,j} + (\alpha/4)[w_{i,j+1} + w_{i,j-1} + w_{i+1,j} + w_{i-1,j} - 4w_{i,j}] \quad (14.48)$$

where the $w_{i,j}$ values are the $(M-1)^2$ components of the corresponding eigenvector. [Note that in Eq. (14.47) we have in effect replaced $w_{i,j}^{(\text{old})}$ by $w_{i,j}$, and $w_{i,j}^{(\text{new})}$ by $\lambda w_{i,j}$.] Using $w_{i,j} = 0$ on the boundary, we can easily solve Eqs. (14.48) via $w_{i,j}^{(m,n)} = \sin(m\pi i/M) \sin(n\pi j/M)$, where m and n are integers; corresponding to all choices of m and n in the range 1 to $(M-1)$, there are $(M-1)^2$ such eigenvectors. The corresponding eigenvalues are

$$\lambda^{(m,n)} = 1 - \alpha[\sin^2(m\pi/2M) + \sin^2(n\pi/2M)]$$

At the start of the iteration process, the $w_{i,j}^{(\text{old})}$ values constitute an original error vector, and this vector may be expressed as a linear combination of the various $w^{(m,n)}$ eigenvectors. Each application of K to this linear combination has the effect of multiplying each coefficient in that combination by the corresponding value $\lambda^{(m,n)}$. Thus for convergence of the $w_{i,j}$ to zero, we must have all $|\lambda^{(m,n)}| < 1$; moreover, the greater the least value of $(1 - |\lambda^{(m,n)}|)$, the more rapid the convergence. An easy calculation shows that this optimal situation is attained for $\alpha = 1$, so that in the Jacobi case there is no advantage to overrelaxation. Since, with $\alpha = 1$, $|\lambda^{(m,n)}|_{\max} \cong 1 - (\pi^2/2M^2)$ for large M, and this is the factor by which the most resistant error component is reduced in magnitude at each application of K, we can deduce that the number N of iteration sweeps to reduce an initial error by a factor of at least 100 is given by

$$[1 - (\pi^2/2M^2)]^N = 1/100$$

so that $N \cong M^2$.

A better relaxation process is to use the new values of $u_{i,j}$ as soon as they are obtained, rather than to wait until the relaxation sweep is complete. This *Gauss–Seidel* procedure, again with a relaxation parameter α, is equivalent to

$$u_{i,j}^{(new)} = u_{i,j}^{(old)} + (\alpha/4)[u_{i+1,j}^{(old)} + u_{i-1,j}^{(new)} + u_{i,j+1}^{(old)} + u_{i,j-1}^{(new)} - 4u_{i,j}^{(old)} - h^2 g_{i,j}]$$

(14.49)

(here we sweep from left to right, and from the bottom up). In Problem 14.11.1 the reader is asked to show that the optimal choice of α is now given by

$$\alpha_{opt} = 2 - 2\sin(\pi/M) \tag{14.50}$$

and that the number of relaxation sweeps necessary to reduce an initial error by a factor of 100 is given by

$$N \cong \tfrac{3}{4}M \tag{14.51}$$

This gain in efficiency is remarkable.

A relaxation method such as this if often termed an *SOR method*, where the initials stand for "successive overrelaxation." If the boundary is not square, then theoretical estimates of convergence speed are more difficult to obtain [see Varga (1962) for an excellent survey of general linear iterative methods], but the results that are available, together with computational experience, indicate that the above results apply—more or less—to nonrectangular regions also, and to situations involving Neumann or mixed conditions as well as Dirichlet conditions. At the boundary of an irregular region, some kind of interpolation (perhaps also iteratively corrected, as the progress of the solution permits one to better estimate derivatives near the boundary) is of course necessary; it may be that the choice of a triangular or other nonrectangular mesh is worthwhile if the boundary shape is a difficult one. If there is a singularity near the boundary (resulting perhaps from discontinuous boundary data), then one can obtain increased accuracy by using a locally fine mesh or by analytically subtracting off the singularity.

Further remarks on iterative methods for elliptic-type problems, together with some useful references, will be found in Birkhoff (1971). In Section 14.12 we turn our attention to some series expansion methods.

14.11 PROBLEMS

14.11.1 Verify results (14.50) and (14.51) and extend the calculations to the three-dimensional case.

14.11.2 Examine the efficacy of superposing on the iterative schemes of Section 14.10 a progressive mesh refinement—i.e., a number of relaxation sweeps is first made with a crude mesh so as to yield a first approximation for functional values (obtained via interpolation) for a finer grid, and so on.

14.11.3 Discuss a numerical approach to the problem

$$u_{xx} + u_{yy} = g(u, x, y)$$

in R, where g is a given function of u, x, and y.

14.11.4 Suppose that in the problem of Eq. (14.43) there are additional linear terms involving u_x, u_y, and u on the left-hand side; moreover, suppose that our main interest is in values of u_x and u_y rather than in values of u. How would you proceed numerically, and why?

14.11.5 Let

$$u_{xxxx} + 2u_{xxyy} + u_{yyyy} = g(x, y)$$

in R, with u and $\partial u/\partial n$ specified on the boundary of R. Discuss various numerical methods applicable to this problem—including the advisability of splitting it into a coupled pair of lower-order equations.

14.12 SERIES; FAST FOURIER TRANSFORM

For problems involving a potential-type equation in a simple region, such as a rectangle or a circle, series expansion methods may be more efficient than iteration methods for solving the associated finite-difference equation. A useful tool in some of these problems is the fast Fourier transform, which is essentially a programming technique for the rapid summation of Fourier series. The efficiency of a series expansion method carries over to some extent to regions made up, say, of rectangles joined together (e.g., an L-shaped region), or to such regions containing small but perhaps irregular holes. In much of the literature, series expansion methods are presented in terms of operations on the coefficient matrix of the finite-difference equations.

Consider, for example, a rectangular region in the (x, y) plane in which we want to solve $\Delta\phi = 0$, where ϕ is specified on the boundary, In finite-

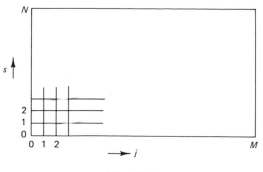

Fig. 14.2

difference terms, we have (cf. Fig. 14.2)

$$\phi_{j-1,s} + \phi_{j+1,s} - 4\phi_{j,s} + \phi_{j,s+1} + \phi_{j,s-1} = 0$$

$$j = 1, 2, \ldots, M - 1 \qquad s = 1, 2, \ldots, N - 1$$

(14.52)

with ϕ_{0s} and ϕ_{Ms} specified for all values of s, and with ϕ_{j0} and ϕ_{jN} specified for all values of j.

Define

$$b_{jr} = \frac{2}{N} \sum_{s=1}^{N-1} \phi_{js} \sin \frac{\pi rs}{N}$$

$$j = 1, 2, \ldots, M - 1; \quad r = 1, 2, \ldots, N - 1$$

$$a_{kr} = \frac{2}{M} \sum_{j=1}^{M-1} b_{jr} \sin \frac{\pi jk}{M}$$

$$k = 1, 2, \ldots, M - 1; \quad r = 1, 2, \ldots, N - 1$$

It then follows [cf. Eq. (14.27)] that

$$\phi_{js} = \sum_{r=1}^{N-1} b_{jr} \sin \frac{\pi rs}{N}$$

$$= \sum_{r=1}^{N-1} \sum_{k=1}^{M-1} a_{kr} \sin \frac{\pi jk}{M} \sin \frac{\pi rs}{N}$$

(14.53)

$$j = 1, 2, \ldots, M - 1; \quad s = 1, 2, \ldots, N - 1$$

If we now multiply Eq. (14.52) by $(2/N) \exp(i\pi rs/N)$, sum over s

from 1 to $N - 1$, and take the imaginary part, we obtain

$$b_{j-1,r} + b_{j+1,r} - 4b_{j,r}$$

$$+ 2b_{jr} \cos(\pi r/N) + (2/N)[\phi_{j0} - (-1)^r \phi_{jN}] \sin(\pi r/N) = 0$$

Multiply this equation by $(2/M) \exp(i\pi jk/M)$, sum over j from 1 to $M - 1$, and take the imaginary part, to give

$$\left(2 \cos \frac{\pi k}{M} + 2 \cos \frac{\pi r}{N} - 4\right) a_{kr} + \frac{4}{MN} \sin \frac{\pi k}{M} \left\{ \sum_{s=1}^{N-1} (\phi_{0s} - [-1]^k \phi_{Ms}) \sin \frac{\pi rs}{N} \right\}$$

$$+ \frac{4}{MN} \sin \frac{\pi r}{N} \left\{ \sum_{j=1}^{M-1} (\phi_{j0} - [-1]^r \phi_{jN}) \sin \frac{\pi jk}{M} \right\} = 0 \qquad (14.54)$$

This is now an explicit equation for a_{kr} for use in Eq. (14.53), and our solution is complete.

In these and similar manipulations, we encounter Fourier sums that are the real or imaginary parts of expressions typified by

$$x_j = \sum_{k=0}^{N-1} A_k z^{jk} \qquad (14.55)$$

where $z = \exp(i2\pi/N)$. The A_k are given, and we want to compute x_j for all values of j. The fast Fourier transform algorithm of Cooley and Tukey[†] does this in an efficient way in those cases in which N is composite.

Suppose first that $N = r_1 r_2$, where r_1 and r_2 are integers. Write $j = j_1 r_1 + j_0$, $k = k_1 r_2 + k_0$, where j_1 and k_0 range from 0 to $r_2 - 1$, and where j_0 and k_1 range from 0 to $r_1 - 1$. Then

$$x_j = \sum_{k_0=0}^{r_2-1} \sum_{k_1=0}^{r_1-1} A_{k_1 r_2 + k_0} z^{j(k_1 r_2 + k_0)}$$

$$= \sum_{k_0=0}^{r_2-1} \left(\sum_{k_1=0}^{r_1-1} A_{k_1 r_2 + k_0} z^{j_0 k_1 r_2} \right) z^{jk_0} \qquad (14.56)$$

where we have used the fact that

$$z^{j_1 r_1 k_1 r_2} = z^{j_1 k_1 N} = 1$$

Let $B_{k_0 j_0}$ denote the quantity in parenthesis in Eq. (14.56). Counting a multiplication and an addition as an operation, to compute $B_{k_0 j_0}$ for each of the r_2 values of k_0 and for each of the r_1 values of j_0 (a total of $r_1 r_2 = N$ combinations) requires r_1 operations. Having computed the

† Cooley and Tukey (1965, p. 297).

$B_{k_0 j_0}$ quantities, each of the N possible values of x_j may be obtained from Eq. (14.56) by a further r_2 operations. The total labor is thus $Nr_1 + Nr_2$ operations, which may be compared with the N^2 operations required in the direct use of Eq. (14.55).

If next r_1 is also composite, say $r_1 = \alpha\beta$ where α and β are integers, then since the formula for $B_{k_0 j_0}$ is of essentially the same form as Eq. (14.55) (where we now treat k_0 as a parameter having r_2 possible values), an extension of the idea permits us to compute the $B_{k_0 j_0}$ quantities by a total of $r_1(\alpha + \beta)$ operations for each value of k_0, and so by a total of $N(\alpha + \beta)$ operations. The second step, as before, requires Nr_2 operations, so the total is now $N(r_2 + \alpha + \beta)$ operations. In general, if $N = r_1 r_2, \ldots, r_n$, the total number of operations that are required by this method will be $N(r_1 + r_2 + \cdots + r_n)$. For example, if $r_1 = r_2 = \cdots = r_n = r$, say, so that $N = r^n$, then the required number of operations is $Nnr = Nr \log_r N$. If N is large, the replacement of $O(N^2)$ operations by $O(N \log N)$ operations is very worthwhile indeed.

14.13 PROBLEMS

14.13.1 (a) Let $N = r^n$ in the fast Fourier transform method just discussed. Show that, for a given order of magnitude for N, the optimal choice for r is 3.

(b) Compare the number of operations required for the direct and fast transform processes in the two-dimensional equivalent of Eq. (14.55).

14.13.2 Repeat the manipulations leading to Eq. (14.54) for the case in which $\Delta\phi = 0$ is replaced by $\Delta\phi = f(x, y)$, where f is a given function. Next, let $f_{ij} = 0$ at all interior mesh points except one, and set $\phi_{ij} = 0$ at all boundary mesh points, so as to obtain a finite-difference Green's function for the problem.

14.13.3 Discuss the possibility of using series expansion methods to solve a potential problem in (a) the interior of a circle, (b) in an L-shaped region, and (c) in a square region from which a central circular region has been excised.†

14.13.4‡ Let u_k, v_k be two sets of complex numbers, defined for $k = 0, \pm1, \pm2, \ldots, \pm K$, where K is some given positive integer. For convenience,

---

† For a discussion of direct methods for such problems, via matrix manipulations, see Buzbee *et al.* (1970, p. 627) and (1971, p. 722).

‡ Problems 14.13.4 and 14.13.5 follow Orszag (1971). Our notation is slightly different.

set $u_k = v_k = 0$ for $|k| > K$. Define

$$w_k = \sum_{p=-K}^{K} u_p v_{k-p}$$

for $k = 0, \pm 1, \ldots, \pm K$. How many operations are necessary to calculate directly all w_k values on a computer?

Consider now the following alternative technique for the calculation of the w_k quantities, and show that it can be much more efficient. Define the transform function

$$U_j = \sum_{k=-N}^{N-1} u_k \exp\left(2\pi i \frac{jk}{2N}\right)$$

for $j = 0, 1, \ldots, 2N - 1$, where N is some integer greater than K; define the transform V_j of v_k similarly. Note the inversion formulas exemplified by

$$2N u_k = \sum_{j=0}^{2N-1} U_j \exp\left(-2\pi i \frac{jk}{2N}\right)$$

Define next $W_j = U_j V_j$, for $j = 0, 1, 2, \ldots, 2N - 1$, and show that the inverse of W_j gives w_k for a suitable (specify) choice of N.

14.13.5 Let $\phi(x, t)$ satisfy the equation

$$\phi_t = a^2 \phi_{xx} + \varepsilon \phi^2 \tag{14.57}$$

in the region $-\infty < x < \infty$, $t > 0$, where a and ε are positive constants. The value of $\phi(x, 0)$ is prescribed, and is periodic in x, with period 2π. The solution $\phi(x, t)$ will clearly also be periodic in x. We can write

$$\phi(x, t) = \sum_{k=-\infty}^{\infty} \psi_k(t) e^{ikx} \tag{14.58}$$

where the ψ_k are complex coefficients depending on t; since ϕ is real, we require of course that ψ_{-k} be the complex conjugate of ψ_k. A feature of the Fourier expansion of a periodic function (possessing derivatives of all orders) is that as $k \to \infty$, $\psi_k \to 0$ more rapidly (asymptotically) than as any power of k^{-1}; thus the convergence of Eq. (14.58) can be expected to be very rapid, and an approximation in which we truncate the series can be very efficient. This leads us to write

$$\Omega(x, t) = \sum_{k=-K}^{K} \Omega_k(t) e^{ikx}$$

where we now try to determine the $\Omega_k(t)$ functions so that $\Omega(x, t)$ is a good approximation to $\phi(x, t)$.

Explain why it is reasonable to determine the $\Omega_k'(t)$ functions by the requirement that

$$\int_0^{2\pi} [\Omega_t - a^2\Omega_{xx} - \varepsilon\Omega^2]^2 \, dx$$

be minimized with respect to each such Ω_k' quantity, and carry out this idea to obtain a set of differential equations for the $\Omega_k(t)$ qantities. How can the method of Problem 14.13.4 be used? Generalize your discussion to the case of a two-dimensional problem of the form

$$\phi_t = a^2 \, \Delta\phi + \varepsilon\phi^2$$

where ϕ is periodic in both x and y. Show that an optimal procedure leads to $O(K^2 \ln K)$ operations, rather than $O(K^4)$ operations, for each time step in the numerical solution of the resulting differential equations. [*Note:* If boundary conditions other than that of periodicity are imposed, expansion in Chebyshev polynomials, rather than in terms of Fourier series, may be useful. See Orszag (1971).]

15

SINGULAR PERTURBATION METHODS

15.1 A BOUNDARY LAYER PROBLEM

One frequently encounters a boundary value problem that contains a small parameter but for which the perturbation methods of Chapter 8 are inadequate. For instance, let $v(x, y)$ satisfy the equation

$$\varepsilon(v_{xx} + v_{yy}) + v_x = y(1 - y^2) \tag{15.1}$$

in the square region R defined by $0 < x < 1, 0 < y < 1$, with the boundary condition that $v = 0$ on all four sides of this square. Here ε is a constant, satisfying $0 < \varepsilon \ll 1$; we note that ε multiplies the most highly differentiated term.

The reader can easily verify that an attempt to describe v by the series

$$v = v^{(0)}(x, y) + \varepsilon v^{(1)}(x, y) + \varepsilon^2 v^{(2)}(x, y) + \cdots$$

is not successful. The underlying reason for this failure is that v is not an analytic function of ε near $\varepsilon = 0$; this fact becomes evident early in the expansion process when one finds that $v^{(0)}$ is governed by the first-order equation $v_x^{(0)} = y(1 - y^2)$, and that none of its solutions can satisfy the conditions $v^{(0)}(0, y) = v^{(0)}(1, y) = 0$.

On the other hand, it does seem plausible that in much of R the behavior of v is governed by the equation obtained by ignoring the seemingly small term $\varepsilon \Delta v$ in Eq. (15.1). This is indeed the case, but before proceeding we remind the reader of the analogous situation in ordinary differential equations. Consider for the moment the problem of finding $w(x)$ so as to satisfy

$$\varepsilon w'' + w' = 1 \tag{15.2}$$

in $0 < x < 1$, with $w(0) = w(1) = 0$, and with $0 < \varepsilon \ll 1$. The solution is

$$w = (x - 1) + \frac{e^{-x/\varepsilon} - e^{-1/\varepsilon}}{1 - e^{-1/\varepsilon}} = \alpha(x) + \beta(x/\varepsilon) \tag{15.3}$$

Thus, for all $x \gg \varepsilon$, w is approximated accurately by $\alpha(x) = x - 1$, which is a solution of $\alpha' = 1$, an equation obtained from Eq. (15.2) by neglecting the term $\varepsilon w''$. The function $\beta(x/\varepsilon)$ plays an important role in the description of w only in that region where w is so "steep" (has large slope, curvature, etc.) that εw_{xx} is no longer small—i.e., in a "boundary layer" region adjacent to $x = 0$.

In an effort to extrapolate these results to the problem of Eq. (15.1), we might write

$$v(x, y, \varepsilon) = h(x, y) + p(x, y, \varepsilon) \tag{15.4}$$

where h satisfies the equation $h_x = y(1 - y^2)$ in R, so that

$$h = y(1 - y^2)[x - A(y)] \tag{15.5}$$

where we have arranged for h to satisfy the boundary conditions $h(x, 0) = h(x, 1) = 0$, for any (finite) choice of the undetermined function $A(y)$. As we have noted before, no choice of $A(y)$ will enable us to satisfy the other two boundary conditions; we thus look to the function p to do this for us by virtue of steepness near one or both of the remaining boundaries.

Substitution of Eq. (15.4) into Eq. (15.1) shows that p must satisfy

$$\varepsilon \Delta p + p_x = -\varepsilon \Delta h \tag{15.6}$$

and, still using the results of the ODE example for guidance, we now try $p = p(\xi, y)$, with $\xi = x/\varepsilon$, to obtain

$$p_{\xi\xi} + \varepsilon^2 p_{yy} + p_\xi = -\varepsilon^2 h_{yy}$$

Neglecting the term in ε^2, we have

$$p \cong B(y)e^{-\xi} + C(y)$$

so that

$$v \cong y(1 - y^2)[x - A(y)] + B(y)e^{-x/\varepsilon}$$

where we have omitted $C(y)$ with no loss in generality, since (if the boundary conditions on $y = 0$, $y = 1$ are to be satisfied) it can always be absorbed into $A(y)$. In order that $v(0, y) = v(1, 0) = 0$, we now set

$$A = \frac{1}{1 - e^{-1/\epsilon}} \cong 1. \qquad B = \frac{y(1 - y^2)}{1 - e^{-1/\epsilon}} \cong y(1 - y^2)$$

and obtain

$$v \cong \phi = y(1 - y^2)[x - 1 + e^{-x/\epsilon}] \tag{15.7}$$

In examining the adequacy of this solution, we might note that ϕ satisfies the condition $\epsilon \, \Delta\phi + \phi_x = y(1 - y^2) + O(\epsilon)$ everywhere, and that ϕ satisfies the boundary conditions except for an error of size $e^{-1/\epsilon}$ at $x = 1$.

Of course, this problem was deliberately constructed to be particularly simple. There was a boundary layer on only one side of R, and the thickness $O(\epsilon)$ of the boundary layer was the same as in the associated ODE problem. In general, one does not know a priori whether there are any boundary layers, where they are, how many there are, or how thick they will be. Accordingly, we must frame the techniques of this chapter to have substantial flexibility if they are to be of general applicability.

15.2 A MORE GENERAL PROCEDURE

To illustrate a more general approach, consider the problem

$$\epsilon \, \Delta v - v_x + 2v = 1 \tag{15.8}$$

in a region R defined by $0 < x < 1, 0 < y < \infty$, with $v = 0$ on the boundary Γ of R. Here $0 < \epsilon \ll 1$.

In any nonsteep region, we approximate $v(x, y)$ by $h(x, y)$ satisfying the equation

$$-h_x + 2h = 1$$

i.e.,

$$h = \tfrac{1}{2}[1 + A(y)e^{2x}] \tag{15.9}$$

where $A(y)$ is an as-yet-undetermined function of y. No matter how we choose $A(y)$, we cannot satisfy all the boundary conditions on Γ. As a result of our experience with the problem of Section 15.1, however, we are led to ask whether $h(x, y)$ can nevertheless adequately represent the solution everywhere except perhaps near one or more boundaries.

Write $v = h + p$; then the equation satisfied by p is

$$\epsilon \, \Delta p - p_x + 2p = -\epsilon \, \Delta h \tag{15.10}$$

in R; on Γ, we require $p = -h$. If now there is a boundary layer near

$x = 0$ (a matter of pure speculation at this stage), then p must vary rapidly with x near $x = 0$. Moreover, p must decay rapidly with increasing x, so that its contribution outside of the boundary layer region is negligible, if h is indeed to be a good approximation to v except near the boundary. Accordingly, we write $p = p(\xi, y)$ where $\xi = x/\varepsilon^\nu$; here ν is a positive constant, as yet undetermined, which characterizes the "thickness" of the boundary layer. The idea is to—hopefully—choose ν so that the new variable ξ is $O(1)$ in the boundary layer region, and moreover so that p, p_ξ, $p_{\xi\xi}$ are all $O(1)$ in that region. Equation (15.10) now becomes

$$\varepsilon^{1-2\nu}p_{\xi\xi} + \varepsilon p_{yy} - \varepsilon^{-\nu}p_\xi + 2p = -\varepsilon\,\Delta h$$

or

$$p_{\xi\xi} + \varepsilon^{2\nu}p_{yy} - \varepsilon^{\nu-1}p_\xi + 2\varepsilon^{2\nu-1}p = -\varepsilon^{2\nu}\,\Delta h \tag{15.11}$$

We have already remarked that ν must be greater than zero if p is to be steep near $x = 0$. If $0 < \nu < 1$, then for small ε the equation would reduce to $p_\xi \cong 0$, which is unsatisfactory since we are now looking beyond first-order derivatives (in any event, we need a second-order equation in order to provide enough flexibility in the solution to fulfill all the requirements). If $\nu > 1$, then Eq. (15.11) would reduce to $p_{\xi\xi} \cong 0$, which has no decaying solutions. We try, therefore, $\nu = 1$, and obtain

$$p_{\xi\xi} - p_\xi \cong 0$$

However, this equation again admits no solution that decays as ξ increases. Thus there can be no boundary layer along $x = 0$. An immediate conclusion is that h as defined by Eq. (15.9) must now represent v right up to the boundary $x = 0$; it follows that $A(y) = -1$, and Eq. (15.9) becomes

$$h = \tfrac{1}{2}(1 - e^{2x}) \tag{15.12}$$

We turn next to the boundary $y = 0$, and ask if there can be a boundary layer adjacent to it. If so, write

$$v = h + w(x, \eta)$$

with $\eta = y/\varepsilon^\beta$, where $\beta > 0$. Substitution into Eq. (15.8) yields

$$\varepsilon w_{xx} + \varepsilon^{1-2\beta}w_{\eta\eta} - w_x + 2w = 2\varepsilon e^{2x} \tag{15.13}$$

The only reasonable choice is $\beta = \tfrac{1}{2}$, which leads to

$$w_{\eta\eta} - w_x + 2w \cong 0 \tag{15.14}$$

The boundary conditions are that $w(x, 0) = -\tfrac{1}{2}(1 - e^{2x})$, that $w(0, \eta) = 0$, and that w decay as η becomes large and positive. Equation (15.14) has the form of a diffusion equation in which x plays the role of the time variable. Its solution subject to the given boundary conditions is easily found (e.g., by a Laplace transform in x); it is

$$w \cong \int_0^x e^{2\gamma} \operatorname{erfc}\left(\frac{\eta}{2\sqrt{\gamma}}\right) d\gamma \qquad (15.15)$$

The composite function $h + w$ satisfies, within terms of $O(\varepsilon)$, Eq. (15.8) and also the boundary conditions along $x = 0$ and $y = 0$. We look finally at the possibility of a boundary layer along $x = 1$. With

$$v = h + w(x, \eta) + u(\sigma, y) \qquad (15.16)$$

where $\sigma = (1 - x)/\varepsilon^\alpha$, $\alpha > 0$, we find

$$\varepsilon^{1-2\alpha} u_{\sigma\sigma} + \varepsilon^{-\alpha} u_\sigma + 2u = O(\varepsilon)$$

and the choice $\alpha = 1$ leads to

$$u_{\sigma\sigma} + u_\sigma \cong 0$$

so that (taking account of the boundary values on $\sigma = 0$ attained by the function $h + w$)

$$u \cong -\left\{ \tfrac{1}{2}(1 - e^2) + \int_0^1 e^{2\gamma} \operatorname{erfc}\left(\frac{\eta}{2\sqrt{\gamma}}\right) d\gamma \right\} e^{-\sigma} \qquad (15.17)$$

With the replacement $\eta = y/\sqrt{\varepsilon}$, $\sigma = (1 - x)/\varepsilon$, we can now use Eqs. (15.15)–(15.17) to obtain an adequate approximation to the desired solution function $v(x, y)$.

15.3 PROBLEMS

15.3.1 Solve anew the problem of Section 15.2 if the boundary condition along $x = 0$ is altered to read $v(0, y) = ye^{-y}$. Introduce next an added modification, in that the sign of the term v_x is reversed; solve the problem once more.

15.3.2 Let $\phi(x, y)$ satisfy the equation

$$\varepsilon \Delta\phi + \phi_x + \phi_y = f(x, y)$$

in the region $0 < x < 1$, $0 < y < 1$, with $\phi = 0$ on the boundary of this square. Here $f(x, y)$ is a given function, and $0 < \varepsilon \ll 1$. Along which sides of the square can there be boundary layers? How thick are they? Repeat for each of

$$\varepsilon \, \Delta\phi + \phi_x - \phi_y = f$$

$$\varepsilon \, \Delta\phi - \phi_x + \phi_y = f$$

$$\varepsilon \, \Delta\phi - \phi_x - \phi_y = f$$

$$\varepsilon \, \Delta\phi - \phi_x = f$$

$$\varepsilon \, \Delta\phi - \phi = f$$

15.3.3 Let $\phi(x, y)$ satisfy the equation

$$\varepsilon\phi_{xx} + \phi_{yy} + \phi_y = 0$$

in the square region $0 < x < 1$, $0 < y < 1$, with $0 < \varepsilon \ll 1$, and with

$$\phi(x, 0) = x, \qquad \phi(x, 1) = 1 - x, \qquad \phi(0, y) = y, \qquad \phi(1, y) = 1 - y$$

Solve this problem exactly (e.g., via a series expansion; the transformation $\phi = x + y - 2xy + \psi$ may be useful), and comment on the presence or absence of boundary layers. Another approach is to write $\phi(x, y) = e^{-y/2}\omega(\xi, y)$ where $\xi = x/\sqrt{\varepsilon}$, and to note that ω satisfies the equation of an elastically restrained membrane, so that ω is easily visualized physically. Would the method of Section 15.2 have suggested the presence of a boundary layer?

15.3.4 In terms of polar coordinates (r, θ), let $u(r, \theta)$ satisfy

$$\varepsilon \, \Delta u - u = 1$$

inside the unit circle, with $u(1, \theta) = 0$, and with $0 < \varepsilon \ll 1$. Obtain the exact solution in terms of a Bessel function, and state whether or not there is a boundary layer region. If so, what is its width?
   What would a boundary layer approach to this problem, as in Section 15.2, yield? [*Note:* with $1 - r = \varepsilon^\alpha\xi$, say, observe that a term like u_r/r becomes $-\varepsilon^{-\alpha}u_\xi/(1 - \varepsilon^\alpha\xi)$, which to a first approximation is just $-\varepsilon^{-\alpha}u_\xi$.]
   Repeat all of the above discussion for the case in which u satisfies the modified equation

$$\varepsilon \, \Delta u + u = 1.$$

15.3.5 In terms of polar coordinates (r, θ), let $u(r, \theta)$ satisfy the equation

$$\varepsilon \, \Delta u - u = e^{2r \, \cos \theta}$$

in $0 < r < 1$, with $u(1, \theta) = \sin \theta$, and with $0 < \varepsilon \ll 1$. Find u by (a) the boundary layer method, and (b) the series expansion method. Compare efficiencies.

15.3.6 As in the case of ordinary perturbations, higher-order contributions in boundary layer problems can often be obtained by an extension of the matching process. To illustrate, let $u(x, t)$ satisfy the wave equation

$$\varepsilon(u_{tt} - u_{xx}) + u_t = 0$$

in $-\infty < x < \infty$, $t > 0$, with $u(x, 0) = f(x)$ and $u_t(x, 0) = g(x)$. Here $f(x)$ and $g(x)$ are prescribed functions of x, and $0 < \varepsilon \ll 1$. (This equation could describe the motion of a heavily damped string.) Since a neglect of the term in ε would preclude the satisfaction of the boundary conditions for $u(x, 0)$ and $u_t(x, 0)$, it is possible that, although the approximate equation $u_t \cong 0$ governs throughout most of the (x, t) plane, there is a boundary layer adjacent to the x-axis. Investigate the consequences of such an assumption, as follows:

(a) In the region outside this boundary layer, write

$$u(x, t) = u^{(0)}(x, t) + \varepsilon u^{(1)}(x, t) + \varepsilon^2 u^{(2)}(x, t) + \cdots$$

and obtain

$$u^{(0)} = \alpha(x)$$

$$u^{(1)} = \alpha''(x) \cdot t + \beta(x)$$

$$u^{(2)} = \tfrac{1}{2}\alpha^{iv}(x) \cdot t^2 + \beta''(x) \cdot t + \gamma(x)$$

$$.$$
$$.$$
$$.$$

by a conventional perturbation approach, where $\alpha(x)$, $\beta(x)$, $\gamma(x)$, ... are as-yet-unknown functions of x.

(b) In the inner region, define $\tau = t/\varepsilon$ and $v(x, \tau) = u(x, \tau\varepsilon)$, so that $v_{\tau\tau} + v_\tau = \varepsilon^2 v_{xx}$, with $v(x, 0) = f(x)$, $v_\tau(x, 0) = \varepsilon g(x)$, and obtain

$$v^{(0)} = f(x)$$

$$v^{(1)} = g(x) \cdot (1 - e^{-\tau})$$

$$v^{(2)} = f''(x) \cdot (\tau - 1 + e^{-\tau})$$

$$v^{(3)} = g''(x) \cdot (\tau + \tau e^{-\tau} - 2 + 2e^{-\tau})$$

$$.$$
$$.$$
$$.$$

(c) Assume now that the expansion of part (b), with $\tau = t/\varepsilon$, matches that of part (a) in some common range of t and ε values. [In particular, let $t \to 0$ as $\varepsilon \to 0$, but at a less-rapid rate, so that $t/\varepsilon \to \infty$; then the exponential terms in the expansion of part (b) can be neglected.] Deduce from the matching process that $\alpha(x) = f(x)$, $\beta(x) = g(x)$, and $\gamma(x) = -f''(x)$; verify that, to the order in ε considered, there is agreement between the two expansions.

15.3.7 Carry out the expansion process as in Problem 15.3.6 with the differential equation modified to read

$$\varepsilon(u_{tt} - u_{xx}) + u_t - \tfrac{1}{2}u_x = 0$$

[Note that the expansion corresponding to part (a) will now involve functions of the variable $(x - \tfrac{1}{2}t)$, and that the Taylor series expansions of these functions, for small t, will occur in the matching process.]

15.4 A TRANSITION SITUATION

We consider now a problem in which a boundary layer alters character as we move along a boundary. Let

$$\varepsilon \, \Delta u + u_x = 0 \tag{15.18}$$

in the interior of the unit circle (with $0 < \varepsilon \ll 1$); thus, in terms of polar coordinates, $u(r, \theta)$ satisfies

$$\varepsilon(u_{rr} + (1/r)u_r + (1/r^2)u_{\theta\theta}) + u_r \cos \theta - u_\theta \sin \theta/r = 0 \tag{15.19}$$

in the region $r < 1$. On the boundary, we require $u(1, \theta) = f(\theta)$, prescribed.

In any nonsteep region, Eq. (15.18) reduces to $u_x \cong 0$, so that u is essentially constant along lines parallel to the x-axis (see Fig. 15.1). The method of Section 15.2 shows that there can be no boundary layer for $-\pi/2 < \theta < \pi/2$, so that, except near the left-hand half of the boundary, we have

$$u(r, \theta) \cong f(\alpha)$$

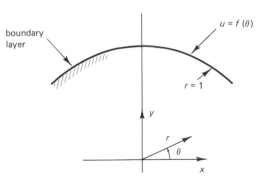

Fig. 15.1 Boundary layer transition.

where $\sin \alpha = r \sin \theta$, $-\pi/2 < \alpha < \pi/2$. Define w by

$$u(r, \theta) = f(\alpha) + w \qquad (15.20)$$

where w is to decay rapidly as we move away from the left-hand half of the boundary (for which $\pi/2 < \theta < 3\pi/2$). Define a boundary layer variable z by $1 - r = \varepsilon z$, so that Eq. (15.19) becomes, for $w = w(z, \theta)$,

$$\frac{1}{\varepsilon} w_{zz} - \frac{w_z}{1 - \varepsilon z} + \frac{\varepsilon w_{\theta\theta}}{(1 - \varepsilon z)^2} - \frac{1}{\varepsilon} w_z \cos \theta - w_\theta \frac{\sin \theta}{1 - \varepsilon z} = O(\varepsilon)$$

so that, for small ε,

$$w_{zz} - w_z \cos \theta \cong 0$$

or

$$w(z, \theta) = [f(\theta) - f(\pi - \theta)]e^{z \cos \theta} \qquad (15.21)$$

where we have made use of the required boundary conditions. Since $\pi/2 < \theta < 3\pi/2$, $\cos \theta$ is negative, so that w does decay as z increases (i.e., as r decreases).

If θ, satisfying $\theta > \pi/2$, is allowed to decrease toward $\pi/2$, then $|\cos \theta|$ becomes smaller, so that the thickness of the boundary layer determined by Eq. (15.21) increases. On the other hand, there is no boundary layer for $\theta < \pi/2$, so that Eq. (15.20), with w as defined by Eq. (15.21), cannot be valid for θ close to $\pi/2$. We need some kind of transition boundary layer in this region. Since our interest will now be in the region $|1 - r|$ small, $|\theta - (\pi/2)|$ small, define ξ and ζ by

$$1 - r = \varepsilon^\nu \zeta, \qquad \theta - (\pi/2) = \varepsilon^\beta \xi$$

and let $\phi(\zeta, \xi)$ represent the behavior of u in this region. Equation (15.19) requires

$$\varepsilon^{1-2\nu} \phi_{\zeta\zeta} - \frac{\varepsilon^{1-\nu}}{1 - \varepsilon^\nu \zeta} \phi_\zeta + \frac{\varepsilon^{1-2\beta}}{(1 - \varepsilon^\nu \zeta)^2} \phi_{\xi\xi} + \varepsilon^{\beta-\nu} \xi \phi_\zeta - \frac{\varepsilon^{-\beta} \phi_\xi}{1 - \varepsilon^\nu \zeta} = 0$$

where $\sin \theta$ has been approximated by 1 and $\cos \theta$ by $(-\varepsilon^\beta \xi)$. We want an equation that can have a solution that melds into the previous boundary layer solution; this suggests that the term in $\phi_{\zeta\zeta}$ will be important. For $\nu > 0$, the first term dominates the second for small ε, so that the only way we can get a term in ϕ_ζ—also a constituent of the boundary layer effect—is if

$$1 - 2\nu = \beta - \nu$$

The equation then reduces to

$$\phi_{\zeta\zeta} + \varepsilon^{4\nu-2} \phi_{\xi\xi} + \xi \phi_\zeta - \varepsilon^{3\nu-2} \phi_\xi \cong 0$$

If $\nu > \frac{2}{3}$, we merely recover the previous boundary layer equation in terms of our new variables. If $\nu < \frac{2}{3}$, the last term would alone occur. Thus we set $\nu = \frac{2}{3}$ (i.e., $\beta = \frac{1}{3}$), and obtain

$$\phi_{\zeta\zeta} + \xi\phi^4 - \phi_\xi \cong 0 \qquad (15.22)$$

For $\zeta = 0$, we require $\phi = f[(\pi/2) + \varepsilon^{1/3}\xi]$, or using the first terms of a Taylor expansion,

$$\phi(0, \xi) \cong f(\pi/2) + \varepsilon^{1/3}\xi f'(\pi/2)$$

For ζ large and positive, ϕ must become asymptotic to the function $f(\alpha) = f(\sin^{-1}[r \sin \theta])$ encountered previously; in terms of ξ and ζ, this condition requires

$$\phi(\zeta, \xi) \sim f[(\pi/2) - \varepsilon^{1/3}(2\zeta + \xi^2)^{1/2}] \sim f(\pi/2) - \varepsilon^{1/3}f'(\pi/2)(2\zeta + \xi^2)^{1/2}$$

for ζ large and positive. This condition holds also for any positive value of ζ, if ξ is large and negative, for we are then well outside any boundary layer region.

Finally, if ξ is large and positive, ϕ must be asymptotic to the previous boundary layer function defined by Eqs. (15.20) and (15.21); a similar calculation leads to

$$\phi(\zeta, \xi) \sim f(\pi/2) - \varepsilon^{1/3}f'(\pi/2)(2\zeta + \xi^2)^{1/2} + 2f'(\pi/2) \cdot \varepsilon^{1/3}\xi e^{-\zeta\xi}$$

Define $\psi(\zeta, \xi)$ via

$$\phi(\zeta, \xi) = f(\pi/2) + \varepsilon^{1/3}f'(\pi/2) \cdot \psi(\zeta, \xi)$$

Then ψ satisfies

$$\psi_{\zeta\zeta} + \xi\psi_\zeta - \psi_\xi \cong 0, \qquad \psi(0, \xi) \cong \xi \qquad (15.23)$$

$$\psi(\zeta, \xi) \sim -(2\zeta + \xi^2)^{1/2} \qquad \text{for} \quad \begin{cases} \zeta \text{ large and positive} \\ \text{or} \\ \xi \text{ large and negative} \end{cases}$$

$$\psi(\zeta, \xi) \sim -\xi + 2\xi e^{-\zeta\xi} \qquad \text{for} \quad \xi \text{ large and positive}$$

where, in the last condition, $(2\zeta + \xi^2)^{1/2}$ has been replaced by ξ for large ξ.

This modified diffusion equation for ψ (think of ξ as a pseudo time variable) is easily solved numerically. The technique is to set $\psi = -(2\zeta + \xi^2)^{1/2}$ for some large negative initial value of ξ, to use the end conditions $\psi(0, \xi) = \xi$, $\psi = -(2\zeta + \xi^2)^{1/2}$ for some suitable large value of ζ, and to solve the equation by taking steps in ξ. With $\zeta_{max} = 5$, $\xi_{initial} = -6$, $\xi_{final} = 10$, and $\Delta\zeta = 0.025$, $\Delta\xi = 0.005$, then the method of Problem 14.3.2, with $A = \frac{1}{2}$, yields the results plotted in Fig. 15.2. For $\xi = 10$, a comparison between ψ and the quantity $(-\xi + 2\xi e^{-\zeta\xi})$ is given in Table 15.1. Thus, this transition solution appears to be quite satisfactory.

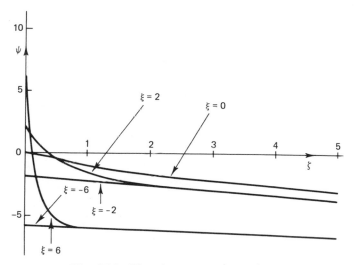

Fig. 15.2 Plot of ψ versus ζ for various ξ.

Table 15.1

ζ	$\psi(\zeta, 10)$	$-10 + 20e^{-10\zeta}$
0	10	10
.025	5.55	5.58
.05	2.10	2.13
.075	$-.594$	$-.527$
.1	-2.69	-2.64
.15	-5.58	-5.54
.2	-7.34	-7.29
.5	-9.91	-9.87
5.0	-10.5	-10.0

15.5 PROBLEMS

15.5.1 This exercise will illustrate a situation in which it is useful to alter the scale of the dependent, as well as of the independent, variable. Let $\phi(x, y)$ satisfy the equation

$$\varepsilon\phi_{yy} - (1 - y^2)\phi_x + 4y^2 = 0$$

in the region $x > 0$, $-1 < y < 1$, with $0 < \varepsilon \ll 1$, and with $\phi(0, y) = 0$, $\phi(x, \pm1) = 1$. (This is the kind of problem that might arise in the flow of a viscous heat-conducting fluid along a channel.) Show first that, if the ε term is neglected, the appropriate solution is given by

$$\phi_0 = 4xy^2/(1 - y^2)$$

and note that ϕ_0 becomes infinite as $y \to \pm 1$. Next, introduce a boundary layer variable near $y = -1$, via $\zeta = (1 + y)/\varepsilon^\alpha$; rewrite ϕ_0 in terms of (x, ζ) so as to show that ϕ_0 is $O(\varepsilon^{-\alpha})$ near the boundary layer. This suggests that we write $\phi = \varepsilon^{-\alpha}\psi(x, \zeta)$ in the boundary layer region; show that

$$\varepsilon^{1-3\alpha}\psi_{\zeta\zeta} - \zeta(2 - \varepsilon^\alpha\zeta)\psi_x + 4(1 - \varepsilon^\alpha\zeta)^2 = 0$$

whence $\alpha = \frac{1}{3}$. Show that ψ satisfies

$$\psi_{\zeta\zeta} - 2\zeta\psi_x + 4 \cong 0$$

$$\psi \sim 2x/\zeta \qquad \text{for large } \zeta$$
$$\text{(in order to merge with } \phi_0)$$

$$\psi(0, \zeta) = 0, \qquad \psi(x, 0) = \varepsilon^{1/3} \cong 0$$

Numerical or approximate methods could now be used to find ψ. Why might it be difficult to obtain directly a numerical solution of the original equation?

15.5.2   In the triangular region of the (x, y) plane defined by the vertex coordinates $A = (0, 0)$, $B = (1, 1)$, $C = (2, 0)$, let $w(x, y)$ satisfy

$$\varepsilon \, \Delta \, \Delta w - w_x = \sin y$$

with $\psi = \psi_x = 0$ on AB and BC, $\psi = \psi_{yy} = 0$ on AC. As before, $0 < \varepsilon \ll 1$. (This problem models a wind-driven ocean circulation situation.) Analyze the boundary layer behavior near AC.

15.5.3   (a)   Invent and solve a boundary layer-type problem in three dimensions.

(b)   Invent a problem in which there is an internal boundary layer, rather than one along a boundary.

(c)   Let $u(r, \theta)$ satisfy $\varepsilon \, \Delta u - u = 0$ in $r < 1$, with $u(1, \theta) = f(\theta)$ prescribed. Here (r, θ) are polar coordinates, and $0 < \varepsilon \ll 1$. Let $\zeta = (r - 1)/\sqrt{\varepsilon}$ be a boundary layer coordinate. In the boundary layer, denote u by $w(\zeta, \theta)$; outside the boundary layer, denote u by $v(r, \theta)$. Expand each as

$$w(\zeta, \theta) = w^{(0)}(\zeta, \theta) + \sqrt{\varepsilon}w^{(1)}(\zeta, \theta) + (\sqrt{\varepsilon})^2w^{(2)}(\zeta, \theta) + \cdots$$

$$v(r, \theta) = v^{(0)}(r, \theta) + \sqrt{\varepsilon}v^{(1)}(r, \theta) + \cdots$$

and solve for the first few of these expansion functions using the fact that w and v must merge at the edge of the boundary layer. Compare your results with an expansion of the exact solution for the case $f(\theta) = \sin \theta$. What would this procedure give when applied to the problem of Section 15.1?

15.6 ASYMPTOTIC ANALYSIS OF WAVE MOTION

We consider solutions of the wave equation

$$\phi_{tt} = c^2(\phi_{xx} + \phi_{yy})$$

that are periodic in time, so that $\phi = \mathrm{Re}[u(x, y)e^{i\omega t}]$. (We take c and ω as constant; the cases of position-dependent c and of three-dimensional motion are left for exercises.) Then

$$\Delta u + k^2 u = 0 \qquad (15.24)$$

where $k = \omega/c$. In physical applications, k is frequently a large number,† and Eq. (15.24) is then reminiscent of boundary layer-type problems.

There are many periodic solutions of Eq. (15.24) that represent propagating wave trains. A simple example is given by $u = e^{ikx}$, which in conjunction with the factor $e^{i\omega t}$, represents a sinusoidal wave pattern propagating to the left with velocity c and without change in amplitude. Another example is $u = H_0^{(2)}(kr)$, as in Problem 9.7.2; here $r^2 = x^2 + y^2$. In this second example, the propagation velocity of the pattern is again c—at least for large values of r for which $H_0^{(2)}(kr) \sim (\mathrm{const})r^{-1/2}e^{-ikr}$.

We now intend to study more general solutions having wave-propagation character, and for this purpose we write

$$u = \alpha(x, y)e^{ik\beta(x,y)} \qquad (15.25)$$

where α and β are real. Thus $\phi = \alpha \cos(k\beta + \omega t)$, so that α is the amplitude and $(k\beta + \omega t)$ the phase of the motion. For a chosen value of time, the phase is determined entirely by the value of β. A substitution into Eq. (15.24) yields

$$k^2(1 - [\beta_x{}^2 + \beta_y{}^2])\alpha + ik(\alpha \Delta\beta + 2\alpha_x\beta_x + 2\alpha_y\beta_y) + \Delta\alpha = 0 \quad (15.26)$$

If now k is indeed large, then the most important term is presumably the first, so that

$$\beta_x{}^2 + \beta_y{}^2 \cong 1 \qquad (15.27)$$

---

† Strictly, of course, we should nondimensionalize; if l is a characteristic length, then the appropriate nondimensional parameter is $(l\omega/c)$. Otherwise, the "largeness" of a parameter is merely a function of the units of measurement.

This means that β satisfies an equation of the same form as the eikonal equation studied in Section 12.7. The characteristics, or rays, of this equation are straight lines, orthogonal to the curves of constant β. If s represents distance along such a ray, then along that ray, $d\beta/ds = \pm 1$, so that a curve of constant phase ($k\beta + \omega t$) moves, normal to itself, with velocity c. This is the same kind of situation as that which orginally led to the eikonal equation of Section 12.7, except that there we discussed a disturbance moving into an undisturbed medium, rather than a moving surface or curve of constant phase.

Choosing β so as to satisfy Eq. (15.27), the next most important term of Eq. (15.26), for large k, is the second one, and we thus choose α so that

$$\alpha \, \Delta\beta + 2(\alpha_x\beta_x + \alpha_y\beta_y) = 0 \qquad (15.28)$$

Because the gradient of β is parallel to a ray direction, Eq. (15.28) is a statement concerning the rate of change of α along a ray. If Eqs. (15.27) and (15.28) are satisfied, then the only error in the satisfaction of Eq. (15.26) [and so of Eq. (15.24)] results from the term $\Delta\alpha$. If the effect of this term is neglected—and the preceding discussion indicates that this may be reasonable for large values of k—then we have obtained a solution of Eq. (15.24) that represents a moving wave train, in which each curve of constant phase appears to propagate with a normal velocity c. This kind of approximate solution of Eq. (15.24) is termed a *WKB approximation*, after Wenzel, Kramers, and Brillouin, who popularized it in wave mechanics.†

With this background, let us now carry out a formal WKB expansion. It is convenient to absorb the amplitude α into the exponent, and so we replace Eq. (15.25) by

$$u = \exp\{ik[\beta^{(0)} + (1/ik)\beta^{(1)} + (1/ik)^2\beta^{(2)} + \cdots]\} \qquad (15.29)$$

where the $\beta^{(j)}$ are functions of x and y. A substitution into Eq. (15.24), followed by a collection of terms of like order in (ik), yields the sequence

---

† It has a long prior history, dating back in fact to Huygens. In his book, *Treatise on Light* (1690), Huygens suggests that the reflection and refraction properties of light can be explained on the basis of a sequence of wave fronts spreading out from a source much as ripples spread out from a stone thrown into water. He lets each point on such a wave front act as a new disturbance source, just as was done in Section 12.7. Although Huygens does not seem to specify exactly what is meant by a wave front (in fact he emphasizes that the spacing between successive wave fronts need not be uniform), his water wave example would indeed be a curve of constant phase.

of equations

$$[\beta_x^{(0)}]^2 + [\beta_y^{(0)}]^2 = 1$$
$$2\beta_x^{(0)}\beta_x^{(1)} + 2\beta_y^{(0)}\beta_y^{(1)} + \Delta\beta^{(0)} = 0$$
$$2\beta_x^{(0)}\beta_x^{(2)} + 2\beta_y^{(0)}\beta_y^{(2)} + [\beta_x^{(1)}]^2 + [\beta_y^{(1)}]^2 + \Delta\beta^{(1)} = 0 \qquad (15.30)$$

.

.

.

A simple example is given by the case of cylindrical waves converging toward the origin of an (r, θ) polar coordinate system. Take $\beta^{(0)} = r$, so that the first of Eqs. (15.30) is satisfied. Then the second of these equations requires that

$$\beta^{(1)} = \ln[\,f(\theta)\cdot r^{-1/2}\,]$$

Similarly,

$$\beta^{(2)} = \frac{1}{r}\left[\frac{1}{8} + \frac{f''(\theta)}{2f(\theta)}\right] + g(\theta)$$

and so on. Here $f(\theta)$ and $g(\theta)$ are functions of θ to be determined from a knowledge of u on some curve. Thus, if $u = e^{ik}$ on $r = 1$, we would obtain $f(\theta) = 1$ and $g(\theta) = -\frac{1}{8}$. Note that this expansion will break down as $r \to 0$; it is in any event asymptotic rather than necessarily convergent.

A large collection of examples of this kind has been given by Keller, Lewis, and Seckler.[†]

In the example just discussed, the origin is a focus of the incoming waves. We have remarked that the WKB method is not valid near this focus; we should not leave this example without digressing a bit to discuss the value of u near a focal point. We will consider a slightly more general problem, in which the incoming waves lie within an angle Γ. As first suggested by Debye,[‡] a fruitful technique is to construct an exact solution of Eq. (15.24) by means of a superposition of plane waves. At the point (r, θ), the value of u resulting from a plane wave coming from the direction ψ is given by

$$u = (\text{const})\,e^{ik[r\,\cos(\theta-\psi)+\text{const}]}$$

A superposition of such waves, coming from directions satisfying $0 < \psi < \Gamma$ where Γ is some aperture angle, therefore has the form

$$u = \int_0^\Gamma m(\psi)\cdot e^{ik[r\,\cos(\theta-\psi)+h(\psi)]}\,d\psi \qquad (15.31)$$

[†] Keller *et al.* (1956, p. 207).
[‡] Debye (1909, p. 755).

Here $m(\psi)$ and $h(\psi)$ denote the distribution of intensity and phase as functions of ψ. By use of the principle of stationary phase,† the value of u at large values of r is given asymptotically by

$$u \sim (2\pi/kr)^{1/2} m(\theta) \cdot e^{ik[h(\theta)+r]-(i\pi/4)} \tag{15.32}$$

for $0 < \theta < \Gamma$, and by

$$u \sim (2\pi/kr)^{1/2} m(\theta - \pi) \cdot e^{ik[h(\theta-\pi)-r]+(i\pi/4)} \tag{15.33}$$

for $\pi < \theta < \Gamma + \pi$, and by $u \sim 0$ (relatively) for other values of θ. Thus there is an apparent phase change of $\pi/2$ along a ray passing through a focal point. We can obtain a specific example by taking $h(\psi) = 0$, $m(\psi) = 1$ for $0 < \psi < \Gamma$. Then the behavior of u for small r may be obtained by expanding the exponential in Eq. (15.31). In particular, at $r = 0$ we have $u = \Gamma$, so that if the amplitude is A at a large distance R, then it is $A\Gamma[kR/(2\pi)]^{1/2}$ at the focus.

15.7 BOUNDARY LAYER NEAR A CAUSTIC

We will now use a boundary layer method to analyze the behavior of the solution of Eq. (15.24) near a caustic—i.e., near a ray envelope. Let the caustic be the arc of a circle of radius R, as shown in Fig. 15.3, and consider, as in Eq. (15.31),

$$u(r, \theta) = \int_0^{\Gamma} m(\gamma) e^{ik[h(\gamma)+r\cos(\theta-\gamma)]} \, d\gamma \tag{15.34}$$

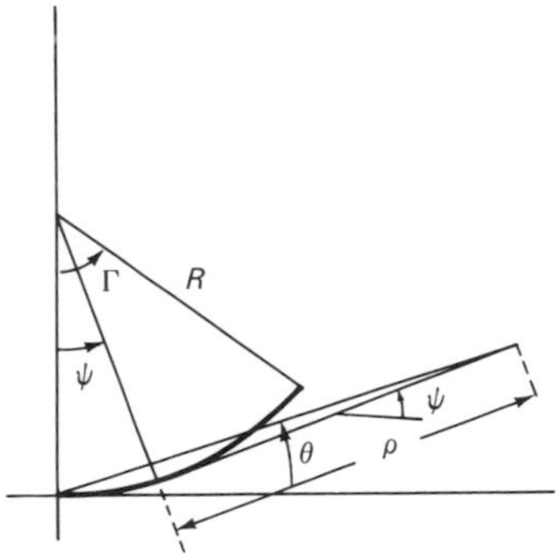

Fig. 15.3 Caustic.

† Carrier *et al.* (1966, p. 272).

where $h(\gamma)$ is to be so chosen that this circular arc is indeed the ray envelope. It is useful to introduce coordinates (ρ, ψ) defined as in Fig. 15.3; along a ray, ψ is constant. We have

$$r \cos \theta = R \sin \psi + \rho \cos \psi, \qquad r \sin \theta = R(1 - \cos \psi) + \rho \sin \psi \quad (15.35)$$

Using Eq. (15.32), we see that at large values of r (and for $0 < \theta < \Gamma$) a curve of constant phase satisfies the equation $r + h(\theta) = $ const. A vector tangential to this curve has x and y components proportional to $(-h'(\theta) \times \cos \theta - r \sin \theta, -h'(\theta) \sin \theta + r \cos \theta)$. A vector in the ray direction has components proportional to $(\cos \psi, \sin \psi)$. Since these vectors are mutually orthogonal,

$$\cos \psi [-h'(\theta) \cos \theta - r \sin \theta] + \sin \psi [-h'(\theta) \sin \theta + r \cos \theta) = 0$$

and using Eqs. (15.35) this condition reduces to

$$-h'(\theta) \cos(\psi - \theta) + R - R \cos \psi = 0$$

But at large r, $\psi \cong \theta$, so we take

$$h'(\theta) = R(1 - \cos \theta) \qquad \text{or} \qquad h(\theta) = R\theta - R \sin \theta$$

within a constant, which we absorb into $m(\theta)$. In order to obtain a specific problem, we must also choose $m(\gamma)$; arbitrarily, we set

$$m(\gamma) = (kA/2\pi)^{1/2} e^{i\pi/4}$$

which has the effect of making

$$u(A, \theta) \cong e^{ik[R(\theta - \sin \theta) + A]}$$

at a large reference distance $r = A$, $0 < \theta < \Gamma$. Then

$$u(r, \theta) = (kA/2\pi)^{1/2} e^{i\pi/4} \int_0^{\Gamma} e^{ik[R(\gamma - \sin \gamma) + r \cos(\theta - \gamma)]} \, d\gamma \quad (15.36)$$

or, in terms of ρ and ψ,

$$u(\rho, \psi) = (kA/2\pi)^{1/2} e^{i\pi/4} \int_0^{\Gamma} e^{ik[R\gamma + R \sin(\psi - \gamma) + \rho \cos(\psi - \gamma)]} \, d\gamma \quad (15.37)$$

We are interested in the behavior of $u(\rho, \psi)$ for small values of ρ. We consider two regions—an outer region in which $\rho = O(R)$, and an inner region in which $\rho = O(k^{-s})$, where s is some positive constant.

In the outer region, write $\rho = R\tau$, and let k become large with τ fixed. Then if τ is not too large, and if ψ is not too close to zero or to Γ, there will be two points of stationary phase† in Eq. (15.37), and a straight-

---

† Corresponding in fact to the two rays through the point in question.

forward calculation yields

$$u \sim (A/R\tau)^{1/2} e^{ikR(\psi+\tau)} \{1 + i e^{ikR[(\gamma_0-\psi)-2\tau]}\} \qquad (15.38)$$

where γ_0 is defined by $\tan[\frac{1}{2}(\gamma_0 - \psi)] = \tau$.

Equation (15.38) fails as $\tau \to 0$, and we need a description of a solution in the inner region that merges with the result of Eq. (15.38) in some intermediate region. Although we could again base our approach on the integral (15.37), it is instructive to return to Eq. (15.24) and use a direct boundary layer calculation.† We start with the substitution (15.25), and set $\beta = R\psi + \rho$ (which satisfies $\beta_x^2 + \beta_y^2 = 1$) in Eq. (15.26). In terms of (ρ, ψ) coordinates, this equation becomes

$$ik\left[2\alpha_\rho + \frac{1}{\rho}\alpha\right] + \left(1 + \frac{R^2}{\rho^2}\right)\alpha_{\rho\rho} - \frac{2R}{\rho^2}\alpha_{\rho\psi} + \frac{1}{\rho^2}\alpha_{\psi\psi}$$

$$+ \left(\frac{1}{\rho} - \frac{R^2}{\rho^3}\right)\alpha_\rho + \frac{R}{\rho^3}\alpha_\psi = 0 \qquad (15.39)$$

If we discarded all terms except the first, for large k, then α would be singular at $\rho = 0$. To avoid this, we stretch ρ via $\rho = k^{-s}\xi$ with $s > 0$; substitution into Eq. (15.39) shows that the appropriate value of s is $s = \frac{1}{3}$. The boundary layer equation then becomes (to lowest order)

$$i\left[2\alpha_\xi + \frac{1}{\xi}\alpha\right] + \frac{R^2}{\xi^2}\alpha_{\xi\xi} - \frac{R^2}{\xi^3}\alpha_\xi = 0$$

The solution of this equation is

$$\alpha = c_1\alpha^{(1)} + c_2\alpha^{(2)} \qquad (15.40)$$

where

$$\alpha^{(1)} = \exp\left(-\frac{i\xi^3}{3R^2}\right) \mathrm{Ai}[2^{-2/3}R^{-4/3}\xi^2 e^{-i\pi/3}]$$

$$\alpha^{(2)} = \exp\left(-\frac{i\xi^3}{3R^2}\right) \mathrm{Ai}[2^{-2/3}R^{-4/3}\xi^2 e^{i\pi/3}]$$

Here c_1 and c_2 are independent of ξ, and Ai is the Airy function of first kind. We determine the c_i by the requirement that the behavior of α for large ξ be the same as the behavior of the term multiplying $e^{ikR(\psi+\tau)}$ in

---

† We here follow the ideas of Buchal and Keller (1960, p. 85).

Eq. (15.38) for small τ; this gives

$$c_1 = e^{-i\pi/12}2^{5/6}(\pi A)^{1/2}R^{-1/3}k^{1/6}$$

$$c_2 = ie^{i\pi/12}2^{5/6}(\pi A)^{1/2}R^{-1/3}k^{1/6}$$

On the caustic itself, $\xi = 0$; Eq. (15.40) then leads to

$$u(0, \psi) = 6^{-1/6}A^{1/2}\pi^{-1/2}R^{-1/3}k^{1/6}\Gamma(\tfrac{1}{3})e^{i\pi/4} \tag{15.41}$$

It is gratifying that a direct calculation of $u(0, \psi)$ from Eq. (15.37) leads to the same result—cf. Problem 15.8.3.

15.8 PROBLEMS

15.8.1 Extend the discussion of Section 15.6 to (a) the case in which $c = c(x, y)$, (b) three dimensions with c constant, and (c) three dimensions with $c = c(x, y, z)$. Proceed as far as you can with the general situation in each of these, and construct some illustrative examples. Show that, in part (c), the equivalent equations to (15.30) imply that the product of the "energy flow rate" $(\alpha^0)^2/c$, where $\alpha^0 = \exp(\beta^{(1)})$, and the cross-sectional area of an infinitesimal ray tube is constant along that tube. [*Hint:* apply the divergence theorem to the vector $(\alpha^0)^2\nabla\beta^{(1)}$.]

15.8.2 Use the fact that $H_0^{(2)}(kr)$ is a solution of Eq. (15-24) to obtain its asymptotic representation via Eqs. (15.30).

15.8.3 Set $\rho = 0$ in Eq. (15.37) and evaluate the result asymptotically for large k to re-derive Eq. (15.41). [*Hint:* in the neighborhood of the stationary phase point, the function in the exponential is a cubic; use a change of variables in the complex plane.]

15.8.4 Discuss the value of u on the nonilluminated side of the caustic of Section 15.7.

15.9 MULTIPLE SCALING

As in the case of an ordinary differential equation, the formal use of two or more naturally occuring scales in a partial differential equation problem may lead to a perturbation expansion valid over a larger range of an independent variable.

To illustrate, consider the signaling problem with cubic damping[†]:

$$u_{tt} - u_{xx} = -\varepsilon u_t^3, \quad u(0, t) = \sin \omega t, \quad u(x, 0) = u_t(x, 0) = 0 \quad (15.42)$$

for the region $(x > 0, t > 0)$, where ω and ε are constants, with $\omega > 0$ and $0 < \varepsilon \ll 1$. In a conventional perturbation procedure, we would substitute

$$u = u^{(0)} + \varepsilon u^{(1)} + \varepsilon^2 u^{(2)} + \cdots \quad (15.43)$$

into Eq. (15.42) and the boundary conditions, and obtain, after some algebra

$$u^{(0)} = \begin{cases} \sin \omega(t - x), & t > x \\ 0, & t < x \end{cases}$$

$$(15.44)$$

$$u^{(1)} = \begin{cases} \frac{1}{8}\omega^2 x[3 \sin\omega(t - x) + \frac{1}{3} \sin 3\omega(t - x)], & t > x \\ 0, & t < x \end{cases}$$

.
.
.

Even without proceeding further, we can see that the expansion (15.43) will involve the factor εx in the second term, and this factor will cease to be small when $\varepsilon x = O(1)$; consequently, the series (15.43) will probably not be usable beyond values of x for which $x \ll (1/\varepsilon)$.

The above calculation suggests that the formal introduction of a second space variable, $\xi = \varepsilon x$ [which quantity, as we have seen, occurs naturally in the expansion (15.43)], might lead to an expansion valid over a greater range of x values. We therefore write $u = u(x, \xi, t)$, and since

$$\frac{\partial}{\partial x} u(x, t) = \frac{\partial}{\partial x} u(x, \xi, t) + \varepsilon \frac{\partial}{\partial \xi} u(x, \xi, t)$$

etc., Eq. (15.42) becomes

$$u_{tt} - u_{xx} - 2\varepsilon u_{x\xi} - \varepsilon^2 u_{\xi\xi} = -\varepsilon u_t^3 \quad (15.45)$$

Substituting the expansion

$$u = u^{(0)}(x, \xi, t) + \varepsilon u^{(1)}(x, \xi, t) + \varepsilon^2 u^{(2)}(x, \xi, t) + \cdots \quad (15.46)$$

[†] This example, and a number of similar ones, have been discussed by Chikwendu and Kevorkian (1972, pp. 235–257).

into Eq. (15.45), we obtain

$$u_{tt}^{(0)} - u_{xx}^{(0)} = 0 \tag{15.47}$$

$$u_{tt}^{(1)} - u_{xx}^{(1)} = 2u_{x\xi}^{(0)} - (u_t^{(0)})^3 \tag{15.48}$$

$$\cdot$$
$$\cdot$$
$$\cdot$$

with the boundary conditions

$$u^{(0)}(0, 0, t) = \sin \omega t$$

$$u^{(1)}(0, 0, t) = 0$$

$$u^{(0)}(x, \xi, 0) = u^{(1)}(x, \xi, 0) = u_t^{(0)}(x, \xi, 0)$$

$$= u_t^{(1)}(x, \xi, 0) = 0$$

$$\cdot$$
$$\cdot$$
$$\cdot$$

The solution for $u^{(0)}$ is

$$u^{(0)} = f(\alpha, \xi) + g(\beta, \xi)$$

where $\alpha = t - x$, $\beta = t + x$; f and g are as yet unknown functions of their two variables. To avoid a wave propagating toward the origin from the right, we set $g = 0$.

The equation for $u^{(1)}$ now reads

$$u_{tt}^{(1)} - u_{xx}^{(1)} = -2f_{\alpha\xi} - f_\alpha^3 \qquad \text{or} \qquad 4u_{\alpha\beta}^{(1)} = -2f_{\alpha\xi} - f_\alpha^3 \tag{15.49}$$

If the right-hand side of this equation were nonzero, then the solution for $u^{(1)}$ would include a term of the form $\beta F(\alpha, \xi)$ which is large compared to $u^{(0)}$ wherever x is large. Thus we take advantage of the flexibility afforded by the additional variable, ξ, and set

$$2f_{\alpha\xi} + f_\alpha^3 = 0$$

An integration with respect to ξ yields

$$f_\alpha^{-2} = \xi + h(\alpha) \tag{15.50}$$

where the function $h(\alpha)$ is determinable from the boundary condition $u(0, 0, t) = \sin \omega t$. In fact, we have

$$u_t^{(0)}(0, 0, t) = \omega \cos \omega t$$

But since $u^{(0)}(x, \xi, t) = f(\alpha, \xi)$,

$$u_t^{(0)}(0, 0, t) = f_\alpha(t, 0)$$

so that, setting $x = \xi = 0$ in Eq. (15.50),

$$h(t) = \frac{1}{\omega^2 \cos^2 \omega t} \qquad \text{for} \quad t > 0$$

whence

$$h(\alpha) = \frac{1}{\omega^2 \cos^2 \omega \alpha} \qquad \text{for} \quad \alpha > 0$$

Thus Eq. (15.50) gives

$$f_\alpha = \frac{\omega \cos \omega \alpha}{(1 + \xi \omega^2 \cos^2 \omega \alpha)^{1/2}} \qquad \text{for} \quad \alpha > 0$$

and so (using the initial conditions)

$$u^{(0)} = \int_0^{t-x} \frac{\omega \cos \omega \alpha \, d\alpha}{(1 + \xi \omega^2 \cos^2 \omega \alpha)^{1/2}} \tag{15.51}$$

for $t > x$, with $u^{(0)} = 0$ for $t < x$.

This zeroth-order term can be written

$$u^{(0)} = \frac{1}{\sqrt{\xi}\omega} \sin^{-1}\left[\left(\frac{\xi\omega^2}{1 + \xi\omega^2}\right)^{1/2} \sin \omega(t - x)\right] \tag{15.52}$$

for $t > x$. The reader should verify [most easily by use of Eq. (15.51)], that a small-ξ expansion of this result agrees with the series obtained in Eqs. (15.44); however, we can now expect Eq. (15.52) to represent the zeroth-order approximation to u even for $\varepsilon x = O(1)$. Although rarely of practical importance, high-order approximations could be found by a continuation of this process. Also, as in the case of ordinary differential equations, it might be possible to make the range of validity $O(1/\varepsilon^2)$, or even greater, by introducing $\eta = \varepsilon x^2$ as a third variable (and similarly $\varepsilon x^3, \ldots$ as further new variables), or equivalently, by modifying the x variable appropriately.

A further discussion of multiple scaling methods will be found in Carrier (1974, Chap. 14), Cole (1968), and Leibovich and Seebass (1974).

15.10 PROBLEMS

15.10.1 Solve the problem of Eq. (15.42), with the same boundary conditions, but with the term $-\varepsilon u_t{}^3$ replaced by $-\varepsilon u_t$ by (a) an exact method, (b) conventional perturbations, and (c) multiple scaling, to at least second order. Draw appropriate conclusions.

15.10.2   Solve Eq. (15.42) for the case $-\infty < x < \infty$, $t > 0$, $u(x, 0) = \sin \omega x$, $u_t(x, 0) = -\omega \cos \omega x$, by multiple scaling. Note that εt rather than εx is now the appropriate second variable.

15.10.3   The temperature $\phi(x, t)$ in a thin rod of slowly varying area satisfies the equation

$$f(\varepsilon x) \cdot \phi_t = f(\varepsilon x) \cdot \phi_{xx} + \varepsilon f'(\varepsilon x) \cdot \phi_x$$

for $t > 0$, $0 < x < \infty$, where $0 < \varepsilon \ll 1$; here $f(\varepsilon x)$ represents the area function. Let $\phi(x, 0) = 0$, $\phi(0, t) = g(t)$. Such special cases as that in which $g(t) = 1, f(\varepsilon x) = (1 + \varepsilon x)^2$, and $\phi(x, t) = (1 + \varepsilon x)^{-1} \mathrm{erfc}(x/(2\sqrt{t}))$ suggest that a multiple scaling procedure might be useful for at least some of these problems. Explore this possibility. Compare your results with that resulting from a consideration of the large-s behavior of a Laplace transform.

15.10.4   Consider the wave propagation problem $u_{tt} - (1 + \varepsilon \alpha)^2 u_{xx} = 0$ in $0 < x < \infty$, $0 < t$, where $u(x, 0) = u_t(x, 0) = 0$, $u(0, t) = \sin \omega t$. Here α is a given function of x and t, with $\alpha(x, t) \geq 0$, and ε a constant satisfying $0 < \varepsilon \ll 1$.

Begin with the special case $\alpha = x$, obtain the exact solution [*hint*: introduce a new independent variable $r = (1/\varepsilon) \ln(1 + \varepsilon x)$], and observe that a conventional perturbation expansion would lead to nonuniformity; verify this latter result for general $\alpha(x, t)$. Returning to the special case $\alpha = x$, sketch the characteristics in the (x, t) plane. One reason for the failure of the conventional perturbation procedure for large values of x is that the characteristics—the lines of signal propagation—have departed significantly from the positions they would occupy for $\varepsilon = 0$. Thus one might expect to obtain a more uniformly valid expansion by using characteristic coordinates (ξ, η) as the basic variables, and permitting x and t, as well as u (or, perhaps more easily, u_t and u_x) to be functions of (ξ, η). Explore this idea,[†] both for the special case $\alpha = x$ and more generally for $\alpha = \alpha(x, t)$.

15.10.5   Let $u(x, t)$ satisfy Burgers' equation (cf. Section 13.3)

$$u_t + u u_x = \varepsilon u_{xx}$$

where $0 < \varepsilon \ll 1$, in $-\infty < x < \infty$, $t > 0$, with

$$u(x, 0) = \begin{cases} 1 & \text{for} \quad -1 < x < 0 \\ 0 & \text{otherwise} \end{cases}$$

---

† See Lesser (1970). Nonlinear problems in two-dimensional supersonic flow are treated similarly by Lin (1954, p. 117). See also Friedrichs (1948, p. 211).

If the ε term is neglected, there will be a shock wave whose equation in the (x, t) plane is given by $t = 2x$ for $t < 2$, and by $t = \frac{1}{2}(x + 1)^2$ for $t > 2$. Assuming that the effect of the εu_{xx} term is to replace the shock wave by a narrow transition region, use a singular perturbation method to extract as much information as you can concerning this transition region. Compare the results with those of Problem 13.4.3.

15.10.6 Solve Problem 13.4.6 anew by using two time scales, t and $\tau = \varepsilon^2 t$, directly in Eq. (13.15). Show in particular that if

$$u(x, t, \tau) = u^{(0)}(x, t, \tau) + \varepsilon u^{(1)}(x, t, \tau) + \cdots$$

with

$$u^{(0)} = A(\tau) \sin k(x - Vt) + B(\tau) \cdot \cos k(x - Vt)$$

where $V = \alpha - \beta k^2$, then A and B must satisfy

$$A' = (B/24\beta k)(A^2 + B^2) \quad \text{and} \quad B' = (-A/24\beta k)(A^2 + B^2)$$

and solve these equations to show finally that the effective wave velocity is the same as that found in Problem 13.4.6.

REFERENCES

Birkhoff, G. (1971). "The Numerical Solution of Elliptic Equations." *SIAM Regional Conference Series No. 1.*

Birkhoff, G., and Rota, G. C. (1962). "Ordinary Differential Equations." Ginn (Blaisdell), Boston, Massachusetts.

Blumenson, L. E. (1970). *J. Theoret. Biol.* **27,** 273.

Boyce, W. E., and DiPrima, R. C. (1969). "Elementary Differential Equations." Wiley, New York.

Buchal, R. N., and Keller, J. B. (1960). Boundary layer problems in diffraction theory, *Comm. Pure Appl. Math.* **13,** 85.

Buzbee, B. L., Golub, G. H., and Nielson, C. W. (1970). On direct methods for solving Poisson's equation, *SIAM J. Numer. Anal.* **7,** 627.

Buzbee, B. L., Golub, G. H., and Nielson, C. W. (1971). *SIAM J. Numer. Anal.* **8,** 722.

Carrier, G. F. (1974). Perturbation techniques. *In* "Handbook of Applied Mathmatics, Selected Results and Methods" (C. Pearson, ed.), Chap. 14. Van Nostrand-Reinhold, Princeton, New Jersey.

Carrier, G. F., and Pearson, C. E. (1968). "Ordinary Differential Equations." Ginn (Blaisdell), Boston, Massachusetts.

Carrier, G. F., Krook, M., and Pearson, C. E. (1966). "Functions of a Complex Variable, Theory and Technique." McGraw-Hill, New York.

Chikwendu, S., and Kevorkian, J. (1972). A perturbation method for hyperbolic equations with small non-linearities, *SIAM J. Appl. Math.* **22,** 235–257.

Coddington, E. A. (1961). "An Introduction to Ordinary Differential Equations." Prentice-Hall, Englewood Cliffs, New Jersey.

Cole, J. (1968). "Perturbation Methods in Applied Mathematics." Ginn (Blaisdell), Boston, Massachusetts.

Collatz, L. (1948). "Eigenwertprobleme." Chelsea, Bronx, New York.

Cooley, J. W., and Tukey, J. W. (1965). An algorithm for the machine calculation of complex Fourier series, *Math. Comp.* **19,** 297.

Courant, R., and Friedrichs, K. O. (1948). "Supersonic Flow and Shock Waves." Wiley (Interscience), New York.

Courant, R., and Hilbert, D. (1953). "Methods of Mathematical Physics," Vol. 1. Wiley (Interscience).

Courant, R., and Hilbert, D. (1962). "Methods of Mathematical Physics," Vol. 2. Wiley (Interscience).

Courant, R., Friedrichs, K., and Lewy, H. (1928). Über die partiellen Differenzgleichungen der mathematischen Physik, *Math. Ann.* **100**, 32. (English trans. *IBM J. Res. Develop.* **11**, 215, 1967.)

Debye, P. (1909). Das Verhalten von Lichtwellen in der Nähe eines Brennpunktes, *Ann. Physik.* **30**, 755.

Embleton, C., and King, C. (1968). "Glacial and Periglacial Geomorphology." St. Martin's, New York.

Erdélyi, A., ed. (1954). "Tables of Integral Transforms," Vol. 1. McGraw-Hill, New York.

Finlayson, B. A. (1972). "The Method of Weighted Residuals and Variational Principles." Academic Press, New York.

Friedrichs, K. O. (1948). Formation and decay of shock waves, *Comm. Pure Appl. Math.* **1**, 211.

Garabedian, P. R. (1964). "Partial Differential Equations." Wiley, New York.

Goldstein, H. (1950). "Classical Mechanics." Addison-Wesley, Reading, Massachusetts.

Greenspan, H. (1968). "The Theory of Rotating Fluids." Cambridge Univ. Press, London and New York.

Huygens, C. (1690). "Treatise on Light" (S. P. Thompson, English trans.). Macmillan, New York, 1912.

Ince, E. L. (1956). "Ordinary Differential Equations." Dover, New York, (orig. ed. Longmans, Green, New York, 1926).

Jeffrey, A., and Kakutani, T. (1972). *SIAM Rev.* **14**, 582.

Kamke, E. (1948). "Differentialgleichungen," Vol. 1. Chelsea, Bronx, New York.

Kantorovich, L., and Krylov, V. (1958). "Approximate Methods in Higher Analysis." Wiley (Interscience), New York.

Keller, E. F., and Segel, L. A. (1970). *J. Theoret. Biol.* **26**, 399.

Keller, J. B., Lewis, R. M., and Seckler, B. D. (1956). Asymptotic solution of some diffraction problems, *Comm. Pure Appl. Math.* **9**, 207.

Kober, H. (1952). "Dictionary of Conformal Representation." Dover, New York.

Kreyszig, E. (1967). "Advanced Engineering Mathematics." Wiley, New York.

Lax, P. D. (1973). "Hyperbolic Systems of Conservation Laws and the Mathematical Theory of Shock Waves." *SIAM Regional Conference Publication No. 11.*

Leibovich, S., and Seebass, A. R., eds. (1974). "Nonlinear Waves." Cornell Univ. Press, Ithaca, New York.

Lesser, M. B. (1970). Uniformly valid perturbation series for wave propagation, *J. Accoust. Soc. Amer.* **47**, 1297.

Lick, W. (1974). Wave propagation. *In* "Handbook of Applied Mathematics" (C. Pearson, ed.), Chap. 15. Van Nostrand-Reinhold, Princeton, New Jersey.

Lighthill, M. J. (1960). "Higher Approximations in Aerodynamic Theory." Princeton Univ. Press, Princeton, New Jersey.

Lin, C. C. (1954). On a perturbation theory based on the method of characteristics, *J. Math. Phys.* **33**, 117.

Lin, C. C. (1955). "Theory of Hydrodynamic Stability." Cambridge Univ. Press, London and New York.

Mikhlin, S. (1964). "Variational Methods in Mathematical Physics." Pergamon, Oxford.

National Bureau of Standards (1964). "Handbook of Mathematical Functions," AMC Ser. 55. US Govt. Printing Office, Washington, D.C.

Orszag, S. (1971). Numerical simulation of incompressible flows within simple boundaries, *Studies in Appl. Math.* **50,** 293.

Pearson, C. (1969). Asymptotic behavior of solutions to the finite-difference wave equation, *Math. Comp.* **23,** 711.

Petrovsky, I. G. (1964). "Lectures on Partial Differential Equations." Wiley (Interscience), New York.

Polya, G., and Szego, G. (1951). "Isoperimetric Inequalities in Mathematical Physics," *Ann. Math. Stud.* 27. Princeton Univ. Press, Princeton, New Jersey.

Richtmyer, R. D., and Morton, K. W. (1967). "Difference Methods for Initial-Value Problems," 2nd ed. Wiley (Interscience), New York.

Stoker, J. J. (1957). "Water Waves." Wiley (Interscience), New York.

Stokes, G. G. (1847). *Trans. Cambridge Phil. Soc.* **8,** 441, (also in "Mathematics and Physics Papers," p. 197. Cambridge Univ. Press, London and New York, 1880).

Strang, G., and Fix, G. (1973). "An Analysis of the Finite Element Method." Prentice-Hall, Englewood Cliffs, New Jersey.

Thomée, V. (1969). Stability theory for partial difference operators, *SIAM Rev.* **11,** 152.

Tricomi, F. G. (1957). "Integral Equations." Wiley (Interscience), New York.

Varga, R. S. (1962). "Matrix Iterative Ananlysis." Prentice-Hall, Englewood Cliffs, New Jersey.

Whitham, G. B. (1974). "Linear and Nonlinear Waves." Wiley (Interscience), New York.

Zienkiewicz, O. C. (1971). "The Finite Element Method in Engineering Science," 2nd ed. McGraw-Hill, New York.

INDEX

A

Acoustic wave, 132
  and Mach number, 134
ADI method, 265
Adjoint operator, 155, 196, 248
Admissible function
  in variational method, 167
Amoebae aggregation, 193
Artificial viscosity method, 273

B

Bagrinovskii–Godunov method, 265
Bicharacteristics, 254
Bilinear function
  in finite elements, 186
Boundary layer method, 285, 287, 292
  near caustic, 300
Burgers' equation, 233
  exact solution, 234
  in singular perturbations, 307
  speed of shock wave, 235
  thickness of shock wave, 237

C

Cauchy data, 75
  on a characteristic, 85, 88, 97, 100, 208
  for elliptic equation, 78, 90
  in fluid motion, 251
  for wave equation, 109, 253
Cauchy–Kowalewski theorem, 76
Caustic, 300
Characteristics, 53, 80, 99, 227
  and Cauchy data, 85, 88, 99, 109
  conoid, 254
  and consistency, 207, 213
  coupled equations, 239, 247
  and discontinuities, 86, 88, 109, 227
  in first-order PDE, 94, 101, 205
  in fluid flow, 239, 242, 247, 250
  in integrated form, 211
  intersection of, 227
  in quasi-linear equation, 89, 95, 101
  Riemann invariants, 247
  and shock waves, 90, 228
  and signal speed, 86, 217
  straight, 244
  strips, 205
Clairaut's equation, 212, 215

Classification of PDEs, 1, 75
Collocation, 178
Complementary error function, 24
  asymptotic expansion for, 25
Complete integral
  and eikonal function, 220
  for first-order PDE, 209
  and Hamilton–Jacobi equation, 223
Complete set, 14
Conformal mapping, 68
Connectivity, 61
Conservation form, of PDE, 231, 274
  for gas flow equations, 236
Consistency
  of finite difference equation, 260
Continuity
  and distribution function, 111
  equation of, 106
Coupled equations, 110, 252
Courant maximum–minimum principle,
  200
Crank–Nicholson method, 262, 265

of pollutants, 9
series solution, 15
in three dimensions, 108
uniqueness, 19, 21, 34
variational approach, 183
see also Finite difference equations
Diffusivity, thermal, 8
Dipole, 149
Dirichlet problem, 57, 108
  for circle, 64
  for conjugate function, 72
  Green's function for, 143
  for half space, 146
  and source distribution, 151
  for sphere, 144
Discriminant, 78
  invariance of, 80
Discretization error, 260
Dispersive medium, 112
Distribution function, 112
Domain of dependence, 242
DuFort–Frankel equation, 263

D

Delta function
  definition, 30
  derivative of, 33
  and Green's functions, 141, 156
  and Laplace transform, 32
  for moving source, 157
  product, 125
  source interpretation, 149
  in three dimensions, 141
Developable surface, 213
  in Legendre transformation, 214
Diffusion equation, 7
  and Cauchy data, 79
  finite elements, 186
  and Green's functions, 139
  in heat flow, 7, 9, 10, 13, 108
  and Laplace transform, 24
  maximum principle for, 19
  method of images, 26
  moving boundary, 26
  and moving source, 159
  nonhomogeneous end conditions, 17
  for phytoplankton, 34

E

Eigenfunctions, 189
  and adjoint operator, 196
  in amoebae aggregation, 193
  expansion in terms of, 195
  Gram–Schmidt process, 192
  and integral equations, 190, 194, 195
  and Legendre functions, 119
  for membrane, 190, 202
  of ODE, 14
  orthogonality, 191
  for PDE, 189
  via perturbations, 197
  reality, 191
  in relaxation method, 276
  for Schrödinger's equation, 192
  in semi-infinite domain, 114
  and series solutions, 15, 113
  unbounded regions, 192
  variational approach, 199
  in variational problem, 171
Eikonal equation, 218, 253
Eikonal function, 220
  and complete integral, 220

Electrostatic fields, 124, 147
  dipoles, 149
  flux integral, 148
  surface distribution, 148
  volume distribution, 148
Elliptic equation, 82
  canonical form, 83
  Green's function for, 155
  in several variables, 110
  see also Potential equation
Envelope, 203, 205
  of characteristics, 227
  and complete integral, 209
  as developable surface, 213
  in disturbance propagation, 217
  of rays, 300
Euler equation
  of variational principle, 165
Evolution, equation of, 261
Expansion wave, 248

F

Fermat's principle, 219
Fast Fourier transform, 278
Finite difference equations, 257
  ADI method, 265, 266
  artificial viscosity, 273
  consistency, 260
  for diffusion equation, 257, 262–264,
    266, 267, 282
  discretization (truncation) error, 260
  error in, 258, 259
  explicit and implicit, 262, 271
  and fast Fourier transform, 278
  generating function for, 268
  in nonlinear case, 272
  for potential equation, 274, 278
  relaxation solution, 275
  series solutions for, 267
  stability, 259, 261, 264, 270
  for wave equation, 270
Finite elements, 180
  for biharmonic operator, 185
  for diffusion equation, 186
  higher order, 184
  linear, 180

First-order PDE, 91, 203
  change in variables, 98
  characteristics, 94, 99, 205
  complete integral, 209
  coupled set, 110, 252
  eikonal equation, 218
  and envelopes, 203
  general form, 205
  general solution, 210
  Hamilton–Jacobi equation, 222, 225
  and Jacobians, 98, 101, 205
  more variables, 101
  quasi-linear, 95, 101
  see also Rays
Fluid motion, 105
  characteristics, 239
  conservation form, 236
  intrinsic coordinates, 240
  Legendre transformations, 216
  perturbation, 130
  potential flow, 111
  Rayleigh–Janzen series, 132
  rotational, 255
  slender body theory, 136
  past sphere, 135
  stability of, 138
  supersonic, 242
  time-dependent, compressible, 247,
    250, 255
  variational principle for, 187
Fluid particle, 106
Focal point, 299

G

Galerkin method, 177
  for eigenfunctions, 201
  and finite elements, 182, 186
Gauss–Seidel relaxation, 277
General solution,
  for first-order PDE, 98, 210
Generating function,
  for finite difference equation, 268
Glacier motion, 99
Gram–Schmidt orthonormalization,
    121, 195
  for eigenfunctions, 192

Green's functions, 139, 147
  in cell diffusion, 160
  and delta function, 30, 141
  for Laplacian, 143
  modified Laplacian, 152
  and perturbations, 160
  and Riemann function 250
  and source distributions, 125, 147
  for sphere, 144
  symmetry, 144, 157
  wave equation, 155, 156
Green's identities, 140, 141
  and electrostatic field, 150
  for surface point, 147
Group velocity, 112

H

Hamilton–Jacobi equation, 222
  in central body motion, 223
  and first-order PDE, 225
Hamiltonian, 222
Harmonic function, 59
  see also Laplace's equation
Harnack inequality, 73, 145
Helmholtz equation, 125
  eigenfunction expansion, 195
  in variational form, 171, 187
Homogeneous
  end conditions, 17
  function, 95
  PDE, 2
Hyperbolic equation, 79
  canonical form, 80, 84
  Riemann function, 249
  in several variables, 110, 247, 256
  see also Wave equation
Hypercircle
  in function space, 187

I

Images, method of
  across plane, 146
  for diffusion equation, 26
  and electrostatics, 148

  for sphere, 143
  for strip, 159
Initial strip, 87
Integral equation
  for harmonic function, 151
  variational approach, 183
Intrinsic coordinates, 240

J

Jacobi relaxation, 275

K

Kelvin inversion, 109
Kirchhoff formula, 156
Kortweg–deVries equation, 237
Kronecker delta, 131

L

Lagrange multiplier, 169
  as function of position, 173
  physical interpretation, 174
Lagrange's equations, 222
Lagrangian, 222
Laplace's equation, 57, 105
  analyticity of solution, 145
  in annular region, 70, 123
  in axially symmetric case, 124
  and Cauchy data, 90
  and conformal mapping, 68
  in finite difference form, 274, 278
  fundamental solution, 63
  and Green's function, 142, 147
  via integral equation, 151
  and Kelvin inversion, 109
  modified, 152
  in perturbed circle, 134
  Poisson integral formula, 64, 65
  polynomial solution, 63, 125
  singularity near boundary, 67, 73
  solution in circle, 64
  solution properties, 59

and source distribution, 125, 147
in sphere, 144
in spherical coordinates, 121
and spherical harmonics, 121
in strip, 67, 158
see also Potential equation
Lax–Wendroff equation, 272
Legendre functions, 116
associated, 120
in axially symmetric problem, 124
as eigenfunctions, 119
orthogonality, 118
polynomials, 117
recursion relations, 117
Rodrigues' formula, 121
and spherical harmonics, 121
Legendre transformation, 214
in fluid flow, 216
Hamilton–Jacobi equation, 222
for minimal surface problem, 216
in thermodynamics, 216
Linearity, of PDE, 2

M

Mach lines, 241
Mach number, 134, 240
in Rayleigh–Janzen series, 132
in slender body theory, 136
Matching, 291
Maximum principle
for diffusion equation, 19
for harmonic functions, 59, 108
Mean value theorem, 60, 108
converse of, 66, 145
for finite difference equation, 275
Mesh, in finite differences, 257
Momentum, equation of, 107
Monge cone, 210
Multiple scaling, 303
Multiply connected region, 61

N

Natural boundary conditions, 167
Neumann problem, 57, 108

and conjugate function, 72
consistency condition, 57
Green's function for, 146
and source distribution, 152
Normal mode, 46, 116, 190
Numerical methods, 257
see also Finite difference equations,
Finite elements, Variational methods

O

Order, of PDE, 2
Orthogonal trajectory, 218
Orthonormality,
of eigenfunctions, 192

P

Parabolic equation, 81
canonical form, 81
in several variables, 110
see also Diffusion equation
Peaceman–Rachford–Douglas method,
265
Perturbations
and acoustics, 131
in amoebae aggregation, 193
boundary shape, 134, 198
for eigenfunctions, 197
flow past sphere, 135
in fluid motion, 131
and fluid stability, 138
of Green's function, 160
matched, 291
Rayleigh–Janzen, 132
regular, 127
slender body approximation, 136
Stokes expansion, 238
see also Singular perturbation methods
Phase velocity, 112
Plateau's problem, 164
Poisson's equation, 55, 105, 149
Poisson integral formula, 64, 65
Potential equation, 55
boundary layer problem, 285, 287, 292
and boundary value problems, 57

Potential equation (cont.)
  and conformal mapping, 68
  discontinuities, 66, 73, 115
  in electrostatics, 58
  finite difference equation for, 274, 278
  and finite elements, 180
  Green's function for, 142
  harmonic functions, 59, 108
  Harnack's inequality, 73
  in heat conduction, 55
  for membrane, 58
  Poisson's equation, 55, 105
  Rayleigh–Ritz approach, 174
  series solutions, 62, 123
  and sources, 125, 147
  uniqueness, 61, 73, 108
  in variational form, 162, 174
  well-posed, 90
  see also Laplace's equation

Q

Quadratic form, 110
Quasi-linear equation, 89, 95, 252
Quintic
  in finite elements, 185

R

Radiation condition, 136, 155, 160
Rayleigh–Janzen series, 132
Rayleigh quotient, 192, 199
  Lagrange multipliers, 200
Rayleigh–Ritz method, 174
  and singularities, 178
Rays, 218
  and bicharacteristics, 254
  and caustics, 300
  Fermat's principle, 219
  in integrated form, 221
  near focus, 299
Relaxation method
  for finite difference equations, 275
  Gauss–Seidel, 277
  Jacobi, 275

overrelaxation, 275
  SOR method, 277
  underrelaxation, 275
Retarded value, 156
Riemann function, 249
  and Green's function, 250
Riemann invariants, 247
Rodrigues' formula, 121

S

Saddle point method, 235
Scalar product
  of functions, 187
Self-adjoint operator, 197
Separation of variables, 11, 108, 113, 122
  and complete integral, 220
  in finite difference equations, 267
Shallow water theory, 37
  hydraulic jump, 239
  Kortweg–deVries equation, 237
Shock waves
  Burgers' equation, 233
  and characteristics, 90, 228
  and weak solutions, 230
Signal propagation
  in acoustics, 130
  and characteristics, 86, 217
  for diffusion equation, 25
  in dispersive medium, 112
  and envelopes, 217
  rays, 218
  see also Wave equation
Simply connected region, 61
Singularity
  near a discontinuity, 66, 73, 74, 115
Singular perturbation methods, 285
  asymptotic wave motion, 297
  for Burgers' equation, 307
  multiple scaling, 303
  transition problem, 292
  WKB method, 298
  see also Boundary layer method
Singular solution
  of ODE, 204
  of PDE, 211, 212
Slender body theory, 136

Soliton, 238
Solution, of PDE, definition, 2, 4
SOR method, 277
Spherical harmonics, *see* Legendre
    functions
Stability
    in amoebae aggregation, 193
    and convergence, 261
    in finite difference equation, 259, 261,
        264, 270, 272
    in fluid motion, 138
Starlike property, 202
Stationary property, 163
Strip relation, 88
Sturm–Liouville problem, 14
Subsidiary condition
    in variational problem, 168
Superposition, of PDE solutions, 2, 13
Supersonic flow, 242

T

Transform
    asymptotic behavior, 27, 51
    diffusion equation, 24, 183
    fast Fourier, 278
    finite, 19
    Fourier, 115, 158
    Laplace, 23, 27
    wave equation, 50, 114
Tricomi equation, 84
Trilinear function
    in finite elements, 186
Trivial solution
    of homogeneous problem, 189
Truncation error, 260

U

Uniqueness
    for diffusion equation, 19, 21, 34
    and discontinuity, 73
    for potential equation, 61, 108
    for wave equation, 52

V

Variational methods, 161
    approximation via, 174
    bounds, 179
    diffusion equation, 183
    direct, 166
    eigenfunctions, 199
    Fermat's principle, 219
    in fluid flow, 187
    Galerkin, 177
    integral equation, 183
    Lagrange multiplier, 169
    maximum–minimum principle, 179, 200
    in mechanics, 224
    natural boundary conditions, 167
    for any PDE problem, 161, 166
    potential equation, 162
    Rayleigh–Ritz, 174
    stationary property, 163
    subsidiary condition, 168
    see also Finite elements

W

Wave equation, 35
    asymptotic analysis of, 297
    caustic, 300
    dispersive medium, 112
    in elastic rod, 38
    for electric cable, 37
    focus of rays, 298
    finite difference equation for, 270
    general solution, 39
    Green's function for, 155
    hanging chain, 47
    and Huygens' principle, 155
    via Laplace transform, 50
    Kortweg–deVries equation, 237
    Kirchhoff formula, 156
    for membrane, 48, 105
    more variables, 253
    and moving source, 157
    multiple scaling, 303
    normal mode for, 46, 116
    radiation condition, 155, 160

Wave equation (cont.)
　radially symmetric, 48, 52
　reflected wave, 44
　Reimann function for, 248
　series solution, 45, 112
　in shallow water, 37
　for string, 35
　uniqueness, 52
　WKB method, 298
　see also Characteristics; Images,
　　method of; Signal propagation

Weak solution, 230
Weierstrass's approximation theorem,
　119
Well-posed problem, 90, 261
WKB method, 298

Z

Zonal harmonic, *see* Legendre functions